Methodologies and Outcomes of Engineering and Technological Pedagogy

Kaushik Kumar
Birla Institute of Technology, Mesra, India

J. Paulo Davim
University of Aveiro, Portugal

A volume in the Advances in Educational Technologies and Instructional Design (AETID) Book Series

IGI Global
DISSEMINATOR OF KNOWLEDGE

Published in the United States of America by
IGI Global
Engineering Science Reference (an imprint of IGI Global)
701 E. Chocolate Avenue
Hershey PA, USA 17033
Tel: 717-533-8845
Fax: 717-533-8661
E-mail: cust@igi-global.com
Web site: http://www.igi-global.com

Library of Congress Cataloging-in-Publication Data

Names: Kumar, K. (Kaushik), 1968- editor. | Davim, J. Paulo, editor.
Title: Methodologies and outcomes of engineering and technological pedagogy / Kaushik Kumar and J Paulo Davim, editors.
Description: Hershey, PA : Engineering Science Reference, 2020. | Includes bibliographical references. | Summary: “This book explores recent developments and advancements in engineering pedagogy”-- Provided by publisher.
Identifiers: LCCN 2019037800 (print) | LCCN 2019037801 (ebook) | ISBN 9781799822455 (h/c) | ISBN 9781799822462 (s/c) | ISBN 9781799822479 (eISBN)
Subjects: LCSH: Engineering--Study and teaching.
Classification: LCC T65 .M574 2020 (print) | LCC T65 (ebook) | DDC 620.0071--dc23
LC record available at https://lccn.loc.gov/2019037800
LC ebook record available at https://lccn.loc.gov/2019037801

This book is published in the IGI Global book series Advances in Educational Technologies and Instructional Design (AETID) (ISSN: 2326-8905; eISSN: 2326-8913)

British Cataloguing in Publication Data
A Cataloguing in Publication record for this book is available from the British Library.

Advances in Educational Technologies and Instructional Design (AETID) Book Series

ISSN:2326-8905
EISSN:2326-8913

Editor-in-Chief: Lawrence A. Tomei, Robert Morris University, USA

MISSION

Education has undergone, and continues to undergo, immense changes in the way it is enacted and distributed to both child and adult learners. In modern education, the traditional classroom learning experience has evolved to include technological resources and to provide online classroom opportunities to students of all ages regardless of their geographical locations. From distance education, Massive-Open-Online-Courses (MOOCs), and electronic tablets in the classroom, technology is now an integral part of learning and is also affecting the way educators communicate information to students.

The **Advances in Educational Technologies & Instructional Design (AETID) Book Series** explores new research and theories for facilitating learning and improving educational performance utilizing technological processes and resources. The series examines technologies that can be integrated into K-12 classrooms to improve skills and learning abilities in all subjects including STEM education and language learning. Additionally, it studies the emergence of fully online classrooms for young and adult learners alike, and the communication and accountability challenges that can arise. Trending topics that are covered include adaptive learning, game-based learning, virtual school environments, and social media effects. School administrators, educators, academicians, researchers, and students will find this series to be an excellent resource for the effective design and implementation of learning technologies in their classes.

COVERAGE

- Higher Education Technologies
- Instructional Design
- Game-Based Learning
- Classroom Response Systems
- Digital Divide in Education
- Hybrid Learning
- Adaptive Learning
- K-12 Educational Technologies
- Collaboration Tools
- Web 2.0 and Education

IGI Global is currently accepting manuscripts for publication within this series. To submit a proposal for a volume in this series, please contact our Acquisition Editors at Acquisitions@igi-global.com or visit: http://www.igi-global.com/publish/.

The Advances in Educational Technologies and Instructional Design (AETID) Book Series (ISSN 2326-8905) is published by IGI Global, 701 E. Chocolate Avenue, Hershey, PA 17033-1240, USA, www.igi-global.com. This series is composed of titles available for purchase individually; each title is edited to be contextually exclusive from any other title within the series. For pricing and ordering information please visit http://www.igi-global.com/book-series/advances-educational-technologies-instructional-design/73678. Postmaster: Send all address changes to above address.

Titles in this Series

For a list of additional titles in this series, please visit:
https://www.igi-global.com/book-series/advances-educational-technologies-instructional-design/73678

Challenges and Opportunities for Transforming From STEM to STEAM Education
Kelli Thomas (University of Kansas, USA) and Douglas Huffman (University of Kansas, USA)
Information Science Reference • © 2020 • 300pp • H/C (ISBN: 9781799825173) • US $195.00

Transforming Music Education in P-12 Schools and the Community
Taichi Akutsu (Okayama Prefectural University, Japan & Seisa University, Japan)
Information Science Reference • © 2020 • 230pp • H/C (ISBN: 9781799820635) • US $175.00

UXD and UCD Approaches for Accessible Education
Ricardo Mendoza-González (Tecnológico Nacional de México, Mexico & Instituto Tecnológico de Aguascalientes, Mexico) Huizilopoztli Luna-García (Universidad Autónoma de Zacatecas, Mexico) and Alfredo Mendoza-González (Universidad Autónoma de Zacatecas, Mexico)
Information Science Reference • © 2020 • 290pp • H/C (ISBN: 9781799823254) • US $185.00

Preparing 21st Century Teachers for Teach Less, Learn More (TLLM) Pedagogies
Pradeep Kumar (Taylor's University, Malaysia) Michael James Keppell (Taylor's University, Malaysia) and Chee Leong Lim (Taylor's University, Malaysia)
Information Science Reference • © 2020 • 332pp • H/C (ISBN: 9781799814351) • US $195.00

Developing Technology Mediation in Learning Environments
Filomena Soares (Polytechnic of Porto, Portugal) Ana Paula Lopes (Polytechnic of Porto, Portugal) Ken Brown (Letterkenny Institute of Technology, Ireland) and Anne Uukkivi (TTK University of Applied Sciences, Estonia)

For an entire list of titles in this series, please visit:
https://www.igi-global.com/book-series/advances-educational-technologies-instructional-design/73678

IGI Global
DISSEMINATOR OF KNOWLEDGE

701 East Chocolate Avenue, Hershey, PA 17033, USA
Tel: 717-533-8845 x100 • Fax: 717-533-8661
E-Mail: cust@igi-global.com • www.igi-global.com

Table of Contents

Section 3
Application and Implementation

Detailed Table of Contents

Quality evaluation is a basic part of education that enables teachers to help learning and to improve instructive programs. Engineering education has been confronting impressive difficulties concerning commendable educating, information organization, and knowledge deployment. Consequently, desires for new teaching methods and learning approaches should be created in the arena. The objective of this chapter is to incorporate various teaching learning methods, educational tools to improve the learning experience of students, and also to fulfil the teaching experience of faculty. The purpose of this research is also to explore the effects of innovative teaching learning strategies based on the performance of student grades. The experiment has been carried out on two courses of electrical and electronics engineering. There are no commendable measures to evaluate the learning outcomes of the student hourly basis in traditional pedagogy. Therefore, this chapter proposed various pedagogical approaches that help to achieve the desirable things.

Evaluation is an essential process for the measurement of transformation that a student attains after a teaching learning process. Outcome-based education (OBE) in academics especially in the field of engineering is an accepted philosophy in recent years. The OBE system departs from the traditional method where assessment of students is based only on grades and/or ranks. Output has been the traditional measurement criterion in education field, which does not address the level of transformation in the learner, whereas outcome is the measurement of level of achievement showing the transformation. Assessment tools are required for the measurement of outcome. These tools could be direct tools for direct assessment or indirect tools for indirect assessment. An assessment can be a formative assessment or summative assessment. Learning is complete only if transformation is observable in all the vital aspects of attitude, skill, and knowledge. It is widely accepted that all these aspects can be measured in OBE.

Teaching is not only a profession. But this is a noble cause of educating and building the society through imparting knowledge. This knowledge makes the students skilled, conceptually and technically, to achieve their goals in life. Present teaching-learning methodology is more concerned with laboratory experiments and hands-on practice with the purpose of learning through real-time simulation exercise around the whole globe. All education authorities are more concerned towards making the students not only educated but employable also. Teachers are playing a major role in adding skills and values for the betterment of students, society, environment, and economy.

of public and environment. Engineers, in their professional quest for an optimal solution, are forced into a dilemma due to clash of values or interests. The explosion of data and its usage is bringing in a lot of concern due to proliferation of unethical practices. Moral values and personal ethics at one end and professional and social ethics at the other end of the spectrum are of points of discussion in academics as well as in society. Engineering programmes strive to offer engineering ethics and professionalism either through direct courses or through embedded capsules in appropriate courses. Promotion of ethics integrating into the engineering profession at all levels could lead to a holistic alternative at universal level, which is self-satisfying, people-friendly, and eco-friendly.

Preface

We, editors, would like to present the book *Methodologies and Outcomes of Engineering and Technological Pedagogy*. Pedagogy is defined as "the study of the methods and activities of teaching". Engineering pedagogy as its title says is closely linked to the technical but also pedagogical sciences. It is also associated with the preparation of teachers who teach in engineering schools and universities. A central role of engineering didactics is primarily dictated by the concept of technology or in other words primary function of the technology is to transform the natural world. In last 100 years the concept of engineering education has moved from Hands-on/ Practical oriented education to education integrating information, computational, and communications technology. With growing environment and consciousness of "Output based Education" importance of this subject has increased many folds. Unfortunately, very few information on Engineering Pedagogy is available, in scattered form, in journal articles, in conference and symposium proceedings, in workshop notes, and in government and company reports. This proliferation of the source material, coupled with the fact that some of the relevant publications are hard to find or are restricted, makes it difficult to identify and obtain the up-to-date knowledge needed to utilize the concept to its full advantage. This book intends to overcome these difficulties by presenting, in a single volume, many of the recent advances in the field of engineering Pedagogy and its recent developments.

An object-related reasoning concerning the designing of learning and teaching processes in academic engineering education illustrates the scientific character of related questions. Verifiable pedagogical and/or psychological qualifications are legally fixed requirements for a teaching career at all levels and in all types of schools of general and vocational education. In contrast, in the sector of higher education, it is assumed that lecturers have teaching abilities due to their high academic qualifications but evaluation results counter this assumption partially. A major reason for this is the complexity of the influence factors and relationships concerning the design of a demand-oriented education in engineering sciences.

Today the task of an engineer is to develop this technology to become qualified in such a manner that they are able to solve technical design problems. In contrast to this, the activity of natural scientists is focused on the discovery of relationships in the world and, thus, solving scientific knowledge problems. Invention and discovery require different ways of thinking and, thus, different methods of academic training. For students in engineering science courses, a future engineer's ability to accomplish social communicative processes in modern structures of production and service has become a focal point of engineering pedagogical considerations.

The chapters in the book have been provided by Researchers and Academicians working in the field and have gained considerable success in the field. The chapters of the book are segregated in four sections, namely Section 1: The State of the Art and Tools; Section 2: Education in the Industry 4.0 Era; Section 3: Application and Implementation; and Section 4: Ethics in Education.

Section 1 contains Chapter 1 to Chapter 6, whereas Section 2 contains Chapter 7 and Chapter 8, Section 3 consists of Chapter 9 to Chapter 13, and Section 4 with Chapter 14.

Section 2 of the book starts with Chapter 1. Here the effects of innovative teaching learning strategies based on the performance of student's grade have been explored and various pedagogical approaches to achieve the same proposed incorporating various teaching learning methods, educational tools to improve the learning experience of students and also to fulfil the teaching experience of faculty. Engineering education has been confronting impressive difficulties concerning commendable educating, information organization and knowledge deployment. Consequently, desires for new teaching methods and learning approaches subsequently is required to be created in the arena. The experiment has carried out on two courses of electrical and electronics engineering.

Chapter 2 intrudes into the study of teaching style among learners with academic accomplishment in engineering. The focuses on learning styles and how learning styles came about in education in highly demanding in this era and many significant things to know about why educators should understand the learning style of a student. Although learning method was one of the significant elements of human existence and students learning styles are among the variables that have been obtained. But, in the present era, extensive research requires a theoretical approach to reviewing appropriate literature on the subject and presents different points of perspective on matching or mismatching learning styles. The chapter concluded with the findings as to how to learn the styles of teaching and learning impressively and effectively, and how an individual understands new information so that they can find the most efficient technique to accumulate, interpret, and present information.

Chapter 3 reviews 42 refereed articles on outsourcing in relation to knowledge management and learning. Among the knowledge and learning potentials of outsourcing in organisations are to focus on core competencies, organisational learning, shorten production cycles, improve quality and enhance innovation.

Outsourcing can also have negative outcomes, such as knowledge loss, competences drainage, organisational learning problems, diminished trust, poorer services, hidden cost and reduction in innovation. The chapter provides summarisation of various hypotheses and two conceptual models which highlight the contribution of outsourcing to either competitive advantage or disadvantage. The chapter can assist managers to seek competitive advantages out of outsourcing of activities while avoiding detrimental outsourcing effects.

Chapter 4 elaborates one of the most important aspect of modern education i.e. Output Based Education (OBE). Evaluation is an essential process for the measurement of transformation that provides student attainment after a teaching learning process. Outcome Based Education (OBE) in academics especially in the field of engineering is an accepted philosophy in recent years world over. OBE system departs from the traditional method where assessment of student is based only on grades and/or ranks. Output has been the traditional measurement criterion in education field which does not address about the level of transformation in the learner whereas outcome is the measurement of level of achievement showing the transformation. Assessment tools are required for the measurement of outcome. These tools could be direct tools for direct assessment or indirect tools for indirect assessment. An assessment can be a formative assessment or summative assessment. Learning is complete only if transformation is observable in all the vital aspects of attitude, skills and knowledge. It is widely accepted that all these aspects can be measured in OBE.

Chapter 5 explains the shift of traditional teaching techniques to the modern one. Teaching is not only a profession but also a noble cause of educating and building the society through imparting knowledge. This knowledge makes the students skilled, conceptually and technically to achieve ultimate objective goal of their life. The current purpose of teaching – learning methodology is more practical oriented with the purpose of learning through real-time simulation exercise considering real life situation and focal point of all types and versions of education are more concerned towards making the students not only educated and knowledgeable but employable as well and in doing so teaching fraternity plays the pivoted role in adding skills and values for the betterment of students, society, environment and economy.

Chapter 6 brings out an introduction and few application courses following the pedagogical engineering approaches. Engineering pedagogical techniques have acquainted with progressive teaching methods and training techniques. The concept of pedagogy has now different dimensions. Along with modern challenges in industries, the teaching approaches have been modified in several respects. Earlier teacher training programs requires to be upgraded with modern pedagogical concepts. In this chapter concepts of technological pedagogical content knowledge, constructive pedagogy is summarized and highlighted.

Section 2 is dedicated to Education in the Industry 4.0 Era and contains two chapters. Chapter 7 talks about preparation of skilled people compatible with Industry 4.0 and the role played by education in achieving the same. Industry 4.0 revolution has changed the thinking of industry owners in terms of technological usage. With the help of modern digital technology, industry can able to fulfil the requirement of customer easily and compete strongly against their competitors. In order to achieve good quality of products in affordable price, industry needs skilled people who aware about autonomous and intelligent components. The chapter starts with which kind of qualifications are needed to fit in smart factory era and also deals with challenges that emerge in education in order to implement skills suitable for Industry 4.0. The chapter concludes with opportunities for education sector as far as smart factory is concerned.

In Chapter 8 an attempt has been made to describe and discuss the current scenario of digital integration in higher educational disciplines. The perception of learning and teaching in the educational universities have been affected by digital technology and this has impacted more with industrial concern over sustainability of resources and efficiency in operation in a digital environment. Hence the need arises to implement digital technologies in the educational setting so that digital competence of the future workforce can be elevated and better industrial output-based education is provided. The issues concerning this integration includes teacher's inability to incorporate digital thinking into student's learning, student's non-adaptability to modern technologies, un-reliability of digital educational resources and lack of infrastructure/power supply in most of the educational institutes. An active digital learning approach in students and extensive training sessions for digital utilization excellence in teacher's and educators are few ways to solve issues regarding the above-mentioned integration.

From here the book starts with Section 3, which groups the application and implementation part.

Chapter 9 elaborates Artificial Intelligence (AI) based systems to adopt Bloom's two-sigma learning performance standard via mastery learning techniques. Modern education is engaged in significant programmatic reforms in response to the business community's call for web-savvy, problem-solving graduates. Web-based intelligent tutors provide a readily accessible vehicle for enhancing business students' learning performance as well as preparing them for the rigors of the global marketplace. Furthermore, intelligent tutors can also be used to identify students at risk, to formulate appropriate intervention plans, and to support team learning. Recent evidence suggests that achieving Bloom's goal may be achievable on a routine basis by 2025. The chapter highlights the growing potential for using intelligent tutors to enhance student and team learning opportunities and outcomes and to outline strategies for implementing this revolutionary process throughout the management education community of practice.

Chapter 10 increases the customer service level by expanding the method of application used by Mystery Customer (MC) and resolves practical and concrete problems concerning the status of the chain stores. The survival of organisations can often be dependent on their customer service level; therefore, there is an immediate need to form a permanent measurement to act as an indicator of that performance. A use case of MC method applied in order to observe the level of quality service in a large retail network placed in Greece. The results suggested an explicit determination of the factors of satisfied customers through the application of MC method. Action diagrams used as a performance-importance map to indicate the strong and weak points in terms of criteria and to define the required improvement efforts. MC data showed the points or issues that service activities performance is likely to cause dissatisfaction and enables proactive measures taken on employee performance and classifies the improvement areas that determined by the analysis and used as a benchmarking tool.

Chapter 11 deals with few innovative learning methods like think pair share (TPS) activity, flipped classroom, online education, virtual classroom techniques, project-based learning, activity-based learning etc. etc. to overcome the difficulties of slow and inactive learners for attaining the course outcome and also to renovate the teaching methodology to provoke the inactive or slow learners to be more active in learning the course. The chapter discusses the significance and impact of using ICT tools over conventional teaching methods. In the present scenario, the innovation in teaching is necessary to engage the students for the course of 36 to 40 hours and at the end of the course, the average student's strength will excel in performance and attains the course outcome with in-depth knowledge. From studies it has been proven that the current generation students are adapted to the modern technology and feel monotonous by conventional teaching methods like black-board and PowerPoint presentations.

Chapter 12 presents a case study at Dedan Kimathi University of Technology, DeKUT, Kenya undertaken during 2015 – 2016 for a subject Machine Design. A feasible approach was implemented in the subject to train engineering students in a developing country to utilize local resources to solve practical societal problems. The approach involved grouping students of different interests and capabilities and tasking them to identify and study various problems in society. The groups were then required to propose machine design solutions to the identified problems. Finally, the groups were tasked to undertake the theoretical design and build CAD models for their projects. The students were monitored through individual weekly presentations to the instructor. The approach was seen to be successful to facilitate training in machine design.

Outcome-Based Education (OBE) is a learning theory based on objectives that derive outcomes for each portion of an instructional scheme. Every student has to accomplish the objectives at the end of the instructional experience. OBE is not limited to well-defined teaching or direct assessment strategies but involves indirect assessments to assist the learners in the attainment of defined outcomes. Chapter 13, the last chapter of the section, provides a case study on a course describing in all aspects of direct and indirect assessments. The chapter starts Initially with development of a correlation between Programme Outcomes (POs) and Course Outcomes (COs) duly analysing the impact of CO on PO. The evaluation of COs using assessment tools are well-defined and attainment percentage of the same is evaluated using statistical methodologies and the same is categorized to High, Medium and Low-level attainments. The achieved level of attainments is correlated to POs and the same is adapted for all the courses in order to initiate the corrective action for further improvements in successive years.

From here the book starts with Section 4, which proposes ethics in education.

The first and only chapter of the section, Chapter 14, proposes that promotion of ethics integrating into the engineering profession at all levels could lead to a holistic alternative at universal level which is self-satisfying, people-friendly and eco-friendly. Engineering as a profession distinguishes itself by having and enforcing code of ethics. Aberrations can lead to commercial considerations to dominate converting the profession into a business rather than promoting safety, health and welfare of public and environment. Engineers, in their professional quest for an optimal solution, many a time, are forced into a dilemma due to clash of values or interests. The explosion of data and its usage is bringing in a lot of concern due to proliferation of unethical practices. Moral values and personal ethics at one end and professional and social ethics at the other end of the spectrum are of points of discussion in academics as well as in society. Engineering programmes strive to offer engineering ethics and professionalism either through direct courses or through embedded capsules in appropriate courses.

First and foremost, the editors would like to thank God for providing the power of thinking and means of expression. The Editors would also like to thank all the Chapter Contributors, the Reviewers, the Editorial Advisory Board Members, Book Development Editor and the team of Publisher IGI Global for their availability for work on this editorial project. Also, the editors, have sincere gratitude towards all who directly or indirectly helped them in this project.

Kaushik Kumar
J. Paulo Davim

Section 1

The State of the Art and Tools

Chapter 1
Technology in Engineering Pedagogy to Progress the Excellence of Teaching:
Teaching Learning Process

Durga Prasad Garapati
https://orcid.org/0000-0002-6972-7801
Shri Vishnu Engineering College for Women, India

Padmaja S.M.
Shri Vishnu Engineering College for Women, India

ABSTRACT

Quality evaluation is a basic part of education that enables teachers to help learning and to improve instructive programs. Engineering education has been confronting impressive difficulties concerning commendable educating, information organization, and knowledge deployment. Consequently, desires for new teaching methods and learning approaches should be created in the arena. The objective of this chapter is to incorporate various teaching learning methods, educational tools to improve the learning experience of students, and also to fulfil the teaching experience of faculty. The purpose of this research is also to explore the effects of innovative teaching learning strategies based on the performance of student grades. The experiment has been carried out on two courses of electrical and electronics engineering. There are no commendable measures to evaluate the learning outcomes of the student hourly basis in traditional pedagogy. Therefore, this chapter proposed various pedagogical approaches that help to achieve the desirable things.

DOI: 10.4018/978-1-7998-2245-5.ch001

INTRODUCTION

The pedagogy of engineering has a very long history. In November 1951, Hans Lohmann's foundation of the Institute of Engineering Pedagogy lastly institutionalized the education and research in engineering pedagogy. His work focuses on technology and technical education. Lohmann's, set the groundwork for an understanding of engineering pedagogy, whose aim is the particular technical and technological design.

As teachers understand, teaching is a complex practice that needs many types of specific expertise to be interweaved. Teaching is thus an instance of an ill-structured discipline which really requires instructors to apply intricate structures of experience across distinct instances and backgrounds (Mishra, Spiro, & Feltovich, 1996; Spiro & Jehng, 1990).

According to studies by (Eggen, Kauchak, 2006) at least four distinct types of knowledge are crucial to understanding specialist teaching:

Content knowledge – we can't teach what we don't know, a comprehensive knowledge of the subjects we teach is crucial in all content fields for all educators.

Knowledge of pedagogical material – the capacity to generate examples, understanding methods of depicting the topic that make it understandable to others, and comprehension what makes it simple or hard to learn particular subjects.

General pedagogical expertise – includes an understanding of overall instructional values and governance of the classroom that defies individual subjects or subject regions.

Knowledge of students and learning–the most significant awareness a professor can possibly have is crucial for efficient teaching. By constantly telling us that we don't teach material, we educate learners, it affects the way we educate. The capacity of teachers to adapt their training based on the knowledge of learners is crucial for efficient learning.

Most conventional teaching and learning techniques are described by particularity (pencil for writing, microscope for viewing tiny items) (Simon, 1969). These techniques attain transparency of interpretation over time (Bruce & Hogan, 1998); they get to be widespread and were not even perceived as innovations in most instances. In comparison, digital technologies such as computers, mobile devices, and application software are impish.

It can be challenging to acquire a fresh knowledge base and skill set, especially if it is a time-consuming activity that needs to fit into a busy timetable. Moreover, it is unlikely that this knowledge will be used unless educators can conceive of technology uses consistent with their current pedagogical views (Ertmer, 2005). In addition, educators were often given insufficient training for this assignment. Many approaches to professional development of educators give a one-size-fits-all approach to inclusion of technology when educators actually work in various teaching and learning environments.

How can teachers include technology in their teaching in the face of these issues? An approach is needed that treats learning as an interaction between what the educator knows and what he understands in his or her classrooms.

Three central elements are at the center of technological excellent learning: content, pedagogy and technology, and the links between them. The heart of the technological, pedagogical and content-based (TPAK) structure are those three knowledge bases (content, pedagogy, and technology).

The efficiency of a lecturer shows the capacity to generate the anticipated results for the teacher in the learning operations. Effectiveness of teachers is the production of predicted results (James E.Ogbu, 2012). In all engineering colleges, electrical discipline curriculum consists of power electronics and Digital Electronics subjects where students will learn about different converters and logical design using synchronous and asynchronous circuits. Today in all industries power electronics converters and digital electronics plays major role in conversion of power and to control. Hence it is necessity for instructor to introduce different education tools like simulation tools, available labs in the department, virtual laboratories etc., for better understanding of the concepts (S. Dormido, 2005; J. Sanchez, 2004; W. G. Hurley, 2005; V. Fedak, 2005).

Project based learning (PBL) is a technique based on science concepts, encouraging learners to explore learning, enhancing perceptual lifestyle strategies, leading to realistic products following genuine issues and themes. PBL can be described as an experimental activity that creates cognitive constructions and reconstitutes understanding according to Piaget's co-operative view. PBL can be described as an experimental activity that creates cognitive structures and reconstructs understanding according to the brain development view of Piaget. It involves actions such as metacognitive thinking, produces an initial product, enables language skills to be used in a team, class or community and provides the finished products (Kimonen, and Nevalainen, 2000; Mergendoller, 2006; Carr, 2012).

For learners referred to as 7Cs, educators recognized seven core competencies. These seven abilities are the following: 1. Critical thinking and the solution of problems., 2. Innovation and creativity., 3. Cooperation, collaboration and management., 4. Comprehension of interculture., 5. Communication fluidity and data., 6. Technology of computers and communication., 7. Career and development of oneself.

Shulman and Tamir, in the Second Handbook of Research on Teaching (Travers, ed., 1973), listed five types of objectives that may be achieved through the use of the Lab based learning (LBL):

- *Skills* - manipulative, inquiry, investigative, organizational, communicative
- *Concepts* - for example, hypothesis, theoretical model, taxonomic category
- *Cognitive abilities* - critical thinking, problem solving, application, analysis, synthesis
- *Understanding of the nature of science*- scientific enterprise, scientists and how they work, existence of a multiplicity of scientific methods, interrelationships between science and technology and among the various disciplines of science
- *Attitudes* - for example, curiosity, interest, risk taking, objectivity, precision, confidence, perseverance, satisfaction, responsibility, consensus, collaboration.

TECHNOLOGY IN EDUCATION

Now-a-days, technology plays a vital role in students' lives in out of the classroom. Technology helps students to involve actively in learning the complex concepts and to engage collaboratively among peers. Hence, it is suggested to implement the technology based learning practices in the classroom based on the availability of the resources. So that, instructor can attract the student towards the lecture and can make the student to concentrate and also to participate interestingly than traditional teaching.

TPACK framework which consists technological knowledge (TK), pedagogical knowledge (PK), and content knowledge (CK), this three offers a prolific approach to deal with a large number of the predicaments that educators face in actualizing instructive innovation in their classrooms. By separating among these three kinds of learning, the TPACK structure outlines how content is being educated and how the educator impacts that content through different pedagogical methods and it must shape the establishment for any viable educational technology combination.

Figures 1 & 2 give a conceptual over view on the TPACK frame work how TK, CK and PK comprises to get the TPACK. TK refers to an understanding of the way that technologies are used in a specific content domain. CK comprises conceptual frameworks and how one can implement proven concept or theory in a superior way. Finally, PK contains common knowledge about teaching and learning process of different theories and concepts.

Right now, innovation is treated as though it is isolated from educating and learning. Mishra and Koehler point to this as a present negative effect of not using the available technologies or applications in regular teaching process. They guarantee that the absence of familiarity with TPACK keeps innovation isolated and prompts four issues with utilizing innovation in the classroom.

Figure 1. Venn diagram representation of TPACK

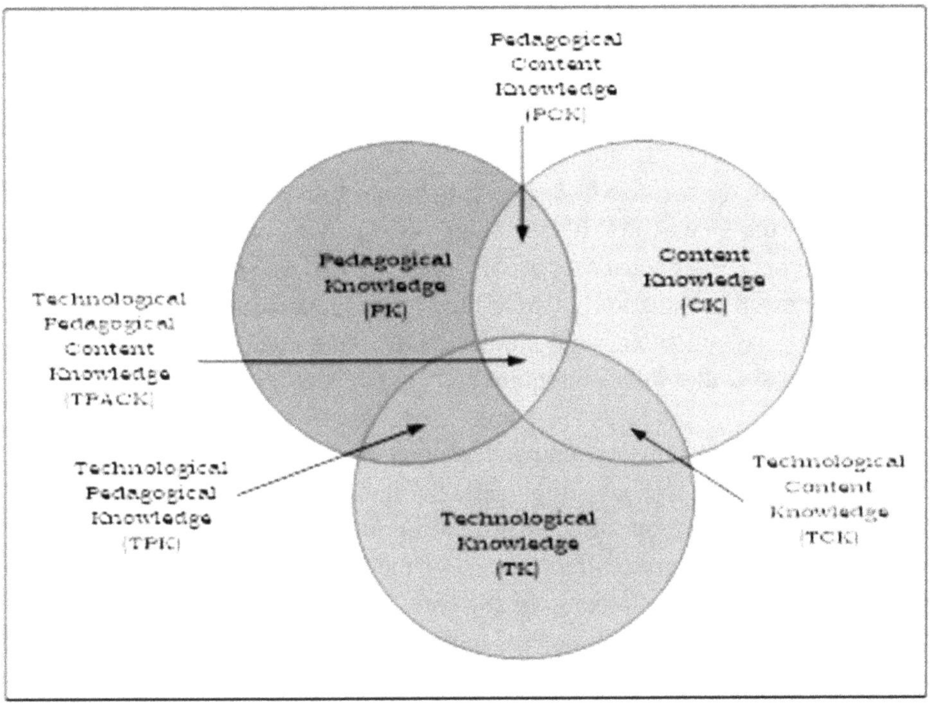

Figure 2. TPACK framework

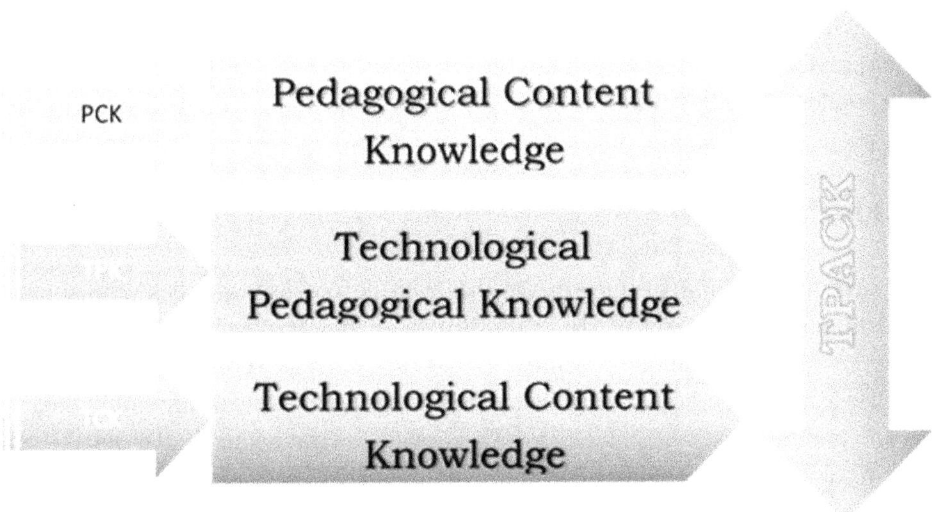

In the first place, there are such quick changes in innovation that it is very hard to stay aware of all the most recent progresses and applications. The second issue is that product is intended for business, not for Teaching. This regularly implies students are figuring out how to utilize the program and not learning the actual content of the class. The third issue with keeping technology separate is the situational idea of the classroom. An instructor

Can alter and do exercise to ensure the content to reach a particular group of audience, but the video can't. It's a similar video playing each time in classroom and the content is never being exercised. Finally, fourth issue is keeping innovation separate places an accentuation on "what" not "how." From the instructor's viewpoint the exercise winds up about what innovation are we going to utilize today, what does it say, what abilities does it require, rather than how might instructor can how to students.

Technology in teaching learning and evaluation process will give good impact. Technological innovative teaching methodology always outperforms than the conventional classroom teaching. It fulfills the individual learning requirements and also enhances the concentration level of the students. This also increases or builds the level of thinking among the students. Students will better understand the concept when they solve the problems in the classroom. An ultimate outcome of this effort is to give high-quality engineering graduate to the society.

VARIOUS TEACHING METHODOLOGIES

Higher educational institutions in the country respond to political, financial, social and technological pressures, to react better to the requirements of learners and to the willingness of learners to play future role in society. Faculties are already under pressure to teach less, increase the interactiveness of learning environments, incorporate technology into learning experience. Below are described some of the most prolific approaches.

Lecture Method

The traditional teaching (Lecture) method has been the most used in college classrooms for many years. In the early 1970s, around 80 percent of all colleges globally reported using the teaching technique (Cashin, 1990). While currently considering the utility of other learning methods, the lecture is still a significant way of providing data.

The traditional lecture can be an efficient way to attain educational objectives, used in combination with active education strategy. The benefit of the lecture strategy is that it offers a way of providing many viewers with a wide variety of data and does not threaten learners. The disadvantages are that teaching minimizes student feedback, presumes an unrealistic level of student understanding and understanding and often diverts learners from the teaching process so that data is forgotten rapidly.

Case Method

It has proved an efficient way to disseminate and integrate understanding for learners to apply what they have learned in the classroom to real life experiences. The case is an educational approach which involves learners in active discussions about practical issues and problems. It can highlight basic dilemmas or critical problems and format ambiguous or contentious scenarios.

Discussion Based Learning

This type of learning helps the students to deeply engage in the topic and to experience the style of learning. Discussion based learning empowers the student's contribution through the faculty thrown questions. This learning practice is much useful when the learning outcomes are complex.

Massive Open Online Courses (MOOCs)

In this digital era teachers and students have plenty scope to learn and update their knowledge and MOOCs is one which offers better learning opportunities to the learners across the globe. MOOCs is the online portal provides platform for discussions through webinars. This improves confidence in the learners, progresses the self-study, can collaborate with others, can maintain interpersonal connectivity, can learn various subjects other than curriculum.

Flipped Classroom

This is one type of delivery method where instructor will provide the data off the classroom to the students and make them learn before they are attending the class. In classroom, instructor will create discussions by questioning and can access the involvement of the students easily. Flipped classroom is the method which is far different from traditional teaching. The flipped classroom deliberately transforms the training to a learner-centered model in which the time is used in the classroom, to study issues more thoroughly and to generate meaningful teaching opportunities (Abeysekera, 2015; Marco Ronchetti, 2010; Greg Topp., 2011).

Active Learning

This the learning strategy making students to learn on their own through proper teaching methods. By this method of teaching students will improve their self-learning capabilities. (Weltman, 2010) Active learning is a way to learn, where learners participate in the learning process actively or experience and where there are distinct levels of active learning depending on the participation of the learners.

Project Based Learning

Students will be learning through projects. Project-based learning provides learners and educators with a wide spectrum of advantages. Increasing educational research promotes the use of collective, project-based learning to involve learners, decrease absence from college, enhance collaborative learning abilities and enhance academic achievement.

Lab Based Learning

This approach involves students to learn through their academic labs or any special labs available in the college or through virtual labs. This approach stretches better understanding and also provides practical exposure on the topics learned theoretically.

Blended Learning

This is an education strategy that combines instructional materials online with possibilities for internet communication with classroom techniques based on traditional location. The physical presence of the student and the instructor needs certain aspects of student control over moment, location, trajectory or speed. The phrases "blended learning," "hybrid learning" are commonly used in identical literature (Martyn, 2003; Bonk, 2006; Bersin, 2004). Although the concepts of the blended learning originally evolved in the 1960s, it wasn't until the late 1990s that official terminology was used to define them. In blended learning, different software's or apps can be used to make students involve among peers and also to evaluate the student performance instantly and provide feedback to students.

The above discussed are few learning methods. In this chapter authors concentrated on PBL and LBL for digital electronics and power electronics subjects.

PROJECT BASED LEARNING

Learning based on projects is known as motivating and problematic learning, which not only enables learners to obtain basic understanding but also helps them transfer their technical expertise into an in-house implementation in real time.

The demand for qualified engineers with technical expertise arises from the rapid changes in society. Training is based on lessons and associated practices. The evolving world and rapid advancement demand change in the engineering education. Here the above issues are addressed by Project Based Learning (PBL). The additional prominence of this type of learning is to inbuilt the challenging spirit among the students. PBL is full of vibrant and committed teaching, which inspires learners to learn the topics they study more thoroughly (G.G.Karady, 2000).

By adopting PBL in classroom the following advantages can be marked: student engagement in class, reduce absentees in the classroom, improve cooperative learning skills, and get better academic performance. PBL encourages learning from experience, allowing students to use and organize what has been learnt to understand problems and also it helps to develop teamwork and communication skills. The PBL at Shri Vishnu Engineering College for Women (SVECW) was introduce in order to inculcate the active learning practices and partially to replace conventional teaching methodology.

The lecturers in the topic gave them the form of more particular goals, taking into account the primary objective and the motivation underlying that little experience. They were elucidated within the demands of the course and developed as skills that must have been obtained by the end of the course by the student.

- Providing learners with the basic of power electronic ideas and preparing them for sophisticated study and research in this region: basic topologies, power supply systems, passive components, semiconductor devices, etc.
- To give students the capacity to suggest alternatives to challenges and to improve the critical thinking required to select the suitable solution according to particular criteria
- To complement the theory of power electronics with experimental experience and to encourage the implementation of conceptual frameworks.

The lecturing team had to pick the most appropriate approach to accomplish that goals, having identified the subject's objectives. In SVECW we chose PBL because this methodology provokes learners to meet a profession's key ideas and values while handling a particular assignment (project), thus allowing the implementation of gained knowledge (D. R. Woods, (1994) & Moursund D, 1999). In addition, PBL extends beyond the connection between understanding and thinking, assisting both "know" and "do" learners This methodology is actually focused on "doing something" and "learning how to do it." PBL is centered on students and depends on their primary skills. The method is designed by students to reach a solution. Their primary issues and abilities are therefore focused on the assignment. The final product is, in reality, a reflection of itself. For this purpose, there is no issue for learners to spend more time executing their initiatives.

PBL enables learners solve issues on their own: improving self-management, project management and critical understanding. PBL thus allows for regular reviews and coherent possibilities for learners to learn better as they program their own job. Throughout development projects, self-assessment occurs continually.

The process involved in PBL is shown in figure and described as follows for power electronics subject, major concepts in the subject has been divided in to categories like, AC- DC conversion, DC-AC Conversion, AC-AC conversion, DC-Dc conversion, Control Techniques etc., Initially the students are trained in conventional methods and provide necessary inputs. After completing classroom teaching, instructor will divide entire strength into different groups and assign a problem to the student groups. Now students will sit with the groups and analyze the assignment and carry out the work in lab and submit the report. By this the PBL is implemented in the classroom and students are actively participating in PBL (M. Undeland, 2003).

The PBL learning process is illustrated in the following algorithm:

Step1: Instructor will introduce the problem specifications
Step 2: Student group analyze the given problem
Step 3: First criteria; proper power semiconductor switches selection.
Step 4: Check different power semiconductor switches and analyze the characteristics of the switch by collecting data sheets and generate report.
Step 5: Get the feedback from the instructor, formulate the solution.
Step 6: Second criteria: topology; discuss and workout for different circuit configurations where the results are effective, layout size, cost etc., submit the report
Step 7: Get the feedback from the instructor, formulate the solution.
Step8: Third criteria; control technique derive the switching pulses based on the operation of the selected topology and submit the report
Step 9: Get the feedback from the instructor, formulate the solution.
Step 10: Same procedure will be repeated for other criteria's also.
Step 11: Simulate the system using proper simulation tool
Step 12: Prepare final documentation and present individually.

Figure 3 shows the students involved in different projects during their study.

Figure 3. Students practicing and doing the assignments

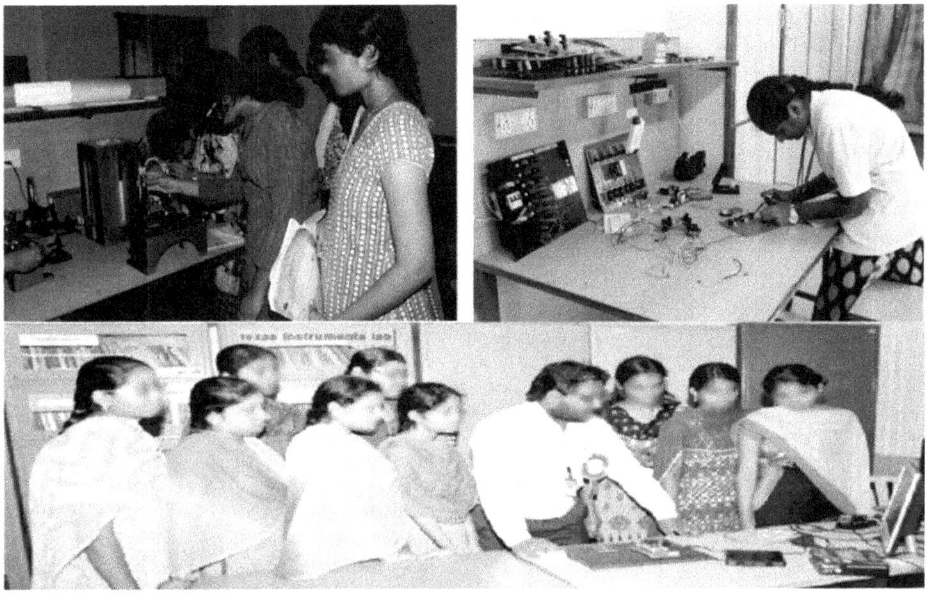

Students involvement in doing the hardware projects in learning the theoretical concepts and practicing the available equipment in the laboratories to strengthen their learnt theoretical knowledge is shown in Figure 3.

By adopting project based learning in addition to classroom teaching students show enthusiasm to learn the concepts and they can relate learned theoretical knowledge to practical approach.

LAB BASED LEARNING

Lab practices in science and technology to teach practical skills are essentially a type of experiential learning. In the theoretical tradition of engineering education, experiential learning is a basic paradigm. Learning by doing' is a constructivist pedagogy approach; it is a process of learning and acquiring skills and expertise using experience in order to increase knowledge as a teaching pedagogy (M. Abdulwahed, 2010 & J. M. Applefield, R. Huber, and M. Moallem, 2001). Students learn by consciously communicating with and experiencing their environment directly through actual experience compared to theoretical learning. Kolb Professor of Organizational Behavior at Case Western Reserve University's Weatherhead School of Management describes teaching "Learning is the method by which knowledge is produced as the conversion of experience," so knowledge outcomes from the mixture of theory and

Figure 4. Objectives of the lab-based learning

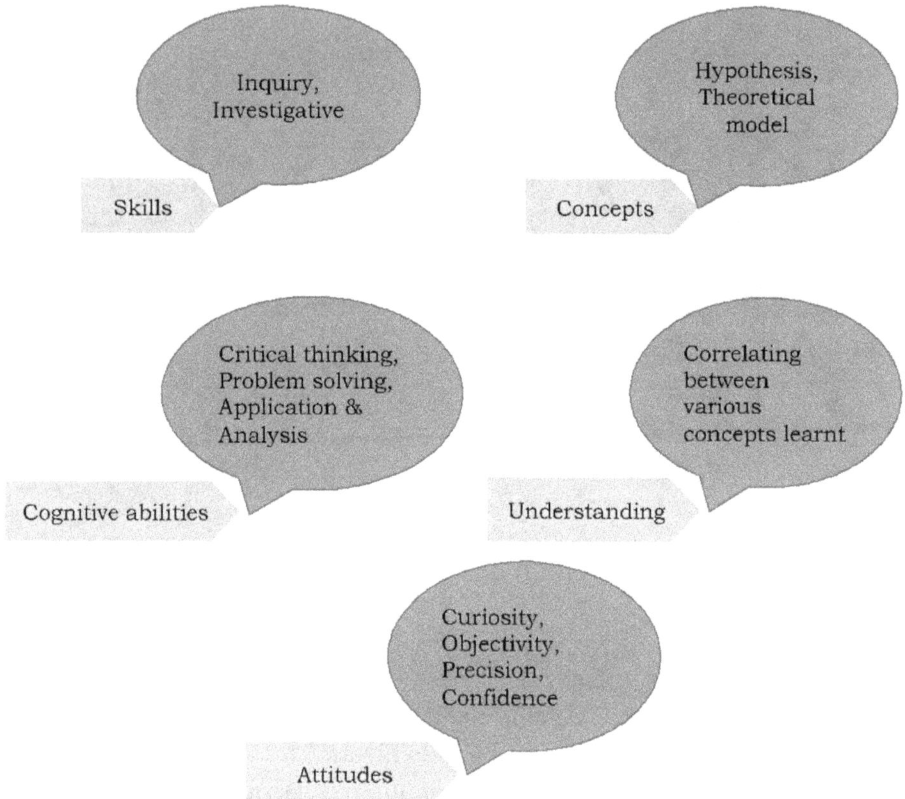

practice (D. A. Kolb, 1984).Laboratories in academic institutions are great settings for teaching and learning engineering courses. They provide students with great prospects to think about, discuss, and solve real time problems.

The five different objectives shown in figure 4 can be achieved by using LBL.

In shri Vishnu Engineering College for Women (SVECW), LBL is practiced for digital electronics course, digital electronics course is thought using physical laboratory available in college and the virtual laboratory available in web. Student after learning concepts in classroom they verify the designed logic circuit in the lab by using Logic IC trainer and verify the output for different Boolean functions, combinational sequential circuits e.t.c. Student have possibility in checking the output for different combinations of inputs and also logic design. Virtual labs are an initiative of ministry of human resource development under the National Mission on Education. Digital electronics virtual lab is designed by IIT Kharagpur, where students will perform various experiments virtually for better theoretical understanding of the course.

ASSESMENT OF TWO TEACHING METHODOLOGIES

This methodology evaluation was scheduled differently from a conventional methodology. Rather than the finished product, the professor wished to evaluate the learning process. At the end of each practical session, the professor is accountable for practical meetings prepared a document on each student. Continuous evaluation was therefore carried out. Students give a detailed explanation of the document. The concept of encouraging oral expression and public participation in the final meeting was also performed an experimentation analysis and validation of the advanced job. They described their solution to the professors and responded with the concept of evaluating the design method they had introduced to answer their questions. Additional competencies were also evaluated in the presentation and defense of the job to promote oral expression and present in public. The assessment may vary depending on the issue, the strategy they have chosen, and the students ' job performed. The evaluation is considered as the internal evaluation (D. R. Woods, (1994) & Moursund D, 1999).

STUDENT SURVEYS

The professor performs a study at the end of the course to ask learners about the implemented PBL methodology in this topic. Figure 5 shows the outcomes of this study. The primary subjects of this study were the development of the primary student abilities that the lecturing team wanted to enhance through this experience, the improvement of other abilities and a general evaluation.

The sample questions of the survey are as follows:

- Evaluate your capacity to solve power electronics issues
- Evaluate your capacity to apply class-acquired understanding
- Evaluate your decision making enhancement.
- Evaluate your capacity to search and assimilate your own data.
- Evaluate your other skills enhancement.
- Evaluate your satisfaction with the job you have accomplished.
- Evaluate the subject's scheduling time

Figure 5. Student survey results

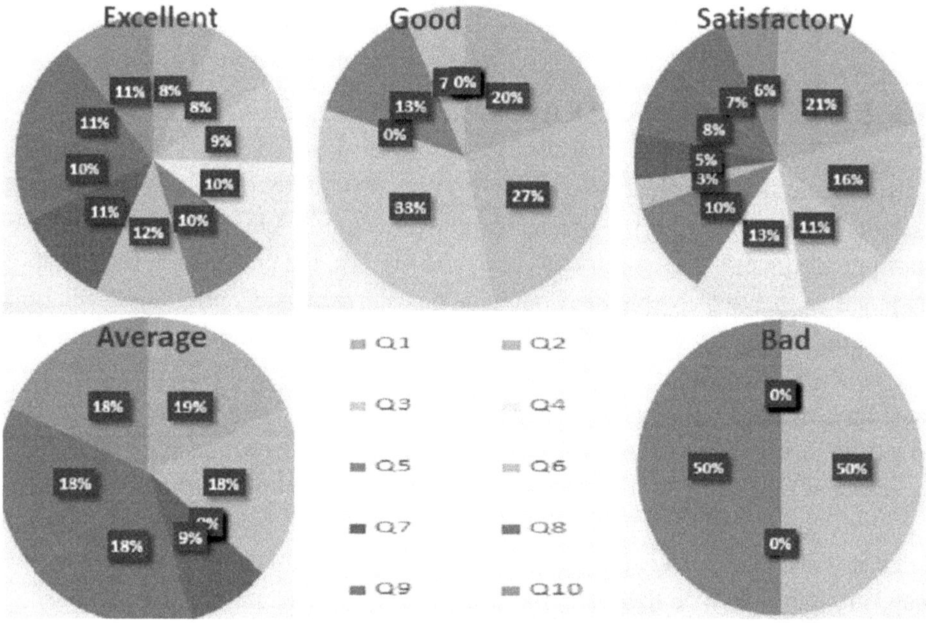

Figure 5 shows the feedback of the students through questionnaire, from the results it is understand that students positively received the change in teaching methodologies and build in confidence in gathering the technical information through many sources to take the decision in solving the practical issues over theoretical concepts. Indeed, the students ‘ view can be summarized in a remark by the students' voice:

Student 1: Encouraging the competitive spirit and teamwork to inculcate interest among the peers

Student 2: Exemplifying day to day examples and small assignments in relation to the course and build the confidence to solve the practical issues.

Student 3: Creates more interest and enthusiasm towards the course by using new teaching aids in classroom

DISCUSSION AND BENEFITS

Students ‘ feedback was positive and empowering. Numerous claimed that after using distinct learning methodologies, they had better understand of the courses. Students grouping was discovered to have a beneficial impact in learning from the study. Most learners liked working in teams and from perception throughout the course, learners who were allocated individual duties or assignments also decided to speak to other student colleagues and speak to each other to complete their assignments.

They enjoyed working exclusively for fast learning students as they could work faster, and with the help of the instructor, they could endeavor more confused assignments in the light of their enthusiasm. The teacher needs to learn about student abilities to make the topic more efficient and interactive for all learners and should recognize the quick learning and slow learning students, set up additional tasks with a greater difficulty level for quick learning learners who have completed their tasks early. The assignment level should be minimized for slow learning learners and the professor should observe it more commonly. While the topic data provided by hands-on methodology should be comparable to that provided by traditional lecture-based instructional programs, hands-on methodology varied in two important respects:

1. Students should effectively take part in their own particular instruction, with the accentuation being on learning.
2. Member's prompt hands-on practice that takes after a theoretical lecture will give sensible representation to that new information and urge them to end up self-directed learners.
3. Students can design their own circuits with digital controller kits.
4. students feel better in understanding the various power electronic circuits and also can design different converters by using discrete components.
5. by introducing various teaching methodologies students are inspired for latter classes
6. students are confident enough to perform lab Viva and to face interviewer.
7. PBL and LBL widen the thinking capabilities of the student towards technical concepts.

Finally, the teaching and learning process is evaluated based on the performance of the students in semester end examinations. After efficiently implementing different learning methodologies in power electronics and digital electronics course for 2018-19 academic year, the outcomes are contrasted with prior learners in the 2017-18 academic year and the outcomes are discovered to be high. Figure 6 and 7 showcases the results of two different subjects in two academic years and it tells that result has been improved by implement project based learning and lab based learning.

Figure 6. Result analysis of power electronics course

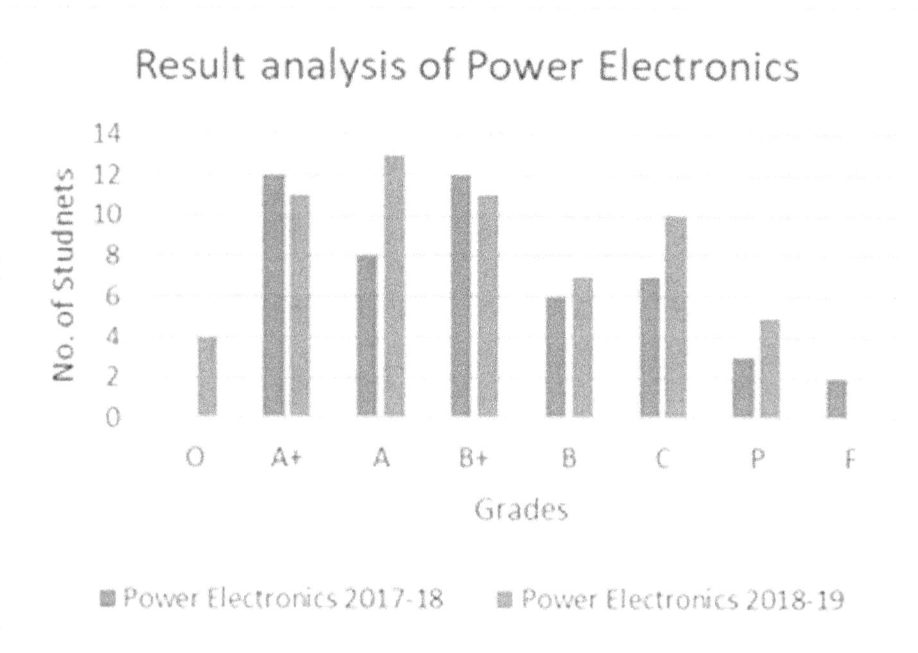

Figure 7. Result analysis of the digital electronics course

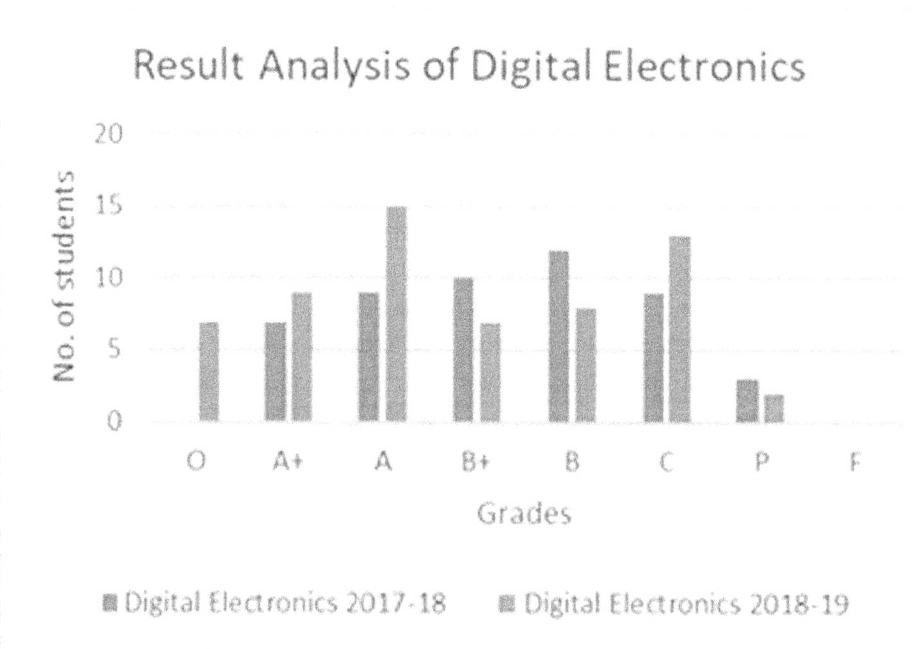

CONCLUSION

Many engineering students ' learning styles and many teachers of engineering teaching styles do not always match. Visual, inductive and sequential learners are many engineering students. On the other side, many teachers of engineering and engineering programs are auditory, logical and spontaneous. This contributes to the instructor's foiling and the student's unplanned study. By adopting different learning styles in the class room instructor may attract the students towards the lecture and make the class more interactive. By using different technology based tools students will gain more practically information than theoretical and also they concentrate what instructor is going to say in the next class then also it is easy for them to solve the assignments and in doing projects.

ACKNOWLEDGMENT

Authors wish to thank Shri Vishnu Engineering College for Women, and VEDIC for supporting in preparing this paper

REFERENCES

Abeysekera, L., & Dawson, P. (2015). Motivation and cognitive load in the flipped classroom: Definition, rationale and a call for research. *Higher Education Research & Development*, *34*(1), 1–14. doi:10.1080/07294360.2014.934336

Bersin, J. (2004). How Did We Get Here? The History of Blended Learning. In *The Blended Learning Book: Best Practices, Proven Methodologies, and Lessons Learned*. Wiley.

Bonk, C. J., & Graham, C. R. (2006). *The handbook of blended learning environments: Global perspectives, local designs*. San Francisco: Jossey-Bass/Pfeiffer.

Bruce, B. C., & Hogan, M. C. (1998). The disappearance of technology: Toward an ecological model of literacy. In D. Reinking, M. McKenna, L. Labbo, & R. Kieffer (Eds.), Handbook of literacy and technology: Transformations in a post-typographic world (pp. 269–281). Academic Press.

Carr. (2012). Project-based learning: increasing social studies interest and engagement. *Rising Tide, 5*.

Cashin, W. E. (1995). Student Ratings of Teaching: The Research Revisited. IDEA Paper No. 32. Kansas State Univ.

Dormido, S., Dormido, S., Dormido, R., S'anchez, J., & Duro, N. (2005). The role of interactivity in control learning. *International Journal of Engineering Education*, *21*(6), 1122–1133.

Eggen, P. D., & Kauchak, D. P. (2006). *Strategies and Models for Teachers Teaching Content and Thinking Skills* (5th ed.). Pearson Education Inc.

Ertmer, P. A. (2005). Teacher pedagogical beliefs: The final frontier in our quest for technology integration. *Educational Technology Research and Development*, *53*(4), 25–39. doi:10.1007/BF02504683

Fed'ak, V., & Bauer, P. (2005). E-learning in education of electrical drives and power electronics: Opportunities and challenges. *International Conference on Electrical Drives and Power Electronics*.

Hurley, W. G., & Lee, C. K. (2005). Development, implementation, and assessment of a web-based power electronics laboratory. *IEEE Transactions on Education*, *48*(4), 567–573. doi:10.1109/TE.2005.856147

Karady, G. G., & Heydt, G. T. (2000). Increasing student interest and comprehension in power engineering at the graduate and undergraduate levels. *IEEE Transactions on Power Systems*, *15*(1), 16–21. doi:10.1109/59.852095

Kimonen, E., & Nevalainen, R. (2000). Active learning in the process of educational change. *Teaching and Teacher Education*, *21*(6), 623–635. doi:10.1016/j.tate.2005.05.003

Lohmann, H. (1954). The Technique and its Teaching-A research sub -program for a scientific engineering pedagogy. *Scientific Journal of the TH Dresden*, *3*(4), 602–629.

Martyn, M. (2003). The hybrid online model: Good practice. *EDUCAUSE Quarterly*, 18–23.

Mergendoller, J. (2006). *Project Based Learning Handbook* (2nd ed.). Novatto, CA: Buck Institute for Education.

Mishra, P., Spiro, R. J., & Feltovich, P. J. (1996). Technology, representation, and cognition: The prefiguring of knowledge in cognitive flexibility hypertexts. In H. van Oostendorp & A. de Mul (Eds.), *Cognitive aspects of electronic text processing* (pp. 287–305). Norwood, NJ: Ablex.

Ogbu. (2012). Benchmark on the Teaching Effectiveness of Basic Electricity Teachers in South Eastern Nigeria. European Scientific Journal, 8(22), 299-326.

Ronchetti. (2010). Using video lectures to make teaching more interactive. *International Journal of Emerging Technologies in Learning.*

S'anchez, J., Dormido, S., & Esquembre, F. (2004). The learning of control concepts using interactive tools. *Computer Applications in Engineering Education, 13*(1), 84–98. doi:10.1002/cae.20033

Simon, H. (1969). *Sciences of the artificial.* Cambridge, MA: MIT Press.

Spiro, R. J., & Jehng, J.-Ch. (1990). Cognitive flexibility and hypertext: Theory and technology for the nonlinear and mutlidimensional traversal of complex subject matter. In D. Nix & R. Spiro (Eds.), Cognition, education, and multimedia: Exploring ideas in high technology (pp. 163–204). Academic Press.

Topp, G. (2011, Oct. 6). Flipped classrooms take advantage of technology. *USA Today.*

Undeland, M., Panaitescu, R. C., Jain, A. K., Jose, P., & Begalke, T. (2003). Restructuring of first courses in power electronics and electric drives that integrates digital control. IEEE Transactions on Power Electronics, 18(1), 429–437.

Weltman, M. W., & Whiteside, M. (2010). Comparing the Effectiveness of Traditional and ActiveLearning Methods in Business Statistics: Convergence to the Mean. *Journal of Statistics Education: An International Journal on the Teaching and Learning of Statistics, 18*(1). doi:10.1080/10691898.2010.11889480

ADDITIONAL READING

Abdulwahed, M. (2010). *Towards Enhancing Laboratory Education by the Development and Evaluation of the 'Trilab' -- a Triple Access Mode (Virtual, Hands-On and Remote).* Laboratory.

Applefield, J. M., Huber, R., & Moallem, M. (2001). Constructivism in Practice and Theory: Toward a Better Understanding. *High School Journal, 84*(2), 35–53.

Moursund, D. (1999). *Project based-learning: using information technology.* Iste publications.

Woods, D. R. (1994). *Problem-based learning: how to gain the most from PB.* Mcmaster University.

Chapter 2
Teaching and Learning:
Pedagogy With Purpose

Archana Dixit
Institute of Applied Science and Humanities, GLA University, Mathura, India

Kuldeep Kumar Saxena
Institute of Engineering and Technology, GLA University, Mathura, India

Jitendra Kumar Dixit
Institute of Business and Management, GLA University, Mathura, India

ABSTRACT

The chapter focuses on learning styles and how learning styles came about in education. There are many significant things to know about why educators should understand the learning style of a student. The learning method was one of the significant elements of human existence. Student learning styles are among the variables that have been obtained. The primary aim of this research was to explore the study of teaching style among learners with academic accomplishment in engineering. The research requires a theoretical approach to reviewing appropriate literature on the subject and presents different points of perspective on matching or mismatching learning styles. The conclusion of the chapter is how to learn the styles of teaching and learning impressively and effectively, and how an individual understands new information so that they can find the most efficient technique to accumulate, interpret, and present information.

DOI: 10.4018/978-1-7998-2245-5.ch002

INTRODUCTION

Teaching and learning pedagogy plays significant role in education, whether it is in engineering, management, medical or any other educational domain. The sole purpose of teaching is to instruct knowledge, extend understanding and transmit skills. This is also defined as a process of inculcating moral etiquettes, capabilities, skills in order to make sure positive transformation in actions functional in developing oneself and the society. It shapes one's thinking and action through giving directions and performing practices that construct a novel behavior and competence. Teaching is an art of disseminating knowledge and understanding that can be achieved through consistent and regular practices.

In Engineering, "teaching" can be defined as "the ideas of a particular person or group, especially about principles, concept and their application in real life, society and religion. These ideas are taught to learners and other people to make their understanding related to realistic application.Teaching is about sparking one's inquisitiveness so that one may care about asking critical questions to gain knowledge of something. It is the act of steering students/ learners into being imaginative scholar's as students are priceless individuals away from their warehouse of specifics (The Oxford Learner's Dictionaries, 2019).

Teacher - Student: Two Sides of the Same Coin

An individual can image himself as a passionate teacher in technical education if he carries integrity and professional skill and whose first priority is to learn and correct self, pupils and colleagues every day (Day and Christopher, 2004).Trust, self-motivation and enthusiasm can be nurtured by these individual's. In teaching, compassion and care are essential features to remain connected with students and colleagues. Teachers and students work well when they are cared 'about': an expression of teachers' personal beliefs and emotional commitment that goes beyond the contractual obligation of caring. It is right to say that teaching is the most effective and impressive form of imparting knowledge to others. Teachers should be enthusiastic in saying that nothing is impossible if students believe in hard work and move in the right direction. Teacher must be problem solvers all the way throughout their lives, so teaching has to engross the expansion of personal perseverance otherwise the confused person will be liable to just give up when the going gets coarse.

Education is widely recognized to be at the heart of international development (Vrasidas, Charalambos et al. 2009). There are multiple challenges in achieving educational quality and that will always be in future too but to achieve educational development goals meaningfully,consistent and continuous effort are need from the teacher and institution both (Brinkhurst et al., 2011; Moore, 2005; Scott et al., 2012).

Teaching and Learning Style: A Conceptual Inception

Learning styles play a very important role in a learner's life. When a student is able to identify his learning style, it will help him to learn fast. As a result, students' confidence will increase and teachers' work load to have control over the class will lessen. Teaching styles adopted by teachers must match with students' learning styles. Matching between learning styles and teaching styles provide a healthy class room environment to both teachers and learners. Thus, it becomes clear that learning is more influential and successful when teachers are aware of why and how their students imagine and learn.

Undoubtedly, learners and teachers are different in various manners but generally, it is supposed that understanding of students' preferences and learning style can be advantageous for both students and teachers (O'Brien, 2013; Courtenay-Hall & Rogers, 2002). Students gain knowledge through various modes in their different phases of life, so it is not easy to modify style of learning of the students in the class. To be more consistent, teachers are required to adopt and customized their learning style according to the categories of the students. In addition, it is recommended that teachers should obtain a balanced approach to teaching styles so that students may make themselves comfortable with various learning styles.

Teaching –Learning: Way to Impart Knowledge

Teaching is a process of imparting knowledge. It's a two way communication process, where teacher is at sender end and students is at receiver end. Teaching will be successful when both the teacher and the student take participation in interactions actively. A student's learning style refers to his preferential way in which he comprehends the information. For example, a student can get the knowledge of building a bridge by his teacher's verbal instruction while the other needs the physical manipulation of building the bridge.

During teaching a lesson, teacher's are required to include different learning styles as many as possible to make the learning more effective and students must feel themselves engaged (Allen, Kelli et al. rightly, 2010). Application of multiple learning activities not only engages the students, but also facilitates the teacher in achieving learning objectives. It helps the students to know and understand the world. It is teaching where a teacher holds the hand of his students and says them to move in an unknown territory as he is with them at every step. Students believe in his teacher's words and proceed in their lives with confidence. A teacher can never forget that he can change his students' lives in a better way and their students may build a better society to live in.

Learning styles leaves impact on the students and provide a way to gain knowledge. It is very important for a teacher to know the student learning styles so that the effectiveness of student learning can be improved. It is well said that students are the future of the nation. Thus, a teacher's task becomes big. He has to make not only his student' future but have a responsibility of building a strong foundation for the nation too. Teaching is not only a job or career or a profession. It is a passion with the sense of responsibility of educating the society and making the student's employable (William, 2001).

Educational institutions are facing new challenges in the present era. The institutions have become business-driven entities. Its result can be seen easily as the students have become clients and teachers have become service providers for the institutions. Hence, teaching – learning process is hampered. Teachers are not paying attention to students to know their learning methodology. They are just providing the service and in return, they stand for their remuneration. It has affected the overall educational structure very badly. Lectures and learning styles are frequently conducted in a way that does not go with the students' requirements. The effectiveness of teaching and productivity of learner's can be improved through self-evaluation by the teachers for themselves and through sharing the outcomes and experience with colleagues and students (Sims et al.1995).Through this instructors can understand what inhibits, frustrates, or promotes learning. Knowledge of learning styles can assist teachers or any individuals in becoming more effective and flexible in teaching, both inside and outside of the classroom.

A change is required in teaching – learning pedagogy to make it progressive and object oriented in near future. It can only be achieved when teachers are aware of students' learning styles and ready to perform their duties in a well-systematic manner. A teacher should always be a learner to know the world and ready to adopt the new challenges of the society. With the passage of time, the ways of teaching are required to be updated. According to Christopher's (2004) point of view, teaching and teacher is a term that distinguishes those who are dedicated, caring and who took their job seriously than those who put their own interests first. Commitment towards their profession gives satisfaction to some teachers, while others treat like a burden and try to escape from their responsibilities saying that teaching is more absorbing and could take over their lives. These thoughts often place limits for teachers on their commitment as a means of survival. In these cases, it is preferable to leave the profession altogether. Commitment of a teacher is directly associated with their satisfaction towards job, motivation, morale and identity. These factors are predictor of a teachers' work performance, absenteeism, burn-out, and turn over and have important influence on students' achievement and attitudes toward the teacher and institution.

Teaching and Learning: In Engineering Education

Students can learn in many ways — listening and seeing, acting and reflecting, intuitively and reasoning, visualizing and memorizing, constructing mathematical models and drawing analogies. Teaching techniques differs and can be customized as per the audiences (Sipos et al., 2008). The student and the professor both suffer when there is a mismatch in the teaching methodology and become disaster in case of technical education where more focus is on experiments and lab rather than theory only.

Richard M. Felder (1988) attempted to show what they have accomplished to match the students learning styles with the teaching style so that the experience is useful for the both professor and the student. He looked at the following issues to do this.

1. Which elements of teaching style are particularly important in engineering education?
2. Which styles of learning are preferred by most learners and which are favoured by the teaching styles of most teachers?
3. What can be achieved to reach learners whose learning styles are not addressed by conventional engineering education techniques?

Thus, the study of teaching and learning styles clearly exhibits that both plays a crucial and significant role not only in building up the career of a student but also in the progress of a nation. They will help the teachers to know their students and support them in achieving their goals of life. Teachers can identify how and when their students are ready to learn and the knowledge of teaching and learning styles to help them. The conclusion of understanding teaching and learning styles is to impressively and effectively learn how an individual understands new information so that they may find out the most efficient technique to accumulate, interpret, and present information.

Technology: A Change Agent in Teaching and Learning

If there is a talk about teaching methods and methodology, it is essential to include the use of technology in the education system. Technology is highly used outside of education but its use in the teaching and interactions will increase the interest of students also. They will pay more attention to their teachers' instructions and lectures. ICT (Information and Communication Technology) plays a significant role in the education system today. It provides a number of creative activities in which students take part willingly and prove themselves better than their previous image in the classroom (McWhirter & Shealy, 2018). Technology is one of the most effective tools to increase the students' awareness in any of their favorite fields. ICT has played substantial role by offering new teaching-learning environment and provoked new educational paradigms (Nikolova, 2001). Teachers are required to adapt these changes, even more to work-out as a "change agents" and helping in shaping new institutional/educationalculture. The use of ICT in education adds worth to teaching – learning process as it increases the effectiveness of classrooms. It makes the teachers' lectures impressive and adds a height to learning that was not earlier accessible. It will increase the confidence of the students at global level and provide them a number of opportunities for getting their livelihood. Definitely, the role of ICT in education is inescapable. As the world is changing with the advancement of technologies, this change indicates that the role of ICT in future will develop enormously in the education system.

ICT also pays its focus to the role of teachers in the classrooms. They will proceed as practical torchbearers for students who want to use electronic media. Ultimately, the exercise of ICT will improve the learning abilities of students. It will help them to integrate with the external world, environment, industry and society and motivates them think freely and communicate innovatively (Brown and Lambert, 2013; Hughes et al., 2013; van der Leeuw et al., 2012). It also makes students capable for building successful careers and lives, in aprogressively technological world.Education is about learning – social, cognitive, emotional, moral, spiritual and physical –teachers must consider the best resources, experiences and environments in which they are required to be supported in all ways (Loveless and Avril, 2003). ICT has supported in inducing these changes in teaching and learning pedagogy. The potential role of ICT is to enhance cognitive learning and develop problem-solving skills with the extension of physical and mental abilities as a new tool used in all ways. Teachers are supposed to explore the ways through which ICT affects the observation, learning and experiences of children at a social and practical level not only through the study material but also through games, source of entertainment and simulation exercises in their day-to-day life.

In today's scenario, technology is a major component of the classroom teaching and it seems that it will go on to be for years to get nearer. However, technology must be utilized and adapted in a manner that will be advantageous to all students. Technology has changed the teaching-learning methods and methodologies. It has made the students life easier through integrating with the external world .Students are more updated about the changing trends, new technologies, industries demand from prospect employees and many more. Technology has changed the educational system. Teachers' are required to make the teaching more innovative, creative and effective through customizing their teaching styles as one solution fit to all philosophy will not works longer. The sole objective of teaching is to prepare students to match with the changing pace of industrial, economical and societal environment for ensuring their survival, growth and achievement of goal of their life. The abilities of students to transmit the information what they have learned to new circumstances gives an index of adaptive and flexible learning.

REFERENCES

Allen, K., Sheve, J., & Nieter, V. (2010). *Understanding Learning Styles: Making a Difference for Diverse Learners*. Shell Education.

Brinkhurst, M., Rose, P., Maurice, G., & Ackerman, J. D. (2011). Achieving campus sustainability: Top-down, bottom-up, or neither? *International Journal of Sustainability in Higher Education*, *12*(4), 338–354. doi:10.1108/14676371111168269

Brown, V. A., & Lambert, J. A. (2013). *Collective Learning for Transformational Change: A Guide to Collaborative Action*. New York, NY: Routledge.

Courtenay-Hall, P., & Rogers, L. (2002). Gaps in mind: Problems in environmental Knowledge-behavior modeling research. *Environmental Education Research*, *8*(3), 283–297. doi:10.1080/13504620220145438

Day, C. (2004). *A Passion for Teaching*". Psychology Press. doi:10.4324/9780203464342

Felder. (1988). Learning and Teaching Styles: In Engineering Education. Energy Education, 78(7), 674–681.

Hughes, T. P., Carpenter, S., Rockström, J., Scheffer, M., & Walker, B. (2013). Multiscale regime shifts and planetary boundaries. *Trends in Ecology & Evolution*, *28*(7), 389–395. doi:10.1016/j.tree.2013.05.019 PMID:23769417

Loveless, A. (2003). *Role of ICT*. A&C Black.

McWhirter, N. D., & Shealy, T. (2018). Teaching decision-making for sustainable Infrastructure: A wind energy case study module. *International Journal of Sustainability in Higher Education*, *19*(5), 893–911. doi:10.1108/IJSHE-10-2017-0183

Moore, J. (2005). Seven recommendations for creating sustainability education at the university level: A guide for change agents. *International Journal of Sustainability in Higher Education*, *6*(4), 326–339. doi:10.1108/14676370510623829

Nikolova, I. (2001). Teacher Development in ICT: Vision and Implementation. In *Information and Communication Technologies in Education, The School of the Future. IFIP TC3/WG3.1 International Conference on The Bookmark of the School of the Future*. Springer Science & Business Media. 10.1007/978-0-387-35403-3_6

O'Brien, K. (2013). Global environmental change III: Closing the gap between knowledge and action. *Progress in Human Geography*, *37*(4), 587–596. doi:10.1177/0309132512469589

Scott, G., Tilbury, D., Sharp, L., & Deane, E. (2012). Turnaround leadership for sustainability in higher education. Final report for the Australian Government Office for Learning and Teaching, Commonwealth of Australia.

Sims, R. R., & Sims, S. J. (1995). *The Importance of Learning Styles: Understanding the Implications for Learning, Course Design, and Education*. Greenwood Publishing Group.

Sipos, Y., Battisti, B., & Grimm, K. (2008). Achieving transformative sustainability learning: Engaging head, hands and heart. *International Journal of Sustainability in Higher Education*, *9*(1), 68–86. doi:10.1108/14676370810842193

Teaching Oxford Learner's Dictionaries. (n.d.). Retrieved from oxfordlearnersdictionaries.com/definition/english/teaching

Van der Leeuw, S., Wiek, A., Harlow, J., & Buizer, J. (2012). How much time do we have? Urgency and rhetoric in sustainability science. *Sustainability Science*, *7*(S1), 115–120. doi:10.100711625-011-0153-1

Vrasidas, Charalambos, Zembylas, Michalinos, & Glass. (2009). *ICT for Education, Development, and Social Justice*. IAP.

William. (2001). *To Teach: The Journey of a Teacher*. Teachers College Press.

Chapter 3
The Knowledge and Learning Potential of Outsourcing

Ingi Runar Edvardsson
https://orcid.org/0000-0002-1167-3994
University of Iceland, Iceland

Susanne Durst
Department of Business Administration, Tallinn University of Technology, Estonia

ABSTRACT

This chapter consists of a review of 42 refereed articles on outsourcing in relation to knowledge management and learning. Among the knowledge and learning potentials of outsourcing in organisations are to focus on core competencies, organisational learning, shorten production cycles, improve quality, and enhance innovation. Outsourcing can also have negative outcomes, such as knowledge loss, competences drainage, organisational learning problems, diminished trust, poorer services, hidden cost, and reduction in innovation. The findings of the review were summarised in a number of hypotheses and two conceptual models that highlight the contribution of outsourcing to either competitive advantage or disadvantage. The chapter can assist managers to seek competitive advantages out of outsourcing of activities while avoiding detrimental outsourcing effects.

INTRODUCTION

Outsourcing of manufacturing and service activities is an accelerating trend (Bustinza *et al.*, 2010; Iqbal and Dad, 2013; Kumar *et al.*, 2010). Outsourcing usually refers to a process of transferring certain activities of a firm to third party providers within

DOI: 10.4018/978-1-7998-2245-5.ch003

a country or offshore location (Iqbal and Dad, 2013). According to the literature, frequent rationality for outsourcing is cost reduction, such as labour costs, less-restrictive work rules abroad, transport, travel, reduced capital expenditure, cheaper raw materials, and other costs. This also refers to indirect costs which might be decreased, for example by reducing staff, simplifying infrastructure, and applying stricter cost control (see Kremic *et al.*, 2006; Lacity *et al.*, 2008; Di Gregorio *et al.*, 2009; Quinn, 1999; van Gorp *et al.*, 2007). Cost driven outsourcing is relevant to transaction cost economics (Williamson, 1979) which focuses on the costs related to monitoring performance, managing contractual obligations, and supervising staff. As a result of specialization and economy of scale, outsourcing can earn a company sufficient savings to perform business functions more economically than other companies carrying out the same functions.

Another reason for outsourcing is more strategic in nature. Here the focus is on the core competencies of organisations other firms have difficulty in copying. This direction is due to the intense competition which forces organisations to reassess and redirect their scarce resources towards their core business functions in order to enable them to expand successfully to worldwide reach. The emphasis on a strategy-driven outsourcing model is linked to the resource-based view of the firm. This examines those resources and capabilities of companies that enable them to generate above-normal rates of return, as well as a sustainable competitive advantage (Barney, 2001). Rare and precious resources, not easily copied, are regarded as core competencies. Such resources should be supported, while others are considered more suitable for outsourcing (Barney, 1991). Quinn (1999) maintains that these critical resources or capabilities tend to be knowledge-based service activities or systems at which the company excels and which are of particular importance to the customers. Another widely cited rationale for outsourcing is to gain access to unique resources, skills, talents, and capabilities possessed by other firms, such as the latest technology and infrastructure. Greater flexibility in managing demand swings, and reducing company risk by sharing it with suppliers are also mentioned as motives for outsourcing (see Kremic *et al.*, 2006; Di Gregorio *et al.*, 2009; Lacity *et al.*, 2008; Quinn, 1999; Tam *et al.*, 2007; Vietor and Veytsman, 2005).

The decisions behind outsourcing tend to change the value chain, production processes, and organisational design. Alterations in the production process or the organisation of a business, in which other actors take part, can result in both new challenges and novel opportunities. Although the outsourcing process has been extensively studied (see Edvardsson and Durst, 2014; Iqbal and Dad, 2013; Lacity *et al.*, 2011; Williams and Durst, 2018), there is still only a limited body of knowledge on the consequences of outsourcing on corporate knowledge management and organisational learning.

The main focus of knowledge management research has been on processes within and across organisations, such as knowledge creation and knowledge transfer, development of a suitable organisational culture and learning, as well as on technologies for knowledge storage and sharing in order to enhance a firm's productivity and sales, lower cost, or increase innovation and quality and thus competitiveness (Durst and Edvardsson, 2012; Edvardsson, 2011; Jashapara, 2011; Kluge *et al.*, 2001; O'Dell *et al.*, 2003). The classification of knowledge into tacit and explicit knowledge (Polanyi, 2009) may be helpful in the context of knowledge transfer, as the specific nature of the two types of knowledge is likely to influence the success of the transfer process (Grant, 1996). With regard to the contribution of different types of knowledge to the competitive advantage of firms, it is usually tacit knowledge that is emphasized, since it resides in the minds of individuals (Currie *et al.,* 2008). Consequently, with a view of the nature of outsourcing, it can have a substantial impact on knowledge and its management within organisations. In the process of outsourcing activities, a great deal of tacit knowledge has to be converted into explicit knowledge in order to transfer the business function to the vendor. This gives rise to the need for new rules, routines and procedures to manage the outsourced activity (Yakhlef, 2009). Moreover, if tacit knowledge is outsourced, the outsourcing relationship becomes more difficult to manage, requires greater interaction between the parties and results in higher costs. Additional challenges also tend to arise during the transfer of tacit knowledge, making the overall process difficult to control (Bustinza *et al.,* 2010).

Finally, Lam and Chua (2009) argue for knowledge outsourcing as a promising alternative strategy of knowledge management. In particular, they recommend knowledge outsourcing when in-house expertise is lacking, external knowledge providers are available, and such outsourcing is beneficial from an economic perspective.

Core competencies comprise the collective learning in an organisation (Kumar *et al.,* 2010). According to Barney (1991), the building blocks of organisational core competencies are, on the one hand, human capital such as training, experience, judgement, intelligence, relationships, and insights of individual managers and workers. On the other hand, those building blocks include the organisational capital that comprises formal reporting structure, formal and informal planning, control and coordination systems and information relations. Firms that outsource their core competencies might lose touch with their technological, market and organisational knowhow which offers opportunities for product and process innovation (Martínez-Sánchez *et al.*, 2007). Thus, outsourcing undermines the knowledge base of companies (Xia and Tang, 2011). As a consequence, many firms in the US and Europe "no longer retained the ability to react to change or even to fully understand their own markets and business" due to their downsizing and outsourcing in the 1980s and 1990s (Quintas, 2002, p. 5).

In this context, it is of interest to look at knowledge-based firms (KBFs) and their outsourcing patterns. In most cases, KBFs produce an intangible output, rather than a tangible product. Consequently, as these firms often produce a service, further characteristics are that 1) consumption and production happen simultaneously, and 2) the customer is highly involved in the production of services, which makes a crucial difference between KBFs and manufacturing companies. Finally, the production of services tends to be more labour and knowledge intensive than the production of manufacturing goods (Daft, 2007; Targowski, 2009). All these aspects, and, in particular, the strong involvement of customers, are highly relevant to understand the motives and realisation of outsourcing in service firms.

Lee (1999) argues that KBFs tend to outsource their non-core processes in order to concentrate on developing their capabilities in their core competencies. Based on a theoretical analysis, Lee reasons that KBFs will outsource only those work activities, or HR requirements, that focus on narrow, explicit knowledge. Other work activities will be kept in-house. Additionally, Lee argues that KBFs will focus 1) on training and development activities addressing those work processes or HR requirements that require broad, explicit knowledge, 2) on informal, group-oriented activities for work processes or HR requirements that require narrow, tacit knowledge, and 3) on research and development activities for work processes or HR requirements that require broad, tacit knowledge. In a similar vein, Currie et al. (2008) concluded from their case study of knowledge process outsourcing in financial services that most of these firms tend to outsource only activities with more defined inputs and outputs. Also here, the more knowledge intensive forms of work are kept in-house. However, they found that some high-end financial services, e.g., financial analytics, portfolio valuation, or risk assessment, were outsourced, but in quite a different manner depending on the company. Thus, while many large organisations established branches of their operation in low-cost countries to carry out the relevant knowledge work, pure play KPOs (primarily start-up firms) are involved in securing and managing customers in developed countries, whose processes will be handed over to offshore service centres. Existing Business Process outsourcers have extended their services into the specialized market of knowledge process outsourcing, whereas professional knowledge taskers provide products and services to specific professional and functional groups, which requires them to become niche-market players, such as in health-care. Finally, knowledge builders focus on the low end of the knowledge process, such as seeking and gathering information as well as analysing it. Murray et al. (2009) developed a theoretical framework for a global sourcing strategy relating to knowledge-intensive business services (KIBS). These authors claim that the more variation in the services' standards or nature, the higher the inseparability of production and consumption, and the more tacit the activity of KIBS, the more they tend to rely on onshore insourcing; that is, to keep their services in-house. However, Murray et al. (2009) also assert that the more innovative the sourcing firms, the more they rely on outsourcing from best-in-class suppliers worldwide.

It is, therefore, evident that outsourcing, seen from the standpoints of knowledge management and learning, is an important topic for both academic research and organisations alike. As will be outlined later in the chapter, a limited number of articles have been written on the consequences of outsourcing on knowledge management and learning. Hence, there is a gap in our understanding of the subject. The aim of this chapter is twofold: Firstly, the purpose is to present a review of the opportunities and risks of outsourcing in relation to learning and organisational knowledge. Thereby the paper seeks to answer the following research questions:

1. How does outsourcing affect knowledge within and between the organisations involved (tacit-explicit, internal-external)?
2. How is organisational learning affected by outsourcing?
3. How does outsourcing affect the organisation's capabilities of maintaining its core competency?
4. How does outsourcing affect the innovation potential of the organisations (client and vendor)?
5. How do the knowledge and learning that outsourcing brings contribute to the competitive advantage of client organisations?

Secondly, the chapter intends, based on the literature review, to contribute to the theory of outsourcing by proposing a number of hypotheses and two conceptual models on the competitive advantages and disadvantages of outsourcing. One of the contributing factors of this study is to name and explain the benefits and challenges of outsourcing from the point of view of learning and knowledge management.

The chapter is structured as follows: The next section presents the methodology of the literature review, the results are presented in the following section, and the paper ends on conclusions.

METHODOLOGY

The methodology of the paper is based on a traditional literature review (Jesson *et al.*, 2011). Originally, the authors had decided to do a systematic literature review and looked for special keywords found in the literature related to the topic of interest such as knowledge externalisation, access to external knowledge, learning, risk related to outsourcing, knowledge loss, unlearning, competency drainage, product development, and lack of control. For this, the databases ABI/Inform and Web of Science were used. This procedure resulted in very few papers. Consequently, the authors turned to the approach of a traditional review and also included prior literature reviews; e.g., a literature review on knowledge process outsourcing (Edvardsson and Durst, 2014), a review on outsourcing trends (Iqbal and Dad, 2013), a review on business process outsourcing (Lacity *et al.*, 2011), and a critical review on outsourcing (Kakabadse and Kakabadse, 2000).

In total 42 papers were identified (see Table 1). The papers were then read and thematically grouped by the authors. As a next step, the authors discussed their findings, synthesised them and assigned them to two broad themes: advantages and disadvantages of outsourcing. As Table 1 indicates many themes related to these two topics were identified.

RESULTS

Among the 42 papers that formed the basis for our analysis, the earliest publication is from 1999 and the most recent ones are from 2018. Most of the papers are of recent origin, which underlines that the topic continues to be of significant interest to researchers. The papers were published in general management, knowledge management, and supply chain management journals.

In the sections below, we present our analysis according to the themes identified. The analysis is divided into learning and knowledge potentials (in the sense of advantages) of outsourcing and disadvantages of outsourcing regarding learning and knowledge.

This is followed by a list of future research directions (i.e., a number of hypotheses and two conceptual models).

Knowledge and Learning Potentials

Focus on Core Competencies

Strategy driven outsourcing is related to enhancing the core competencies of firms. Many studies have stressed that the outsourcing of non-core processes is applied in order to focus more on the organisations´ core competencies (see Edvardsson and Teitsdottir, 2015; Iqbal and Dad, 2013; Kakabadse and Kakabadse, 2005; Kremic *et al.*, 2006). This suggests that firms should specialise in certain knowledge areas that eventually generate innovation, better customer services, and organisational knowhow to meet environmental changes (Quinn, 1999). Core competencies, however,

Table 1. Outsourcing effect on knowledge and learning in organisations

Outsourcing effect	Studies
Advantages	
Focus on core competencies	Edvardsson et al. (2011) Edvardsson & Teitsdottir (2015) Kremic et al. (2006) Kakabadse & Kakabadse (2005) Mehta et al. (2006) Quinn (1999)
Organisational learning	
Knowledge externalisation	Edvardsson et al. (2011) Gottschalk & Solli-Sæther (2007) Yakhlef (2009)
Access to external knowledge/global talent	Bustinza et al. (2010) Di Gregorio et al. (2009) Edvardsson et al. (2011) Ghodeswar & Vaidyanathan (2008) Kremic et al. (2006) Luvison & Bendixen (2010) Sharma & Loh (2009) Simon et al. (2009)
Organisational learning enhances Innovation	Presbitero, Roxas & Chadee (2017)
Shorter product/process design cycle time	Kakabadse & Kakabadse (2005)
Increased innovation	Corral de Zubielqui et al. (2015)
Increased innovativeness when standardised services are outsourced	Martínez-Noya & García-Canal (2015)
Quality improvement	Iqbal & Dad (2013) Braun et al. (2011) Gupa et al. (2005) Quinn (1999)
Disadvantages	
Knowledge loss/loss of critical skills	Agndal & Nordin (2009) Claver et al. (2002) Kakumanu & Portanova (2006) Kakabadse & Kakabadse (2000) Lowman et al. (2012) Quintas (2002) Hanafizadeh & Ravasan (2018)
Loss of cross-functional skills	Kakabadse & Kakabadse (2000)
Loss of corporate memory	Kakabadse & Kakabadse (2000)
Role of time on knowledge at risk	Williams & Durst (2018)
Competency drainage	Kotabe, Mol & Ketkar (2008) Rundquist & Haila (2008)
Organisational learning problems	Schlosser et al. (2006) Becker & Zirpoli (2017) Martínez-Noya & García-Canal (2015) Williams and Durst (2018)
Reduction in trust	Kumar et al. (2010) Qi & Chau (2013) Lee et al. (2008)
Increase in cost/hidden cost	Barbieri et al. (2018) El Fadil & St-Pierre (2016) Becker & Zirpoli (2017) Hanafizadeh & Ravasan (2018)
Poor service/manufacturing quality	El Fadil & St-Pierre (2016) Gorla & Lau (2010) Kumar et al. (2010) Young (2008) Becker & Zirpoli (2017)
Higher turnover rate	Kumar et al. (2010)
Decreased innovation	Kakabadse & Kakabadse (2000) Lacity et al. (2011) Mahmoodzadeh et al. (2009) Miozzo & Grishaw (2005)

are difficult to define and protect, and they can change over time (Iqbal and Bad, 2013). Independent of the attitude toward them, both knowledge management and organisational learning are considered as central in this respect.

Organisational Learning

Outsourcing can enhance organisational learning through access to *external knowledge, increased understanding of cost and processes,* and *externalisation of tacit knowledge.* Bustinza *et al.* (2010) referred to access to external knowledge as an organisational learning potential of outsourcing. Know-how can also be gained through collaborative relationships between partners in the outsourcing relationship. Furthermore, Luvison and Bendixen (2010, p. 34) point out the learning potential of outsourcing:

In other words, functions moved outside the locus of the firm create the need to learn complex behaviors within the locus of the firm. This second paradox of outsourcing therefore implies that outsourcing requires firms to learn and employ behaviors that they do no typically possess... Such routines are often tacit, requiring repeated exchanges that are highly dependent on communication and relationship factors.

Former studies have shown that outsourcing has been used as a tool to access top class knowledge. For example, in their study on outsourcing of new product development among medium sized firms in four Swedish industries, Rundquist and Halila (2010) found a pattern among best performing firms. These companies chose outsourcing partners based on the quality of their technology knowledge. A study by Howells *et al.* (2008) listed, amongst other things, access to expertise which is not available in the firm as the main reason for R&D outsourcing. Gaining access to global talents in low cost countries is another reason (Ghodeswar and Vaidyanathan, 2006; Shama and Loh, 2009). Edvardsson *et al.* (2011) showed that also SMEs (Icelandic service firms) highlighted access to external knowledge as one of the most prominent motives for outsourcing. Pointing in the same direction, Di Gregorio *et al.* (2009) argue that gaining access to external knowledge is one of the benefits of international offshoring for SMEs.

The outsourcing process can be a means of inter-organisational learning initiated through the externalisation of knowledge. This can apply to all partners involved. Yakhlef (2009), for example, argues that IT outsourcing implies the externalization of IT activities – these are activities which are learned over a number of years, often through learning by doing, and are thus fairly tacit and not fully documented in an organisation. When these activities are outsourced, the firms are forced to codify and transform the knowledge underlying those activities into explicitly communicable specifications. The firms' dealings with their suppliers would lead to intense (face-to-face, telephone, and document based) interactions and exchanges of information from different sources. From the present perspective, externalization would subsequently result in the internalization of knowledge, or organisational learning, which takes the form of organisational rules, routines, and procedures, becoming part and parcel of firms' corporate memory. Yakhlef (2009) also emphasises that outsourcing leads to new knowledge regarding cost and processes within firms. Similarly, three case studies on IT outsourcing showed that both vendor and service organisations discovered the importance of knowledge transfer after the implementation of the IT outsourcing arrangement (Gottschalk and Solli-Sæther, 2007). In their study on information technology service providers in the Philippines, Presbitero et al. (2017) show that knowledge sharing capability is positively related to innovation and that organisational learning capability fully mediates the effects of knowledge sharing on innovation. Becker and Zirpoli (2017) stress that learning and knowledge sharing do not exist automatically but require certain conditions, e.g. possibilities for exchange.

Shorter Product/Process Design Cycle Time

Outsourcing can decrease the product/process design cycle time when best-in-class suppliers are involved in the process. In that way, each supplier can provide more sophisticated knowledge in specialised fields, and, accordingly, offer far better inputs than individual suppliers (Kakabadse and Kakabadse, 2000).

Quality Improvement

In their review on outsourcing trends, Iqbal and Dad (2013) mentioned that quality improvement is one effect of outsourcing. Accordingly, Braun *et al.* (2011) show that human resource services are generally of a higher standard when provided by outside vendors compared to in-house services. In a similar manner, Gupa *et al.* (2005) argue that improvements in service quality and staff are among the motivations for outsourcing in higher education institutions in three states in the USA. Quinn (1999), moreover, emphasises that outsourcing is considered to improve the organisations' quality management through cooperation with best-in-class suppliers.

Increased Innovativeness

In the modern knowledge economy, innovation is essential for firms. In their study on South-Australian SMEs, Corral de Zubielqui et al. (2016) found that despite the fact that very few SMEs source knowledge from localised learning networks, such as universities and research institutions, regression analyses indicate that the pattern of outsourcing R&D to these networks increased innovativeness. Martínez-Noya & García-Canal (2015) found out in their study that a firm's innovativeness increases when standardized services are outsourced. Yet, this is not the case when the transfer of tacit and firm-specific knowledge is required.

Knowledge and Learning Disadvantages

Knowledge Loss: Loss of Critical Skills

Agndal and Nordin (2009) point out the negative consequences firms are likely to experience, for example, that the ability to perform the outsourced function is partly or fully lost and/or, as a consequence of interdependencies between functions, the outsourcing can also negatively affect the performance of other related business capacities. Their findings clarify that a firm has to pass through a learning process before suitable procedures become available in order to handle negative consequences related to outsourcing. Hanafizadeh and Ravasan (2018) revealed in their study on outsourcing decision in e-banking services in Iran, that the perceived loss of organisational knowledge was negatively related to the IT outsourcing decisions. Similarly, Lowman *et al.* (2012) highlight the negative consequences of outsourcing. In the context of the pharmaceutical industry, their study's findings indicate the necessity of maintaining integrative mechanisms that ensure absorptive capacity, knowledge diffusion, and interactive learning. Claver *et al.* (2002) observed that the loss of critical skills was seen as one of the negative aspects of IT outsourcing in Spanish universities. Kakabadse and Kakabadse (2000) suggest, in addition to the loss of critical skills, the loss of cross-functional skills and the impairment of company memory as possible further negative consequences. And, as already outlined in the introduction, Quintas (2002) argues that outsourcing can mean reduced knowledge of markets and customers. Still another risk of knowledge loss is through higher turnover rates associated with outsourcing (Kumar *et al.*, 2008). Studying knowledge at risk over the period in which knowledge-intensive work is 'handed over' from a client firm to a vendor firm, Williams and Durst (2018) showed that „the knowledge at risk varies according to the amount of codified knowledge that the client attempts to pass over to the vendor in a given time period, relative to the vendor's readiness and ability to absorb and utilize this knowledge as the client intended“ (pp. 9-10).

Competency Drainage

In their longitudinal analysis of three major consumer electronics manufacturers (Emerson Radion in the US, Sony in Japan and Philips in Europe), Kotabe *et al.* (2008) found that outsourcing and offshoring lead to competency destruction which they coin “the vicious outsourcing cycle”. The destruction of core competencies is manifested, among other things, in the fact that Emerson and Philips chose to exit from the consumer industries and thereby admit that the competency loss is too large to overcome, while Sony has increased in-house production. This loss can be related to a sacrifice of intellectual property rights to suppliers; the suppliers can compete with the outsourcing firms, with a resulting loss of knowledge as the firms no longer engage in learning-by doing. In a similar manner, Rundquist and Halila (2010) found strategic decisions from the board and risk of competency drainage as main reasons not to outsource.

Organisational Learning Problems

The central values of organisational learning are a shared vision and experience, commitment to learning and open-mindedness (Schlosser *et al.*, 2006). Schlosser *et al.* (2006) further argue that outsourcing weakens organisational learning as its temporary nature undermines the traditions and routines associated with strong organisational culture. This is especially the case with the interpretation of a shared perspective and using established communication channels and routines. However, outsourcing can improve learning from the experience of others, particularly from highly specialised firms. Martínez-Noya & García-Canal (2015) provided empirical evidence why firms are reluctant in sharing information in strategic alliances/ outsourcing activities and as a consequence let them follow strategies that reduce optimal learning outcomes. With regard to the outsourcing of innovation tasks, Becker and Zirpoli (2017) stressed that there is a “risk of losing the learning opportunities required for maintaining integrative and monitoring capabilities that are needed in integrating the inputs of external parties (p. 27). Williams and Durst (2018) showed that during the transition period in information system offshore outsourcing, the learning in both the client and vendor teams will not remain constant.

Reduction in Trust

Trust is essential for knowledge sharing, especially of complex knowledge (Hislop, 2013; Lee *et al.* 2008; Qi and Chau, 2013). Kumar *et al.* (2010) point out that outsourcing tends to lead to a struggle for trust and power with new leadership. Qi and Chau (2013) stress that many failures of IT outsourcing are related to trust issues or relationship building between the client and the provider. Their findings show that interpersonal trust is more important for knowledge sharing and outsourcing success than inter-organisational trust. Lee *et al.* (2008) found that mutual trust between partners is essential for knowledge sharing and outsourcing success.

Increase in Cost/Hidden Cost

While cost reduction is often the aim of offshoring and outsourcing, El Fadil and St-Pierre (2016) found out in their study of manufacturing offshoring to China, that cost was far higher than anticipated in the beginning. Also, Hanafizadeh and Ravasan (2018) came to the conclusion that perceived cost was an obstacle for e-banking. In their literature review on reshoring of manufacturing activities, Barbieri et al. (2018) notice that many studies identify miscalculation of the actual cost as a reason for resourcing. In the same vain, Becker and Zirpoli (2017) stressed that the negative consequences of outsourcing might produce additional costs.

Poor Service/Manufacturing Quality

In her analysis of the failure of large outsourcing projects in the public health sector in Australia, Young (2008) concludes that the outsourcing contract did reduce the quality of services. It is, however, unclear whether this was due to unclear specifications by the hospital, or a result of the low contract price. Examples of poor quality are inadequate cleanliness of the hospital, poor service quality, poor food quality, etc. In addition, communication problems arose between patients, their families, and staff wards after low skilled workers were transferred to the vendor. Kumar *et al.* (2010) also mention unreliable quality as a further risk related to outsourcing, and primarily to offshoring; that is, to low cost locations. Gorla and Lau (2010) observed that the negative experience of IT outsourcing can affect re-outsourcing decisions. Their findings indicate that problems with vendor competence in relation to infrastructure and vendor coordination have a direct effect on outsourcing decision making. El Fadil and St-Pierre (2016) findings on offshoring show that products made in China can be defective or non-compliant, and product delays are common. Becker and Zirpoli (2017), by using Fiat for showing the consequences of R&D outsourcing, showed that extreme outsourcing resulted in not meeting the desired quality levels.

Diminished Innovation

Evidence suggests that innovation, as organised through outsourcing, may be reduced due to the conflict between the short-term cost-based nature of the contract and the flexibility and freedom required for innovation (Kakabadse and Kakabadse, 2000; Miozzo and Grishaw, 2005). In their extended review, Lacity *et al.* (2011) point out that business process outsourcers (BPO) are increasingly expecting their BPO suppliers to innovate as the latter have the core knowledge in the field. Furthermore, the low prices charged for outsourcing seem to produce another challenge regarding the provision of new solutions and technologies. Mahmoodzadeh *et al.* (2009), who studied outsourcing in the context of business process management and knowledge management, found there is a risk of firms losing touch with new technological opportunities for product and process innovations.

FUTURE RESEARCH AVENUES

As outlined above, the second aim of this chapter is to contribute to the theory of outsourcing in relation to knowledge management and learning. In order to do so, we put forward hypotheses to be tested in further research and we propose two conceptual models which highlight the advantages and disadvantages of outsourcing.

Competitive Advantage of Outsourcing

In the findings presented above, some papers have pointed out the advantages of outsourcing for knowledge and learning. Based on this, we propose the following hypotheses:

H1: Strategically selected vendors facilitate a company's focus on core competencies.
H2: Knowledge specialisation with regard to core competencies increases the quality of services and products.
H3: Knowledge specialisation with regard to core competencies increase incremental innovation in key business area(s).
H4: Highly competitive vendors provide access to critical external expert knowledge.
H5: Through the combination of documentation and visualisation outsourcing can bring about the transfer of tacit knowledge to explicit knowledge.
H6: An emphasis on core competencies, access to critical external knowledge, new technologies and innovation lead to quality improvements as regards processes and products.
H7: Outsourcing helps reduce product/process design cycle time.
H8: Innovation increases when standardised services are outsourced.

H1 to H8 contribute to competitive advantage through outsourcing.

From these hypotheses, we have drawn up a conceptual model of the competitive advantage of outsourcing in relation to knowledge creation and learning (Figure 1).

Figure 1. Competitive advantage through outsourcing

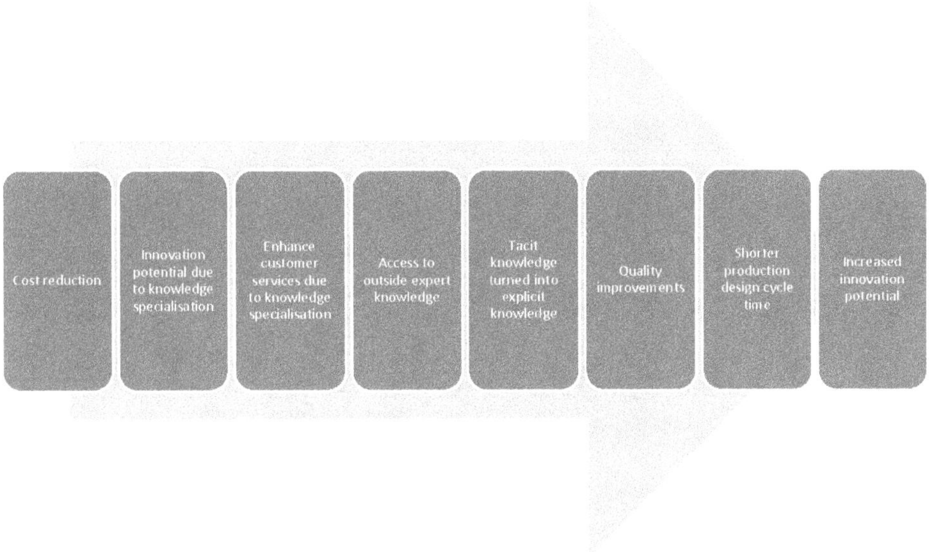

Disadvantages of Outsourcing

The majority of articles cited in this paper have addressed the dangers of outsourcing in relation to knowledge and learning. We propose the following hypotheses related to the findings:

H9: Outsourcing leads to the loss/attrition of knowledge and skills in relation to design, production, and markets.
H10: Outsourcing leads to increased staff turnover.
H11: Outsourcing leads to competency drainage.
H12: Outsourcing decreases organisational learning.
H13: Outsourcing reduces trust in organisations, thus making knowledge sharing even more complicated.
H14: Outsourcing limits innovation through short contracts and low-cost strategies.
H15: Outsourcing leads to hidden costs.
H16: H9 to H15 contribute to poorer services and products.
H17: H9 to H15 contribute to a competitive disadvantage through outsourcing.

As before, from these hypotheses, we have drawn up a conceptual model of competitive disadvantage through outsourcing in relation to knowledge and learning (see Figure 2).

Figure 2. Competitive disadvantages of outsourcing

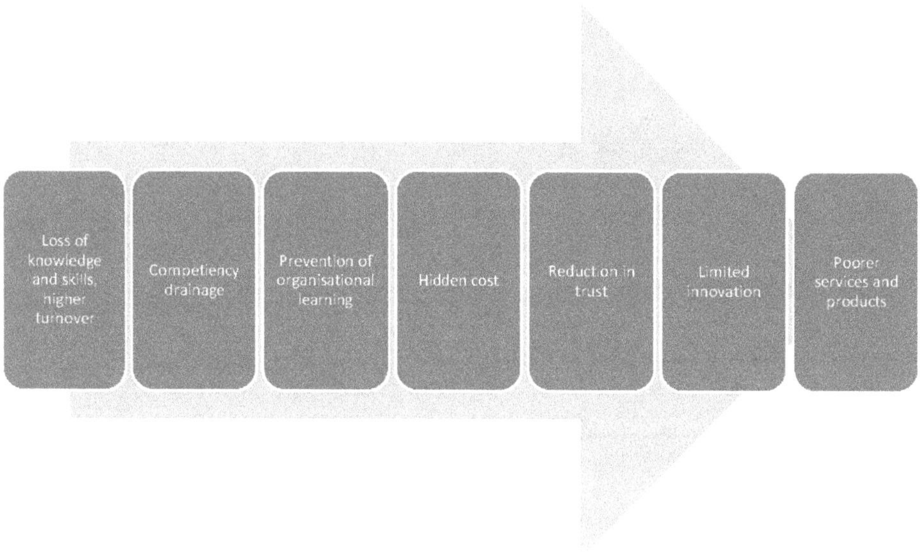

CONCLUSION

In this chapter, we have reviewed articles on the relationship between outsourcing and knowledge and learning. In total, 42 articles were identified and analysed. The focus has been on knowledge in and between organisations involved in outsourcing, organisational learning, innovation, core competency of organisations and their competitive advantage. We identified two main themes in the literature: First, competitive advantages of outsourcing in relation to knowledge creation and learning; and second, competitive disadvantages of outsourcing. The former tends to emphasise that through outsourcing firms are likely to focus on their core competencies. This gives them the opportunity to specialise in key knowledge areas which also provides access to best-in-class knowledge, that can enhance innovation, improve customer services and shorten product design cycle time. These findings are closely related to theories of core competencies and resource-based theory of the firm (Barney, 2001; Quinn, 1999), and mainstream outsourcing research (Kremic *et al.*, 2006; Iqbal and Dad, 2013; Di Gregorio *et al.*, 2009). These research paradigms share the characteristic of being economic/operational in nature and looking upon cost, specialisation and inter-organisational cooperation as a means to gain competitive advantages.

According to the literature reviewed, outsourcing has many disadvantages relating to knowledge and learning. The most important defects are potential loss/attrition of knowledge and skills in relation to design, production, and markets, as well as competency drainage. Organisational learning can also be affected in a negative manner, as well as innovation, even leading to poorer services and products, and increase in cost. The research into disadvantages of outsourcing is critical in nature (Agndal and Nordin, 2009; Kotabe *et al.*, 2008; Young, 2008), and relates to knowledge management (Durst and Edvardsson, 2014; Jashapara, 2011; Polanyi, 2009), and organisational learning studies (Ludvison and Bendixen, 2010, Schlosser *et al.*, 2006). The topics of analysis in the critical studies, knowledge management literature and organisational learning research are the social, cultural and organisational factors affecting knowledge sharing and learning within and between organisations. Hence, they tend to point out critical areas of managing intangible assets.

One of the contributing factors of this study is naming and explaining the benefits and challenges of outsourcing from the point of view of learning and knowledge management. By presenting 17 hypotheses and two conceptual models we have attempted to present a holistic view of outsourcing in relation to knowledge management and learning. In that respect, the model incorporates both economic/operational and socio-cultural aspects of knowledge and learning associated with outsourcing.

This review highlights that there still is a need for more research about the link between outsourcing, knowledge, and learning. Despite recent research, however, suggests the relevance of taking into account the trade-off between short- and long-term advantages of outsourcing. The former being access to new competencies and knowledge while the latter addressing the risk of missed learning opportunities or knowledge decay (Becker & and Zirpoli, 2017). We, therefore, call for further research in this area and urge researchers to test our hypotheses. In the global economy, strategic outsourcing can enhance the competitiveness of organisations; at the same time, opportunistic outsourcing can be a disaster where the loss of knowledge and competency and, in the worst scenario, the closing down of the company can be the end result.

The paper has theoretical and practical implications. As regards the theory of outsourcing there is a gap in our understanding; this concerns, in particular, the disadvantages of outsourcing. More empirical research is therefore needed in order to enhance our understanding of the actual benefits of outsourcing in manufacturing and service firms.

As regards practitioners (e.g., managers), the paper provides useful insights as it points out both the advantages and disadvantages of outsourcing. Given the former, managers should take notice of the elements that can enhance the competitive advantages of the firm when developing outsourcing strategies. At the same time, and thus addressing the latter, managers also have to consider the many pitfalls associated with outsourcing, such as loss of knowledge and learning that can eventually lead to poorer performance.

Outsourcing represents an industry many countries have specialized in, for example, China, India, Costa Rica or the Czech Republic; thus the findings of our paper also have implications for society. Having a better understanding of the advantages and disadvantages of outsourcing can help policy makers contribute to the development of best practice recommendations, thereby supporting entire industries, as well as individual companies and their staff.

This paper is not without limitations. It is likely that we have missed some articles in our search, even though our effort covered all the leading journals in the field. In sum, we believe that the findings presented in this paper contribute to the theory of outsourcing by taking into consideration the topics of learning and knowledge management. Nevertheless, further research is needed to test the quality of the review.

REFERENCES

Agndal, H., & Nordin, F. (2009). Consequences of outsourcing for organizational capabilities: Some experiences from best practice. *Benchmarking: An International Journal, 16*(3), 316–334. doi:10.1108/14635770910961353

Barbieri, P., Ciabuschi, F., Fratocchi, F., & Vignoli, M. (2018). What do we know about manufacturing reshoring? *Journal of Global Operations and Strategic Sourcing, 11*(1), 79–122. doi:10.1108/JGOSS-02-2017-0004

Barney, J. B. (1991). Firm resources and sustained competitive advantage. *Journal of Management, 17*(1), 99–110. doi:10.1177/014920639101700108

Barney, J. B. (2001). Resource-based theories of competitive advantage: A ten-year retrospective on the resource-based view. *Journal of Management, 27*(6), 643–650. doi:10.1177/014920630102700602

Becker, M. C., & Zirpoli, F. (2017). How to Avoid Innovation Competence Loss in R&D Outsourcing. *California Management Review, 59*(2), 24–44. doi:10.1177/0008125617697941

Braun, I., Pull, K., Alewell, D., Störmer, S., & Thommes, K. (2011). HR outsourcing and service quality: Theoretical framework and empirical evidence. *Personnel Review, 40*(3), 364–382. doi:10.1108/00483481111118667

Bustinza, O. F., Molina, L. M., & Gutierrez-Guiterrez, L. J. (2010). Outsourcing as seen from the perspective of knowledge management. *Supply Chain Management, 46*(3), 23–39. doi:10.1111/j.1745-493X.2010.03196.x

Claver, E., González, R., Gascó, J., & Llopis, J. (2002). Information systems outsourcing: Reasons, reservations and success factors. *Logistics Information Management, 15*(4), 294–308. doi:10.1108/09576050210436138

Corral de Zubielqui, G., Jones, J., & Statsenko, L. (2015). Managing Innovation Networks for Knowledge Mobility and Appropriability: A Complexity Perspective. *Entrepreneurship Research Journal, 6*(1), 75–109.

Currie, V. L., Michell, W., & Abanishe, O. (2008). Knowledge process outsourcing in financial services: The vendor perspective. *European Management Journal, 26*(2), 94–104. doi:10.1016/j.emj.2007.11.002

Daft, R. L. (2007). *Understanding the Theory and Design of Organizations.* Mason, OH: Thompson South-Western Publishing.

Di Gregorio, D., Musteen, M., & Thomas, D. E. (2009). Offshore outsourcing as a source of international competitiveness for SMEs. *Journal of International Business Studies, 40*(4), 969–988. doi:10.1057/jibs.2008.90

Durst, S., & Edvardsson, I. R. (2012). Knowledge management in SMEs: A literature review. *Journal of Knowledge Management, 16*(6), 879–903. doi:10.1108/13673271211276173

Edvardsson, I. R. (2011). The impact of outsourcing strategies on companies' intellectual capital. *Knowledge Management Research and Practice, 9*(4), 286–292. doi:10.1057/kmrp.2011.33

Edvardsson, I. R., & Durst, S. (2014). Outsourcing of knowledge processes: A literature review. *Journal of Knowledge Management, 18*(4), 795–811. doi:10.1108/JKM-01-2014-0033

Edvardsson, I. R., & Teitsdóttir, U. D. (2015). Outsourcing and financial crisis: Evidence from Icelandic service SMEs. *Employee Relations, 37*(1), 30–47. doi:10.1108/ER-11-2013-0168

El Fadil, J., & St-Pierre, J. (2016). Analysis and management of risks associated with outsourcing in China: The experience of seven Canadian firms. *Strategic Outsourcing, 9*(2), 218–242. doi:10.1108/SO-08-2015-0018

Ghodeswar, B., & Vaidyanathan, J. (2008). Business process outsourcing: An approach to gain access to world-class capabilities. *Business Process Management Journal, 14*(1), 23–28. doi:10.1108/14637150810849382

Gorla, N., & Lau, M. B. (2010). Will negative experience impact future IT outsourcing? *Journal of Computer Information Systems, 50*(3), 91–101.

Gottschalk, P., & Solli-Sæther, H. (2007). Knowledge transfer in IT outsourcing relationships: Three international case studies. *International Journal of Innovation and Learning, 4*(2), 103–111. doi:10.1504/IJIL.2007.011687

Grant, R.M. (1996). Towards a knowledge-based theory of the firm. *Strategic Management Journal, 17*, 109-122.

Gupta, A., Herath, S. K., & Mikouiza, N. C. (2005). Outsourcing in higher education: An empirical examination. *International Journal of Educational Management, 19*(4), 396–412. doi:10.1108/09513540510607734

Hanafizadeh, P., & Ravasan, A. Z. (2018). An empirical analysis on outsourcing decision: The case of e-banking services. *Journal of Enterprise Information Management, 31*(1), 146–172. doi:10.1108/JEIM-11-2016-0182

Hislop, D. (2013). *Knowledge Management in Organizations: A Critical Introduction* (3rd ed.). Oxford, UK: Oxford University Press.

Howells, J., Gagliardi, D., & Malik, K. (2008). The growth and management of R&D outsourcing: Evidence from UK pharmaceuticals. *R & D Management, 38*(2), 205–219. doi:10.1111/j.1467-9310.2008.00508.x

Iqbal, Z., & Dad, A. M. (2013). Outsourcing: A review of trends. Winners & losers and future directions. *International Journal of Business and Social Science, 4*(8), 91–107.

Jashapara, A. (2004). *Knowledge management: An integrated approach*. Harlow: Prentice Hall.

Jesson, J. K., Matheson, L., & Lacey, F. M. (2011). *Doing Your Literature Review: Traditional and Systematic Techniques*. Los Angeles, CA: Sage.

Kakabadse, A., & Kakabadse, N. (2000). Critical review – outsourcing: A paradigm shift. *Journal of Management Development, 19*(8), 670–728. doi:10.1108/02621710010377508

Kakabadse, A., & Kakabadse, N. (2005). Outsourcing: Current and future trends. *Thunderbird International Business Review, 47*(2), 183–204. doi:10.1002/tie.20048

Kakumanu, P., & Portanova, A. (2006). Outsourcing: Its benefits, drawbacks and other related issues. *The Journal of American Academy of Business, Cambridge*, *9*(2), 1–8.

Kluge, J., Wolfram, S. & Licht, T. (2001). *Knowledge Unplugged. The McKinsey & Company Global Survey on Knowledge Management*. Houndsmills: Palgrave.

Kotabe, M., Mol, M. J., & Ketkar, S. (2008). An evolutionary stage model of outsourcing and competence destruction: A triad comparison of the consumer electronics industry. *Management International Review*, *48*(1), 65–93. doi:10.100711575-008-0004-1

Kremic, T., Tukel, I. G., & Rom, O. W. (2006). Outsourcing decision support: A survey of benefits, risks, and decision factors. *Supply Chain Management*, *11*(6), 467–482. doi:10.1108/13598540610703864

Kumar, S., Deivasigamani, A. J., & Omer, W. W. (2010). Knowledge for sale – the benefits and effects of off-shoring knowledge-based jobs in engineering, design, and R&D – a case study. *Knowledge Management Research and Practice*, *8*(4), 351–368. doi:10.1057/kmrp.2010.24

Lacity, M., Solomon, S., Yan, A., & Willcocks, L. (2011). Business process outsourcing studies: A critical review and research directions. *Journal of Information Technology*, *26*(4), 221–258. doi:10.1057/jit.2011.25

Lam, W., & Chua, A. Y. K. (2009). Knowledge outsourcing: An alternative strategy for knowledge management. *Journal of Knowledge Management*, *13*(3), 28–43. doi:10.1108/13673270910962851

Lee, J., Huynh, M. Q., & Hirschheim, R. (2008). An integrative model of trust on IT outsourcing: Examining a bilateral perspective. *Information Systems Frontiers*, *10*(2), 145–163. doi:10.100710796-008-9066-7

Lee, S.-H. (1999). Taxonomy of HRM policies of knowledge-based firms: A resource-based explanation. *International Journal of Innovation Management*, *3*(4), 379–395. doi:10.1142/S1363919699000190

Lowman, M., Trott, P., Hoecht, A., & Sellam, Z. (2012). Innovation risks of outsourcing in pharmaceutical new product development. *Technovation*, *32*(2), 99–109. doi:10.1016/j.technovation.2011.11.004

Luvison, D., & Bendixen, M. (2010). The behavioral consequences of outsourcing: Looking through the lens of paradox. *The Journal of Applied Management and Entrepreneurship*, *15*(4), 28–52.

Mahmoodzadeh, E., Jalalinia, S., & Nekui Yazdi, F. (2009). A business process outsourcing framework based on business process management and knowledge management. *Business Process Management Journal, 15*(6), 845–864. doi:10.1108/14637150911003748

Martínez-Noya, A., & García-Canal, E. (2015). The framing of knowledge transfers to shared R&D suppliers and its impact on innovation performance: A regulatory focus perspective. *R & D Management, 46*(2), 354–368. doi:10.1111/radm.12191

Mehta, A., Armenakis, A., Mehta, N., & Irani, F. (2006). Challenges and opportunities of business process outsourcing in India. *Journal of Labor Research, 27*(3), 324–338. doi:10.100712122-006-1026-8

Miozzo, M., & Grimshaw, D. (2005). Modularity and innovation in knowledge-intensive business services: IT outsourcing in Germany and the UK. *Research Policy, 34*(9), 1419–1439. doi:10.1016/j.respol.2005.06.005

Murray, J. Y., Kotabe, M., & Westjohn, S. A. (2009). Global sourcing strategy and performance of knowledge-intensive business services: A two-stage strategic fit model. *Journal of International Marketing, 17*(4), 90–105. doi:10.1509/jimk.17.4.90

O'Dell, C., Hubert, C., Lopez, K., Odem, P., & Raybourn, C. (2003). Successful KM implementations: a study of best-practice organizations. In C. W. Holsapple (Ed.), *Handbook on Knowledge Management* (Vol. 2, pp. 411–441). Berlin: Springer. doi:10.1007/978-3-540-24748-7_19

Polanyi, M. (2009). *The Tacit Dimension*. Chicago: University of Chicago Press.

Presbitero, A., Roxas, H., & Chadee, D. (2017). Sustaining innovation of information technology service providers: Focus on the role of organisational collectivism. *International Journal of Physical Distribution & Logistics Management, 47*(2-3), 156–174. doi:10.1108/IJPDLM-07-2015-0171

Qi, C., & Chau, P. Y. K. (2013). Investigating the roles of interpersonal and interorganizational trust in IT outsourcing success. *Information Technology & People, 26*(2), 120–145. doi:10.1108/ITP-09-2012-0088

Quinn, J. B. (1999). Strategic outsourcing: Leveraging knowledge capabilities. *Sloan Management Review, 40*(4), 9–21.

Quintas, P. (2002). Managing knowledge in a new century. In S. Littler, P. Quintas, & T. Ray (Eds.), *Managing Knowledge: An essential reader* (pp. 1–14). London: The Open University of Sage Publications.

Rundquist, J., & Halila, F. (2010). Outsourcing of NPD activities: A best practice approach. *European Journal of Innovation Management, 13*(1), 5–23. doi:10.1108/14601061011013203

Schlosser, F., Templer, A., & Ghanam, D. (2006). How human resource outsourcing affects organizational learning in the knowledge economy. *Journal of Labor Research, 27*(3), 291–303. doi:10.100712122-006-1024-x

Sharma, A., & Loh, P. (2009). Emerging trends in sourcing of business services. *Business Process Management Journal, 15*(2), 149–165. doi:10.1108/14637150910949425

Simon, J. C., Poston, R. S., & Kettinger, B. (2009). Creating better governance of offshore services. *Information Systems Management, 26*(2), 110–122. doi:10.1080/10580530902794778

Tam, F. Y., Moon, K. L., Ng, S. F., & Hui, C. L. (2007). Production sourcing strategies and buyer-supplier relationships: A study of the differences between small and large enterprises in the Hong Kong clothing industry. *Journal of Fashion Marketing and Management: An International Journal, 11*(2), 297–306. doi:10.1108/13612020710751446

Targowski, A. (2009). The architecture of service systems as the framework for the definition of service science scope. *International Journal of Information Systems in the Service Sector, 1*(1), 54–77. doi:10.4018/jisss.2009010104

van Gorp, D., Jagersma, K. P., & Livshits, A. (2007). Offshore behaviour of service firms: Policy implications for firms and nations. *Journal of Information Technology Case and Application Research, 9*(1), 7–19. doi:10.1080/15228053.2007.10856104

Vietor, R.H.K., & Veytsman, A. (2005). *American outsourcing*. Harvard Business School Case 705-037.

Williams, C., & Durst, S. (2018). Exploring the Transition Phase in Offshore Outsourcing: Decision Making amidst Knowledge at Risk. *Journal of Business Research*. doi:10.1016/j.jbusres.2018.01.013

Williamson, O. (1979). Transaction-cost economics: The governance of contractual relations. *The Journal of Law & Economics, 22*(2), 233–261. doi:10.1086/466942

Xia, Y., & Tang, T. L.-P. (2011). Sustainability in supply chain management: Suggestions for the auto industry. *Management Decision, 49*(4), 495–512. doi:10.1108/00251741111126459

Yakhlef, A. (2009). Outsourcing as a mode of organizational learning. *Strategic Outsourcing, 2*(1), 37–53. doi:10.1108/17538290910935882

Young, S. (2008). Outsourcing in public health: A case study of contract failure and its aftermath. *Journal of Health Organization and Management, 22*(5), 446–464. doi:10.1108/14777260810898688 PMID:18959298

Chapter 4
Assessment Practices in Outcome-Based Education:
Evaluation Drives Education

Lavanya C.
Gokaraju Rangaraju Institute of Engineering and Technology, India

Jandhyala N. Murthy
Gokaraju Rangaraju Institute of Engineering and Technology, India

Satyanarayana Kosaraju
https://orcid.org/0000-0001-5271-5743
Gokaraju Rangaraju Institute of Engineering and Technology, India

ABSTRACT

Evaluation is an essential process for the measurement of transformation that a student attains after a teaching learning process. Outcome-based education (OBE) in academics especially in the field of engineering is an accepted philosophy in recent years. The OBE system departs from the traditional method where assessment of students is based only on grades and/or ranks. Output has been the traditional measurement criterion in education field, which does not address the level of transformation in the learner, whereas outcome is the measurement of level of achievement showing the transformation. Assessment tools are required for the measurement of outcome. These tools could be direct tools for direct assessment or indirect tools for indirect assessment. An assessment can be a formative assessment or summative assessment. Learning is complete only if transformation is observable in all the vital aspects of attitude, skill, and knowledge. It is widely accepted that all these aspects can be measured in OBE.

DOI: 10.4018/978-1-7998-2245-5.ch004

NEED FOR ASSESSMENT

The three vital things in any teaching learning process are objectives, teaching methodologies and assessment. Assessment by the teacher is to support teaching learning process. It provides significant probability for improving student learning and to develop competency. During this process, teacher needs to assess and evaluate whether the teaching methodologies contribute to behavioral transformation or not. They also need to identify the student performance as an individual and as a group member in all vital aspects of attitude, skills and knowledge.

Assessment is the key to measure performance. Continuous assessment by teachers along with appropriate feedback, leads to the development of essential changes in the way students can be encouraged to express their point of view and the ability to deal with and guide their own learning.

OUTCOME BASED EDUCATION

Outcome based methods have been implemented in education system around the world, at various levels. Outcome Based Education (OBE) policies have been adopted by Australia and South Africa since 1990s (Donnelly Kevin, 2007 and Allais Stephanie, 2007). United States has also had an OBE program over the years, since 1994 (Austin Tammy, 2014).Hong Kong adopted OBE for its universities in 2005 (Kennedy Kerry, 2011).Malaysia in 2008, implemented OBE in all their public schools (Mohayidin, Mohd Ghazali, 2008).

The European Union has proposed an education shift to focus on outcomes, across the EU (European Commission, 2013). As an international effort to accept outcome based education, the Washington Accord was created in 1989. It is an agreement to recognize undergraduate engineering degrees that were obtained using OBE methods. Full signatories as on2018 are Australia, Canada, China, Hong Kong, India, Ireland, Japan, Korea, New Zealand, Malaysia, Pakistan, Peru, Philippines, Russia, Singapore, South Africa, Sri Lanka, Taiwan, Turkey, the United Kingdom and the United States (Washington Accord, 2012 and 2017).

Traditionally the education system all over the world has been relying on the quantitative measurement of the student outputs against an expected standard as a measure for student learning and student gradation. The test scores do not completely represent the actual behavior of the student in real life. Then evolved the new paradigm outcome based education. The basic difference between traditional teaching and outcome based education is mentioned below in Table 1.

Table 1. Difference between output and outcome based education

Category	Output Based Education (Traditional Teaching)	Outcome Based Education (Transformation Teaching)
Methodology	Teacher – centered	Learner – centered
Focus Points	Number of students graduated	Course outcomes; Course Outcome that describes what a student is expected to know and can do at the end of the course.
	Grades and /or ranks of the students	Attitude, Skills and Knowledge of students
	Number of students employed or career placed	Professional achievements of graduates
Role of Instructor	Teacher	Facilitator

Outcome Based Education (OBE) is a learner-centered approach that focuses entirely on student competency in various aspects. Success for each student is the only goal and it is measured by the student ability to meet the pre-defined outcomes (William G Spady, 1994). It consists of setting outcomes, achieving outcomes, measuring success and evaluation.

The Key components of OBE are Vision, Mission, Program Educational Objectives, Graduate Attributes, Program Outcomes, Course Objectives, Course Outcomes, Mapping, Assessment, Evaluation, Rubrics and Grading.

- Vision is the statement that describes an institute aim for future prospect.
- Mission statement describes what an institute is doing now to achieve the vision.
- Program Educational Objectives are broad statements that describe what graduates are expected to attain within a few years of graduation. These should be aligned with the mission of the Institution.
- Graduate Attributes form a set of individually assessable program outcomes that are the components indicative of the graduate potential to acquire competence to practice at the appropriate level. These are the attributes expected of a graduate of an accredited program. The graduate attributes consist of engineering knowledge, problem analysis, design of solutions, conduct investigations of complex problems, modern tool usage, engineer and society, environment and sustainability, ethics, individual and team work, communication, project management and life-long learning (Washington Accord, 2012).

- Program Outcomes describe what students are expected to know and be able to do by the time the student graduates. These are the narrower statements and relate to the attitude, skills and knowledge. These reflect to all the graduate attributes at the end of the program. These are derived from Graduate Attributes.
- Course Objectives are the statements that describe what students are expected to attain in terms of attitude, skills and knowledge after completing the course. This has strong bearing on the syllabus of the course which is teacher-centred.
- Course Outcomes describe what students are expected to know and be able to do at the time of completion of the course. These relate to attitude, skills and knowledge that the student acquires as they progress through the course. These are specific, measurable and student-centred. One useful indicator to identify the level is Blooms Taxonomy. The endeavor is to reach the higher order level of knowledge. (Anderson Lorin & Krathwohl David, 2001).
- Mapping is the process of representing preferably in matrix form, the correlation among the parameters such as Program Educational Objectives, Program Outcomes, Course Outcomes, etc.
- Rubric is a tool that helps to make performance measurements as objective and consistent as possible by defining the criteria. It describes the characteristics of different levels of performance from beginner to exemplary.
- Grading is the process of evaluating students across a value through scale. It focuses on strength and weakness of the students individual learning and performance.
- Evaluation is the process of interpreting the data accumulated by the various assessments. It results to what extent the objectives are achieved and in decisions, action taken regarding the improvement of the program. For example, in test at the end of the semester for a course, the answer sheet of the student is evaluated against the approved key and marks are allotted.
- Assessment is the measurement of student effective involvement during learning and reflecting the effective teaching methodologies. It provides feedback on performance and results the areas of improvement. The selection of appropriate assessments methods plays a key role reflecting course objectives and outcomes. As an illustration, an undergraduate student is assessed as pass if he/ she attain 40% of marks whereas the pass criterion for a post graduation student could be 50%.

An assessment can be a formative assessment or summative assessment. A formative assessment is a technique or a method which measures the outcome instantaneously and recognizes about the student learning and suitability and/or effectiveness of teaching methodology during teaching learning process. This helps to measure outcome attainments and areas that may need further improvement at mid-course stage. Summative assessment is the one which is done after completing the entire teaching learning process for the whole course providing outcome of the completed course. Summative assessment is about final result of the course / academic activity that has been completed whereas formative assessment is carried out at regular intervals of interactions which can be even after each instructional period during the course.

Assessment tools are required for the measurement of each outcome. These tools could be direct tools for direct assessment or indirect tools for indirect assessment. Direct tools are the tools where measurement can be done directly based on the results obtained through midterm, periodicals, semester exams or final exams, projects and whereas, indirect tools are the tools which are used where direct measurement is not possible. An indirect tool for assessment must be done through various surveys from stakeholders or constituents, which may include but are not limited to faculty, staff, alumni, parents, students, employer, foundations, partners, etc. Feedbacks are collected at regular intervals and are analysed for enhancement during the course and at the end of the course.

Direct assessment can be pursued for certain aspects like applying knowledge, development of solutions, research of complex problems, life-long learning, etc. whereas indirect assessment is needed for some of the aspects like societal impact, environment and sustainability, ethics, team work, project management, etc.

FORMATIVE ASSESSMENT

Formative assessment is one of the types of assessment which refers as a continuous assessment. This method provides the required information to improve teaching learning accordingly as per the outcome which is an indicative quality. The main objective of this method is to monitor, provide ongoing feedback and allow teachers to improve and change their teaching methodologies for students to improve their learning. This process can be reviewed frequently and/or continuously.

Formative assessment gives comprehensive feedback to both teacher and student about the performance and learning of the student. It observes students transformation during the learning process (Huhta Ari, 2010).

Formative assessment techniques are quick to use and fit perfectly into the instructional process. Descriptive feedback may go along with formative assessment to let students know whether they have mastered an outcome or whether they need more practice (Resourced Pomethean, 2019). This is a flexible method and informal way of assessing a students progress and their understanding of certain content. For example, plan for a spontaneous and informal question and answer sessions, class activity where students can informally present the results, observations and discussions during class activities, etc.

This also helps students to identify their strengths and weaknesses and target areas that need to work. It helps teachers also to identify where students are struggling and to address their problems instantaneously (Huhta Ari, 2010). At an instructional level, teachers use this information to identify weak areas across the context and develop strategies for improvement. Progressively, further formative assessments indicate whether teaching plans need to be revised to strengthen or extend learning.

Tools for measurement of formative assessment could be an innovative strategy, for example, analysing students through tests and quizzes, surveys, critical thinking, answering Higher Order Thinking, Think Pair Share activity, learning by doing activities, Plickers, etc. The tests can be preannounced/ scheduled or surprise.

According to some teachers, formative assessment can slow down instructional time itself with a requirement to hasten to proceed with assessments and evaluations. The frequency of formative assessment and relative weightage given to formative and summative assessment are also important. If the weightage is relatively less or non-substantial, student may take formative tests less seriously. This could lead to bias to results and teachers may misinterpret the feedback.

SUMMATIVE ASSESSMENT

Summative assessment is one of the types of assessment which is a standard evaluation of student learning at the end of the course duration typically a semester. It measures the student ability. Typical weightage bias exists towards summative assessment which makes this method a high grading value under controlled conditions and therefore has more visibility. Students tend to focus more and put in greater effort for this assessment for obvious high return advantage.

Summative Assessment focuses only on the result of the end exam as by giving the grading to the students. This helps to check the understanding of the students after the end of the course to what extent they have learned in the entire course. This method helps to evaluate the effectiveness of the course with the help of scores or grades attained by the students. It provides an essential benchmark to check the progress of students with respect to a course, program and institute. For example, final examinations, midterm examinations, projects, etc.

As a result, summative assessment contributes largely towards planning and improving the curriculum. The assessment analysis can indicate gaps between student knowledge and learning targets, course can be restructured by improved curriculum, teaching methods, improve pedagogy or even suggest new learning objectives and outcomes.

Figure 1. Comparison of formative assessment and summative assessment

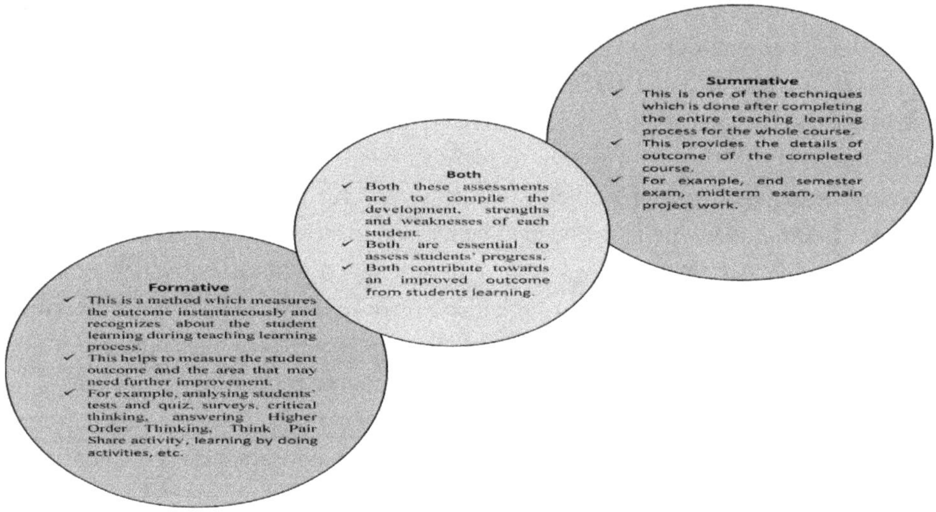

"When the cook tastes the soup, that's formative. When the guests taste the soup, that's summative." By Robert E. Stake, Professor Emeritus of Education at the University of Illinois.

The common goal for both the formative and summative assessment is to assess the student progress in terms of development, strengths and weaknesses of each student. Both the assessment techniques provide different point of views for teachers in terms of teaching methodology and pedagogy. Both contribute towards an improved outcome from students learning and make a better end result in Figure 1

FEEDBACK

Feedback aims at obtaining information on reaction to various academic activities in teaching learning environment. This plays a vital role in the assessment process. Teacher must facilitate for learning rather than teach for curriculum/course coverage. This is the most successful and active technique for the student learning.

Assessment for learning (formative assessment) is assessing with the main aim of providing feedback and guidance to help students to improve. Whereas Assessment of learning (summative assessment) focuses on evaluation of a student's performance, often at the end of a course and provide useful feedback for subsequent assessments (Lorna M. Earl, 2003).

The main objectives of feedback are to guide students to take to improvement, motivate and develop their capability to monitor, evaluate and regulate their own learning. Feedback must be constructive and meaningful. Feedback is valuable only if it is analyzed and proper corrective measures are put in place for further assessment.

Every course should be designed to provide opportunities for students to engage with feedback for their ongoing performance. Feedback becomes essential for improving knowledge and skills and for understanding the significant factor in motivating student learning.

Feedback must be taken at all stages of student learning right from the beginning till the end of the course and program and thereafter also for continuous improvement. Periodic feedbacks from the students become necessary as an indirect measurement for the assessment.

Typical feedbacks from various stake holders with time frames are as given below:

- Orientation Program - At the beginning of the program
- Course Exit Feedback -At the end of the course by the faculty to know whether the course outcomes have been achieved or not.
- Multi-level Survey -Every year by the faculty and students founded on the participation or delegation in co-curricular activities related to technical events, extra-curricular activities related to non-technical dealing with physical education and support services.
- Program Exit Survey - Every Year by final year students for each program grounded on all the aspects of attitude, skill and knowledge.
- Alumni Feedback - Whenever available by the students eliciting information on important positions held, awards and certificates obtained, corporate social responsibility activities undertaken, working as a team leader and with multi-disciplinary teams.
- Employer Feedback - Whenever available by the Employer centered on the students knowledge in technical areas, analyzing and interpreting data, understanding professional and ethical responsibility, etc.
- Parent Feedback - Whenever available by the Parent, on the level of education of their ward, social responsibility, knowledge in environment and sustainability, behavior and attitude, etc.
- Student Feedback on Faculty - More than once in course duration by the students on involving students in discussions, learning environment in class, etc.

Feedback by the stake holders is seldom available and at times not very keen to give it. Institutions should endeavour through ingenious methods to elicit true and unbiased feedback.

RUBRICS

Tests, Quizzes, etc are the direct tools for quantitative measurement of the performance. But there are some academic activities like projects, seminars, presentations, etc for which direct tools may be inadequate because of subjectivity and the number of indicators to judge the level of performance could be varying as per the activity. For such cases, a suitable assessment tool is rubric.

Rubric is a tool for a teacher that helps to make performance measurements as objective and consistent as possible by defining the criteria (Susan Brookhart, 2013). It describes the characteristics of different levels of performance from beginner to exemplary. It may be used by an individual or in multiple to review student work. It is meaningful and useful when shared with students before their work is reviewed so that they better understand the various levels of performance and measurements. It defines what is expected from a student to get a grade on an assignment.

The main purpose of rubrics is to assess performance of the student in different aspects. Rubric can be a general rubric or specific rubric. Rubric describes various levels of quality for each performance criteria. These levels may be written as beginner, developing, acceptable and exemplary for which score could be given as 1, 2, 3 and 4 respectively. For all the performance criteria's, levels are described separately. Sample rubric for is shown in Table 2 below.

Rubrics help the teachers to grade student work and make it easier to explain to students about the grade points they got and what they can improve. To write a rubric, list the performance criteria that will be used in assessing and for each performance criterion, give a distinct description under each level from beginner to exemplary. Now the rubric is ready to be used to evaluate each student.

CONTINUOUS IMPROVEMENT

Assessment is required and considered necessary for improvement. Reevaluation of assessment process with the intention of continuous improvement leads to the quality of education and student learning in Figure 2. Rubrics, Feedback, Assessment tools, etc. play an important role for continuous improvement.

Continuous improvement in education is fundamental. All the processes in the teaching learning environment need to be continuously and constantly monitored keeping in view the changing technology and changing generation gaps for possible course corrections so that the objective achievement is maximized. Improvements can be envisaged in teaching methodologies as well as in assessment tools. The various teaching methodologies help to put forward the improvement not only for teachers but also for students. Presentations and lectures are some of the effective ways of

Table 2. Sample Rubric

Level / Performance Criteria	Beginner (1)	Developing (2)	Acceptable (3)	Exemplary (4)	Score
Data Collection	Doesn't collect any information relating to the topic.	Collects very limited information, some relate to the topic.	Collects some basic information, most refer to the topic.	Collects a great deal of information, all refers to the topic.	4
References of journals	Very few references.	Few references.	Use of references indicates ample research.	Use of references indicates substantial research.	3

Figure 2. Continuous improvement

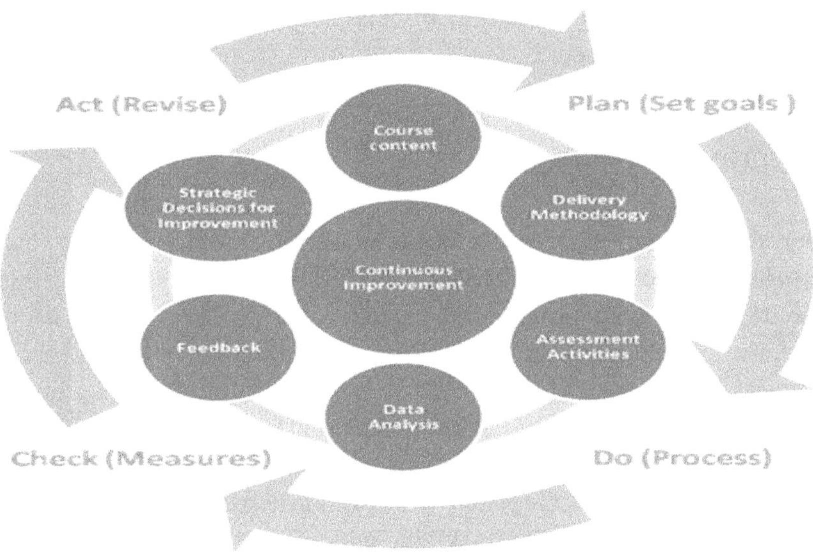

achieving course outcomes. The course objectives and outcomes could not be better achieved without these. Lectures are the best ways to get facts, make students think and get better in their attitudes. The invitation to guest speakers, external resource persons also add value to the course. Seminars and workshops create interest to both students and faculty. These serve as a platform for sharing knowledge and expertise in advanced areas which results in collaborative attempt for further enhancement of the skills and techniques necessary for their improvement.

The extent of self-learning and self-assessment is changing day by day through teaching - learning process from teacher prospective as well as student aspirations. There are multiple techniques or methods to update the traditional teaching to outcome based education. The facilities for self-learning beyond the curriculum help active learning. Massive Open Online Courses (MOOC), Information and Communication Technology (ICT), blended learning, flipped class rooms, etc are changing the active learning environment and have quickly materialized as the support for effective education in the present global scenario of teaching methodologies. One of the main objectives is to create competencies for proper and effective use of various teaching methodologies in education to empower teachers to use technology effectively and to understand the pedagogy associated with online and blended learning.

Teachers are incorporating MOOCs and other blending learning and ICT based tools along with traditional classroom teaching to support various learning preferences. This is one of the ways of integrating traditional teaching with technology effective learning models. Students can benefit from modern technology supplementing to traditional teaching may follow modern teaching tools. Many tools of effective teaching in education are enhanced with online content results in positive impact on teaching learning process.

CONCLUDING REMARKS

Higher education with engineering programs are immensely benefited by following OBE system as it focuses on quality of achievement and student capability. It is student-centered. This system helps in setting outcomes and measuring the transformation that a student has achieved. The measurement of outcome is done through appropriate assessment tools to identify the level of transformation. This transformation is evident in all the vital aspects of attitude, skills and knowledge. Rubrics, Feedback, Assessment tools, etc play an important role in OBE for continuous improvement. Higher education is in midst of paradigm changes in terms of asynchronous learning, MOOCs, quick up-skilling and re-skilling with new technologies. Outcome based education is proving to be ideal for this fast-changing teaching learning environment.

REFERENCES

Allais, S. (2007). Education Service Delivery: The Disastrous Case of Outcomes-Based Qualifications Frameworks. Progress in Development Studies, 7(1), 65–78. doi:10.1177/146499340600700106 doi:10.1177/146499340600700106

Anderson, L. W., & Krathwohl, D. R. (2001). *A taxonomy for learning, teaching, and assessing: A revision of Bloom's taxonomy of educational objectives.* Allyn and Bacon.

Austin, T. (2014). *Goals 2000 - The Clinton Administration Education Program.* Academic Press.

Brookhart, S. M. (2013). How To Create And Use Rubrics For Formative Assessment And Grading. Alexandria, VA: ASCD.

Donnelly, K. (2007). Australia's Adoption of Outcomes Based Education – A Critique. Issues in Educational Research.

Earl, L. M. (2003). Assessment as Learning: Using Classroom Assessment to Maximize Student Learning. Thousand Oaks, CA: Corwin Press, Inc.

European Commission. (2013). *Commission presents new Rethinking Education strategy.*. European Commission.

Huhta, A. (2010). Diagnostic and Formative Assessment. In The Handbook of Educational Linguistics (pp. 469–482). Oxford, UK: Blackwell.

Kennedy, K. (2011). Conceptualising Quality Improvement in Higher Education: Policy, Theory and Practice for Outcomes Based Learning In Hong Kong. Journal of Higher Education Policy and Management, 33(3), 205–218. doi:10.1080/1360080X.2011.564995 doi:10.1080/1360080X.2011.564995

Mohayidin, M. G., Suandi, T., Mustapha, G., Konting, M. M., Kamaruddin, N., Man, N. A., . . . Abdullah, S. N. (2008). Implementation of Outcome-Based Education In Universiti Putra Malaysia: A Focus On Students Learning Outcomes. International Education Studies, 1(4). doi:10.5539/ies.v1n4p147 doi:10.5539/ies.v1n4p147

Promethean, R. (2019). Retrieved from https://Resourced.Prometheanworld.Com/Types-Of-Summative-Formative-Assessment/

Spady. (1994). Outcome-Based Education: Critical Issues and Answers. American Association of School Administrators.

Washington Accord. (2012). International Engineering Alliance.

Washington Accord. (2017). International Engineering Alliance. Retrieved 20 April 2017, http://www.Ieagreements.org/Accords/Washington/Signatories/

Chapter 5
Teaching and Learning:
A Paradigm Shift

Jogindra Nath Sahu
BHEL Bhopal, India

Ankita Awasthi
College of Engineering and Technology, IILM Academy of Higher Learning, Greater Noida, India

Kuldeep Kumar Saxena
Institute of Engineering and Technology, GLA University, Mathura, India

ABSTRACT

Teaching is not only a profession. But this is a noble cause of educating and building the society through imparting knowledge. This knowledge makes the students skilled, conceptually and technically, to achieve their goals in life. Present teaching-learning methodology is more concerned with laboratory experiments and hands-on practice with the purpose of learning through real-time simulation exercise around the whole globe. All education authorities are more concerned towards making the students not only educated but employable also. Teachers are playing a major role in adding skills and values for the betterment of students, society, environment, and economy.

INTRODUCTION

A teacher is a person whose profession is like a farmer. He prepares his land for cultivation, sow the right seed, and feeds with knowledge of water and transforms with his light of support, love, and power. This deed brings blossoming buds, flowers, vegetables, and fruits, which ultimately causes happiness and contentment in the

DOI: 10.4018/978-1-7998-2245-5.ch005

lives of human beings. However, there is a difference in the deeds of a farmer and teacher. A farmer may not get a chance again if he fails in his profession. A good teacher always tries to nurture the minds, ignite the dreams of young buds, and also show them the way to achieve them.

Let's remember those masters who show their talent in the field by putting their hard work at different crucial environment on the cost of their life. The emanate teachers who have contributed honestly to their nation and profession include Aristotle, Pluto, Dr. A.P.J. Abdul Kalam, and many more. Teachers have known by their scholars and remembered after their life also in this universe.

The main aim of this topic is to analyze teaching and its importance for each individual. It looks easy task in its form, but it is not so eased in an actual way (Grattan, 2017). The main question is arising why and why not? The answer is very simple. The art is not measurable.Art is an individual trait and rooted in the inner conscious of the teacher. In the art of teaching, a teacher visualizes in his soul and the student in his body. Of course, the body-soul is moreover integrated to give their best performance in its form. Hence, in teaching style, this art is applied to build a scholar who can prove his identity confidently in this world without any fear.

Whatever it may be, the philosophy of teaching never changes in which someone is ready to deliver his skill and knowledge to a recipient who is in need. But this basic philosophy of teaching changes from the ancient era to the present time, and what is going to happen in the future is still a mystery (Cannella & Reiff, 1994). Let us focus on one of ancient education and teaching mechanism as a compared teaching system. As per the existing evidence and observation, the teaching process has changed remarkably from ancient to a new process, where we have lost the progressive movement of the teaching system and teaching philosophy due to the changing technological scenario. Books have replaced bya virtual study in which the library loses its identity. The existence of teachers and institutions are also on stake (Mary & Thomas, 2000).

TEACHING AND LEARNING CHALLENGES

The rain is poured uniformly on a surface,whereas the accumulation of water is different as per the location and situation. Similar to the teaching methodology that will uniform for all the students, but the accumulation of knowledge remains uneven. The role of a teacher and teaching pedagogy starts to make this accumulation uniform throughout the class. Teaching and learning methodology should adapt in a manner that subject and content become understandable to everyone as per their capacity (Kaufman, 1996). The present and ancient education system in India cannot be overlooked by narrating the teaching and learning system in this chapter.

Figure 1. Education system in India; a) ancient and b) present

The ancient Indian teaching system is entirely different from the present scenario and shown in Fig. 1. The moral base ancient teaching system in India is well known all over the world, and many scholars attracted to India for their careers in different disciplines. The history is full of such evidence for incredible Indian teaching systems where it is Ayurveda, classical music, astronomy, etc.

In the present scenario, the newly developed technology has an important role in teaching methodology. It clears that the involvement of technology makes the teaching system much easier for the teachers and more interesting for the students. The SMART classes with the live videos and the actual test results of the topic with more animation and powerpoint application make ever seen changes in the teaching and learning style by which the understanding capacity of the student by visualization of the subject in a different way for more clarity and understanding (Prasad and Prasad, 2005). Figure 2 shows the smart class teaching.

Various factors are playing a role in the teaching system, which are teachers' availability, well-equipped institute, teaching methodology and guidelines, students, and technology. Still, the system has a drawback. The problem in terms of unskilled, semi-skilled, and unemployable stuff coming out of the technical institution, not ready to be employed in the industry, is causing major challenges and a question mark on teaching-learning method and methodology. Lots of effort is cultivated to get a robust teaching and learning system in this country.

EFFECTIVE TEACHING METHODOLOGIES

Classroom culture is sometimes very volatile because students of different backgrounds with different mindsets, personalities study together. So teachers handling this sort of mixed group have greater responsibility, they need to be more innovative and adaptive to any environment and conditions. To be an effective and efficient teacher,

Figure 2. A presentation of smart class teaching
(https://digitallearning.eletsonline.com/2018/12/schools-digital-infrastructure-speeding-up-learning)

one has to know the basic strategies which make him more accessible to his student's mind. These strategies are as follows:

1. **Visual Approach:** Usage of visual aids like virtual reality, field visits of some experimental sites can help gain some extra knowledge apart from following the mentioned syllabus.
2. **Self-Building Approach:** Even students can contribute to building confidence amongst the fellows. Students can be so friendly and united that they can help a slow learner to become a fast one.
3. **Lifelong Learning:** Teachers should always guide their students whenever they are in need. The teacher-student relationship is permanent integration, which continues throughout life. There should be some critical problem-solving sessions to give a deep understanding of real problems to young learners.
4. **Assessment of Capabilities of Students:** Teachers should be connected with his or her students so that they can assess the capabilities of their students and allocate the roles and responsibility accordingly.
5. **Technology Enables Classroom:**Technology always plays a crucial role in maintaining the bridge between ancient and modern teaching methods. In this fast competitive world, everyone is looking for their survivability and resource learning. Technology provides a platform where students physically engage themselves and also develop their minds towards research.

Figure 3. Education system at present
(https://bdeducationinfo.com)

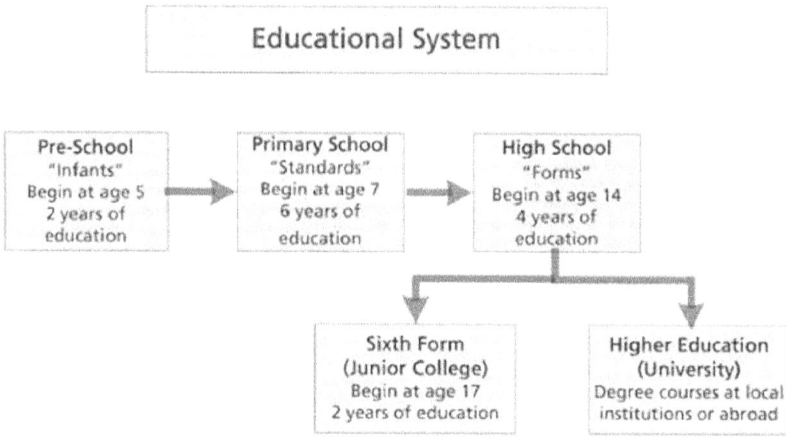

6. **Revival of Educational Policies:** New educational policies should be proposed, which can benefit both stakeholders, and also old policies be reviewed in terms of any improvement.

PRESENT TEACHING/ LEARNING PRACTICES IN INDIA

The Indian teaching system based on a mixture of many countries because of its different languages and cultures, as well as historical values. The teaching methods,as well as the learning system also based on the rules which were ruled this country for a long decade. The present education system and the schooling in India presented in Figure 3.

If we see the present Indian educational scenario, the comment of Dr. A.P.J. Abdul Kalam cannot be ignored. Once he commented, shown in following Figure 4:

The education system of the Indian students depends upon government decisions and policy. When we grow with this thought to educate ourselves, it seems to be more theoretical rather than skill-based education. If we compare our professional/ higher education enrolment with the few developed countries (shown in Figure 5), it shows an amazing situation in our countries. Till now we depend on new technology invented by those countries whose youths are enrolled less for the higher studies as compared to us.

Figure 4. Comment of Dr. A.P.J. Abdul Kalam for education industry

Figure 5. Effect of education system on GDP

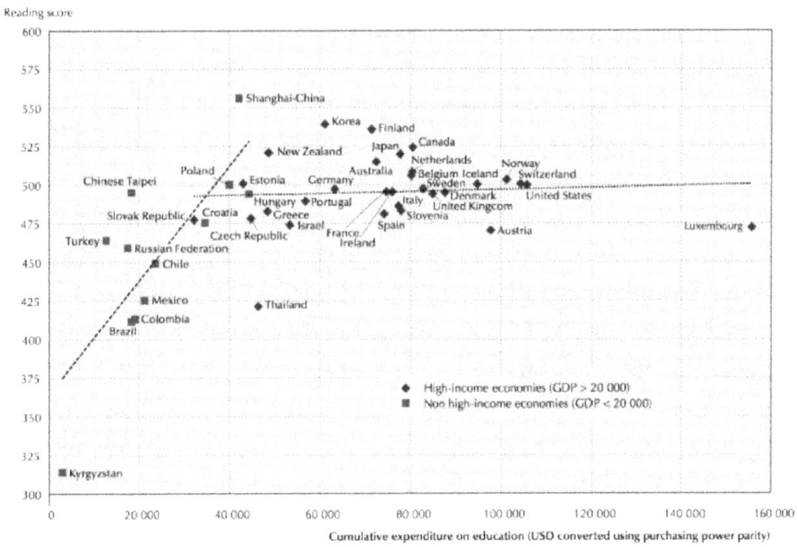

Table 1. Indian higher education enrollments in top 7 receiving countries, 2005-2015

Indian Higher Education Enrollments in Top 7 Receiving Countries, 2005 - 2012								
Year	**USA**	**UK**	**Australia+**	**Canada**	**NZ**	**China**	**Germany**	**Total**
2012	96,754	N/A	12,629	28,929	11,349	10,237	5,745	190,055**
2011	100,270	29,900	15,395	23,601	12,301	9,370	4,825	228,774
2010	103,895	39,009	21,932	17,549	11,616	9,014	3,821	253,743
2009	104,897	38,500	28,020	9,561	9,252	8,468	3,236	247,631
2008	103,260	34,065	28,411	8,325	6,348	8,145	3,217	216,516
2007	94,563	25905	27,078	7,304	3,855	7,190	3,431	205,852
2006	83,833	19,228*	25,497	6,927	2,599	3,245	3,583	158,215
2005	76,503	16,872*	22,529	6,688	N/A	N/A	3,807	N/A

Sources: IIE Open Doors, UK Higher Education Statistics Agency, Australia Education International, Citizenship and Immigration Canada, New Zealand Ministry of Education, China Scholarship Council, DAAD/ HIS (Germany).

*Inferred from percentage of total international student body

**Minus UK total for 2012 which is not currently available

+Higher education (University) enrollments only; no VET

The data in Table 1 shows the scholars of Indian origin enrolled in 7 developed countries till the year 2012, whereas the number is much higher than the later one.

If we see the expenditure on education in India as compared with the world's economically developed countries as well as economically undeveloped countries, it will give a clear idea of how we look after our education system in India.

The purpose of this teaching and learning practice is only to full fill the education calendar of the institute and in turn, convey the message to the educational watchdogs to full fill the official duty. Which, in turn, gives the output in the opposite way, which is noticed by all of us. Lots of youth are in the country running pillar to post for search their bread and butter with well-printed certificates. The newly formed teaching style gives no teaching skill to the society, which is waiting for his scholar solve their problems in different ways, in turn, this new teaching and learning system givesa new problem to the country a huge unemployment situation as well as a scholar without its education values. Now the question arises for anybody ever thought how to solve the so-called situation in the education system and its outcomes? Is anybody voluntarily taking responsibility for this failure? Everybody putting the fault on others' heads and blame games transfer from government to government and institute to institute, person to person. The time-bound policies were missed to guide this teaching system to travel on the right path over time. Now the time everybody should under the grave of the situation and deep brainstorming process carried out for eradicatingthe fear of the learner after acquiring his qualification.

BRIDGE BETWEEN ANCIENT TEACHING AND MODERN TEACHING TECHNIQUES

The flow of knowledge between teachers and taught establishes a strong relationship between the two. And this has remained unchanged since the inception of human civilization. From time to time, new techniques of teachings introduced,but the fundamental issue remains the same. The relationship between the teacher and its students always gets stronger due to faith, trust, duty, and respect. With time commercialization became an important phenomenon. The increasing influence of the capitalist mindset gripped the world. The mindset of people started changing. Now everything has a cost in terms of money. Currency needs circulation; this was the main concept. Therefore everything became commercialized, including packaged water, canned air, taking care of baby, and also education. Nowadays schools take special care to introduce students with modern and expensive gadgets so that they could charge a huge amount of money from their parents. But the question is: Is it worth it? The answer is no. If the quality and reach of education are to be improved, then the basics need to be corrected. The following are the basics that never change:

1. Teacher should treat all his students equal. No biasness or partiality
2. Student should respect their teacher. One should always remember the flow is always from higher potential to lower potential. Hence a student should always look at what new he could learn from his teacher.
3. Student should always revise these lessons before coming to the class. After the teacher has taught, one needs to go through the lessons for better understanding.
4. The more a student reads and thinks, the better the understanding.
5. Education is invaluable.

ROLE OF EDUCATION GOVERNING AUTHORITY IN TEACHING/LEARNING SYSTEM

The educational authorities such as UGC, AICTCE, autonomous university, minority educational society make the guidelines for establishing the teaching and learning system and given the authority to certify the courses as well give the certificate to learners after completing the course. The guidelines, as well as standard operating process (SOP), issued from time to time, but anybody is there doing correct monitoring how much this guide lien used in the so-called teaching and learning system. Lots of institutes affiliated without basic infrastructure. Lots of computations created among the institutions and the institutes become a business center rather than a teaching and learning platform. Due to the deficiency of the infrastructure as well as the skilled experts at the educational institute, outcomes are very poor and not full fill the completeness of the teaching and learning style in the schools and colleges and educational institutes. There is no personal relationship among the teachers, and his scholars exist in the study process. Everybody is busy to complete their syllabus and examination process. There are no moral values that are existing in the students, and no proper guidelines are existing at any level of this education and learning system. It leads to an unsocial activity in the school, college, and institutional compound without any reason. The products are many, but there is no use of products that lead to different types of situations in the upcoming life of the scholars. Overly an unbalanced education system created in the present scenario without any reason. It seems a big challenge for all of us to understand the act according to it. Every sector should work towards this problem and strengthen our backbone for an upcoming developed country.

SOLUTIONS TO OVERCOME

Now the time comes to find the solution to the different problems of the teaching and learning system and adhere to the suggestions strictly to overcome the challenges. As per the author, the following suggestions may consider for its betterment and uplift of the education system. All this could get ignored except knowledge and education. Whatever it may be as per the Vedas "shradhabanlabhatigyanm" means the love and devotion lonely lead to a learner to acquire skill from his master and his trainer. It has seen that urgency is an address rather than important. Which leads to the unbalance situation in every step of life? It shows that till now, it is not clear that what is urgency and which is important in the teaching and learning system. The lawmakers and law followers both should deeply analysis the relationship between urgency and importance in every step of the teaching-learning system.

Time management is life management. Hence the time management in the teaching temples should not be the clock management; rather, it should be skill management. The study syllabus should design such a way that it has to complete within the stipulated time frame, and the unnecessary burden should avoid in the teachers as well as the scholars. The course of work should require more skill and base rather theoretical.

VALUE OF EDUCATION SYSTEM FROM ANCIENT TO PRESENT

Educational is the backbone of the country and civilized society in which teachers are an architect for the reinforcement of this backbone in the educational system. If the material to construct the backbone is not in order and foundation is not strong, then the backbone which is going to bear the total load of the structures should be strong enough then how can one expect the output and safety of the structure. The total word divided intoa three-way as per time and situation.

1. Undeveloped country
2. Under developing country
3. Developed country

This so-called status is considering in the may angels such as economy, education, health, human values, natural resources, and technology. In the pretext, how ape-man comes to a developed digital man a few courses of time and explore the space, land, water to its full extent. The creative name developed, civilized, etc. due to its adaptability to life and his learning style. In the present situation, the world is very small due to technological development and web connection. At the same time, education also crosses its regionalists and becomes a single world platform and can be available for everyone to acknowledge anywhere form this earth. Many watchdogs are coming up to monitoring the education system and teaching process worldwide (Raths et al., 1996). This all effort connecting each person to teach each other and mutual sharing their knowledge among themselves.

The teaching and learning scenario in India and Indians are always ahead to makea needfull contribution from ancient times till now. The Vivekananda's America journey, world yoga day in a live example for new days ahead in this world. There is a saying in India "Gyandanmahadan" means sharing knowledge is auspicious work for a human being on this earth. Where it is ancient India or modern-day teaching, and learning style is unique in India as compared to another part of this earth. This show many scholars from different part of the world moving towards India for acquiring different learning system such as art, culture, music, astronomy, technology, etc. Hence we fill the body is referring to Hardwar and mind is software as compared to the computer system. It is right where it is an ancient or modern era in every situation Indians show the educational light through their excellent teaching methods.

The ancient teacher ignored their contribution to the teaching and learning process for their enormous work. The value has added in this world through Ayurveda, Artha Sastra, and philosophic as well as science and technology. Every subject addressed to its full extent. The main interesting point here to discussing the methodology as well as the learning technology is always entirely different in India as compared to another part of the world few examples have given clarity of the topic showing teacher and learner and their relation. The best examples are BhagwatGeeta in which Lord Krishna teaching and Arju is his dispel and unpunished the Yamraj (king of death) is teacher and Nachiketa is his dispel.

Many talents are born and vanished from this earth, but their contribution to distribution makes a benchmark in all the time and situation. The infrastructure was not upto the mark pre-independence of India as well as another part of the world. People don't know how to preserve their work for innovative ideas and education easily available for all. Literacy level was very poor in the country, and few are rules many due to their educational level. Teachers are all-time do their crucial work to lift the environment in his full extent up.

In the twenty-first century, we are progressed tremendously in every direction of human development. Technology adds value in every sector and recognized as a developed civilized with our new learning system. In education and co-curriculum, teaching and learning styles also take a new step more methods, and the methodology process has developed to the framed learner with its full color. In the process of developing educational institutes, multiplied in a larger way to make the learning system more comfortable. More teachers have created, and infrastructure is provided to meet the challenged a newly formed country for its development. The journey starts with the full color of its form from past to present. Young India and its Indians walk in the path of teaching and learning.

Now we are standing in the mid of the path of progress in the education system and observing both sides of path what we covered and where we are standing and in which direction we are moving. The past shows us a lot of achievement from the value-based education system with extremely less in fracture and facility. The gurukulhad run by the mercy of landlords and the so-called kings of that time. Scholars are moving door to door to collect alms for their solvability. In his difficulty and challenging condition, what we achieve in the teaching and learning process of the teachers and their scholars are a matter of observation. Veda has to memorized to its original form;the medicinal practice was learning from so-called Vaidya's (Unani doctors). Similarly, there was an entirely different cultural maintain between the teacher and the learner. The teaching and learning become heritage from generation to generation to transfer the knowledge from his father to son, and so on.

ROLE OF TEACHERS IN THE TEACHING SYSTEM

Teachers/instructors are the main instrumental component in the teaching and learning system. They should know that teaching philosophy from the core of their hearts. The teaching philosophy says that teachers should have mothers' hearts. Who always think three things in her mind, Such as:

1. Whatever her child eats still she feels that my child is hungry
2. Whatever the precious things she got for herself kept separately for her child by ignoring her requirement.
3. She always gives priority to her child by scarifying her comfort in every step of life and situation.

Similarly, in the teaching process, the teachers should always think of what modification he can do in his/her teaching method so that the requirement of the students can full fill. In the second case,the information should collate from all the sources that should preserve for the students to enhance their knowledge. At last but one the student is the only priority of teacher nothing else in the teaching and learning system and dispels interest isthe above all in every situation. Hence every step the valuation of the mother's hearts should carry out for teachers, and they should have motivated to bring this philosophy in their life and teaching process. One more matter should consider improving the teaching system without any doubt. In the present teaching system, the instructors are employed more in non-teaching jobs rather than their prime duty. The teachers are looking after construction, admission, and other internal and external matters. In this environment, how can we think about the educational value that they are going to impart to the teaching system? If we want to improve the education system and their teaching style, they should free from all extra duties and motivate them to improve their methodology and teaching aids. One more hindrance is arising among the teaching community, not aware of the changing environment of the teaching process in the present scenario. Therefore, they should lake of their up to date knowledge system and not able to stand confidently in front of the fully aware student community. They should also change their mind setup and make themselves more effective by referring the literature as well as the up to date to the new technology and changing environment in the education system.

EVALUATION OF TEACHERS AND THEIR WORK VALUE

Evaluation of teachers and their work measured differently in different places. However, the teaching process is moreover affected by this personal valuation process. The teachers/ instructor have become more self-improve to match with the evaluation system and always try to fight among themself for their promotion. The work cultures in the institutes have changed to athletic race among themselves to achieve their career. In the competitive surrounding, the teaching has been ignoring completely, and hence the teachers' career evaluation process should have modified by the rule makers should work more carefully to give priority the teaching and learning system rather than the personal achievement of the individual. The teaching value and teachers respect should maintain by uplifting their future career as well as the economic and social value at the highest level under any cost. Teachers should have respected by all who have directly or indirectly benefited from the teaching system. Now we should focus on the learner part who also an important component of this chapter. The whole education system designed for them, and they are the only beneficiary of this teaching and learning system.

ANALYSIS OF THE LEARNING SYSTEM THROUGH THE LEARNER'S POINT OF VIEW

The student and student relationsare called Guru sisyaparampara from the stone age to till now, and the relationship will be remaining as it is from 'adikal to anatakal' (ancient era to infinite era). The teaching and learning system is a never-ending process, but its process from time to time. There was a time in which the teachers are the supreme god for the students or a scholar. In the present day, the scenario has changed to a large extent in which the student is controlling the teacher's career. The tremendous changes are over-served from the ancient age of learners to the present scholars. At ancient age, teachers are upper hand, whereas nowadays, students are at a higher place due to changes in rules and guidelines. Lots of evidence are seen form the ancient scripts such as the teaching of Budha,Mahavir, Christ, etc. and their teaching philosophy at that time. The teaching methodology and the learning process show its benchmark till now. There was a time in which all the scholars coming to India from different parts of the glove to acquiring their qualification and enlighten them with Indian masters, whereas at present Indian students are moving different parts of the glove to find their education and career.

The learning system should have modified in such a way so that the students can get maximum benefits out of it. At present, the learner is biased due to a variety of courses and their utilities. In the present time, education became nothing but a business for the students as well as the education providers. It is now very difficult to retain the originality of the learning system in the education market due to its earning visions rather than learning vision. Now the question ariseson how to bring the originality of the learning system where neither learner nor his environment is not in a learning process. Due to the present situation, the learning process loses its vision and moral in totality in our education system.The more focus should give to improve the learning system by all who benefited from it.

ROLE OF LEARNER IN THE TEACHING AND LEARNING SYSTEM

Today's learner is tomorrow's leader and pathfinder of the country. The learning system is also beingthe backbone of the country; hence, its improvement cannot get ignored at any cost. It is the moral responsibility of a learner to find a solution on how to lead himself to get the optimum output by the existing teaching process to improve his skill and knowledge in a particular field. They should come out of their comfort zone and put their whole heartily effort into understanding the subject and curriculum, which designed for them. The social, personal and ethical value among the teachers and the student should be intact under all the conditions in the leaning process. We can say that learning is also an art. The art is not able to teach anyone without his interest. Hence the devotion is required from the learner side to full fill the learning process of this art. Hence everyone should work towards this direction to motivate the learner to furnish a duty to the full extent.

THE OUTCOMES OF LEARNING/ TEACHING SYSTEM

The educational institutes have been producing the learner at the fixed and calendar time. The market has flourished by lots of educated youth with different qualifications and skills. Everyone comes through the education system trying to find his bread and butter as per his capacity in this society and society also have expectation form this learner for its development and service from them. But the scenario is opposite to the expectation and becomes a news headline in the country. The selection of job profile by the educated students in our country becomes a headline in newspapers. Thousands of engineering students applied for the bank clerical post like many new, which are giving the actual position of our education and its outcomes. It is the right time to analyze the reasons for these outcomes carefully while we are having improved learning and teaching system.

CONCLUSION

Every problem has its solution, whereas the question arises who will take the lead to deal with the so-called shortcoming of the present education system. Blaming is not a matter of discussion at this point, but peacefully find a solution to the problem is the main thing whichis affected by this problem. The common realization in the teaching and learning process is opening more educational institutions to educate the number of people to uplift the individual and the education system of the country. In this process, many guidelines and rules for opening the educational institution have overlooked by the government and autonomous bodies who are involved in giving the accreditation and permission for the opening of the institute. How many rules and guidelines have used at the user label are not defined. We will agree at this point that the monitoring comities have constituted at every state of the education system to the proper function of the teaching and learning system. Now a question arises where is the fault at the system, which gives a reversed output beyond our expectation. As per the author, the shortfall of teaching and learning system is well known to all of us, but everybody is speechless due to the physiological and psychological changes of the individual and society. Few suggestions and recommendations may fellow in the present condition to find some solution to this distorted system and bring the teaching and learning system in the main track.

1. There is a proverb "Charity begins at home" let us change his words accordingly in favor of the education system" the teaching and learning process begins at home." Home consist of parent and their children who are only future for them and different bodies with the same soul. The parent should well aware of the limits and the capabilities of their ward without any biased condition. Hence the first correction should be carried out by the parent to bring the learning system in the proper track.
2. The role of education institutes should be motivating the scholars to address important rather than urgent.
3. The rule-makers should make the guidelines not based only on the keep in the files and cupboards. The proper monitoring process should be carried out from time to time.
4. The experts who are designing the education curriculum and the syllabus should have a more impotent to practice orientation rather than theory.
5. The discipline and moral values should be maintained every step of this learning and teaching process by everybody, such as a parent, student, teacher, government etc.

6. Leaning skill of the students should not have ignored while allocating the job for proper utilization of education as well as self-satisfaction of the learner.
7. Time to time brainstorming should carry out by all who have involved this teaching process for its improvement by viewing the changing scenario in the environment.

If we take care of the few life-changing steps honestly towards our improvement, then the result may come in the course of time,which has awaited at this junction for all of us.

The entire chapter puts many ideas and inputs what the actual condition of our teaching and learning system and how to overcome the shortfalls in the system. It also highlights the duties of all who are involved and benefited from this process. The role of watchdogs and rule makers are highlighted to bring the education system in proper track and proper implementation of the rules and guidelines. The chapter explains the glory of our ancient learners and masters who put their innovative methods to build their scholars. The chapter gives a clear picture of past, present, and future aspects of our education system and the prime focuson getting farther improvement on it. At last, the main conclusion of the chapter to get a robust teaching and learning process in the present condition is “important should be addressed rather than urgent.”

REFERENCES

Barksdale-Ladd, M. A., & Thomas, K. F. (2000). What’s at stake in high-stakes testing: Teachers and parents speak out. *Journal of Teacher Education*, *51*(5), 384–397. doi:10.1177/0022487100051005006

Cannella, G. S., & Reiff, J. C. (1994). Individual constructivist teacher education: Teachers as empowered learners. *Teacher Education Quarterly*, 27–38.

Grattan Institute Report No. 2017-01, February 2017.

Kaufman, D. (1996). Constructivist-based experiential learning in teacher education. *Action in Teacher Education*, *18*(2), 40–50. doi:10.1080/01626620.1996.10462832

Prasad, H., & Prasad, D. (2005). Towards professionalism in education. *University News*, *43*(18), 323–337.

Raths, L. E., Harmin, M., & Simon, S. B. (1966). *Values and Teaching*. Columbus, OH: Charles E. Merriu Publishing Company and A. Bell & Howel Company.

Chapter 6
Modern Pedagogy Tools in Engineering Education

J. Srinivas
National Institute of Technology, Rourkela, India

ABSTRACT

Engineering pedagogical techniques have received wide attention in recent times. Various fields of engineering have acquainted with progressive teaching methods and training techniques. The concept of pedagogy now has different dimensions. Along with modern challenges in industries, the teaching approaches have been modified in several respects. Earlier teacher training programs are to be upgraded with modern pedagogical concepts. This chapter brings out an introduction and a few application courses following the pedagogical engineering approaches. The concepts of technological pedagogical content knowledge and constructive pedagogy are summarized.

INTRODUCTION

Pedagogy is the act of teaching. It is a Greek word meaning 'lead' referring to the progress in teaching concepts for future engineers. The pedagogy employed by the teachers convert their actions into the students' needs. In general, the teacher is treated as a knowledge holder and students are considered as the recipients of the knowledge. However, the theories of pedagogy identify the students as agents and the teacher as a facilitator. Pedagogy techniques in science and engineering have witnessed tremendous developments during the past two decades (Wilson, 2001; Melonçon et al., 2019). The practical aspects of future engineer's ability are required

DOI: 10.4018/978-1-7998-2245-5.ch006

for modern engineering courses. Thus, teaching with practical skills improves the understanding ability of the recipients.

An object-related reasoning oriented to the design of learning and teaching processes in academic engineering education is very helpful. Modern engineering education has to consider the updates of industrial practice in its curricula and training methods. The International Society for Engineering Pedagogy (IGIP) was formed in 1972. Currently, several universities are working to provide special modules such as engineering education in theory and practice, psychology and sociology, scientific writing issues etc. Slowly, these lead to the development of a curriculum for an advanced training course in the field of engineering pedagogy. Communication between learners and experts is another issue in pedagogy research (Scherer et al., 2018).

Today, due to the developments in computational technology and internet, the webinars and online training programmes help the teachers in developing certain technical modules. However, the regular teaching courses cannot afford the full-time internet orientation and power point presentations at many places. Some other means are therefore necessary. The demand-oriented advanced training is accredited both by the International Society for Engineering Education (IGIP) and by the Scientific Society for Engineering Education (IPW). The IGIP curriculum follows a classical study structure. First, discipline-oriented foundations (pedagogy, psychology, sociology) are taught, followed by selected applications (project work, communication, etc.). The IPW curriculum also focuses on the target group of teachers of engineering sciences. It is a very open curriculum, which can be adapted to the needs of the respective university and the respective engineering sciences. The decisive positive fact is the intensive strive for the pedagogical qualification of university teachers. The exposure of students to the product life cycle gives the technical skills of automation and the critical thinking skills needed to use the contemporary digital design and manufacturing tools. In teaching computer aided design for example, the parametric and feature based modeling tools are required in parallel to core topics such as surface and solid modeling. Likewise, motion simulations and finite element tools are part of the CAD lectures to create interest and utilize the learned concepts. While explaining the pedagogical techniques of design for non-design students, recently the concepts of design based pedagogy were described (Royalty, 2018). It was defined as an education environment with instructional scaffolds that allow students to solve problems through design practice. Such a concept was proved to result in a robust education environment. Similarly, from production perspective, smart manufacturing and automated process planning technologies can be adapted in learning environments to understand more effectively the concepts of manufacturing processes. In recent discovery, it was found that content, pedagogy and technology form a knowledge frame work. Content knowledge refers to the teachers' knowledge about the subject matter, while the pedagogical knowledge is the teachers' knowledge about the processes and practices or methods of teaching and finally, the awareness of tools and resources related to a technology may be referred as technological knowledge. The agglomeration of all these knowledge bases is found in modern pedagogical practices.

A pedagogical framework for assisting educators to develop the learning modules is needed and hence supports the students in learning sustainable manufacturing technology. Active learning resources improve the participation of engineering students. Educational system requires skilled instructors actively engaging the students in learning. Introducing CNC technology, hydraulics and pneumatics concepts in curricula as active learning sources contribute to learn from commonalities of technology and education. Recent times, the concepts of robotics, artificial intelligence, condition monitoring, and energy harvesting procedures have become more popular and require additional active learning sources such as simulators and small scale training kits during teaching. In addition to tutorials, the live demos and active resources would help the teachers in delivering the lectures perfectly. In the next two sections, content knowledge, constructive pedagogy concept and few application courses are presented.

TECHNOLOGICAL CONTENT KNOWLEDGE AND CONSTRUCTIVE PEDAGOGICAL APPROACHES

Due to rapid advancements in technology in recent times, it is essential to integrate technology into different teaching methods. Teachers need a good understanding of how technology can be coordinated with pedagogy and content knowledge. A framework that includes technological knowledge (referred as technological pedagogical content knowledge framework) was first introduced by Shulman in 1986 and later developed by several authors (e.g., Mishra and Koehler, 2006). It consists of seven cognitive tools derived from a critical synthesis of the knowledge employed by most of the effective teachers. Here, there are three main components of technological knowledge (i) basic and advance technologies, (ii) Pedagogical knowledge referring to instructional method process and practices, (iii) Content knowledge referring to the subject matters of the students (Saengbanchong et al., 2014). The framework has been conceptually developed and assessed to signify it as a perfect guideline for the development of the pre-service teachers. It is the technology utilization in supporting scientific inquiry in the classrooms. Technological pedagogical knowledge is the integration of technology with pedagogical methods like engaging the students with technology-oriented programs. The knowledge of the technological tools used in data collection and analysis come under this line.

Today, technology skills alone are not sufficient. Even teachers have awareness to utilize technology, its use to promote student learning activities is also important. That is a balanced knowledge with respect to technology, pedagogy and content is required. An extension of pedagogical content knowledge is pedagogical content knowledge. This is based on the fact that knowledge about content and knowledge about pedagogy exist independently but their overlap creates a new form of knowledge known as knowledge about how to teach content of a particular subject matter. There are three kinds of knowledge in recent identification: (1) content knowledge (2) pedagogical knowledge and (3) Technology knowledge. Content knowledge is the teacher's understanding about the subject matter to be taught and learned. Pedagogical knowledge is the teacher's knowledge about the processes, practices and approaches to teaching and learning. Knowledge about students assessment and class room management come under this category. The pedagogical knowledge is not static, but changes from time to time due to experiences. The third category is technological knowledge, which is continuously changing (Koehier et al., 2013).

Moreover, the constructive approach of teaching also has become common practice in technology. It is based on combination of researches with cognitive and social psychologies. In such mechanism, students come to learn situations with already formulated knowledge, ideas and understandings. Constructive teaching approach views learners as actively engaged people. To improve efficiency, teachers need to turn as learners/researchers to strive for greater awareness of the environments. Here, teaching allows students to interpret their experiences in class and try to make sense of them. Bruner has given the following principles of constructive teaching:

1. Teaching or instructions should share the experiences and contexts to have a student willing approach.
2. Structured teaching approach is required so as grasp easily by the students.
3. Instructions should allow in facilitating the extrapolation and clarify the doubts.

Constructive teaching relies on collaboration among the students. It promotes social and communication skills. A mutual respect between teacher and students is idea behind it. Such idea is characterized by the mutual respect between the teacher and children. Driscoll defined some constructive characteristics that should form pedagogical foundation for designing learning. Complex learning environments must be provided to incorporate authentic activity. Similarly, social negotiations are to be arranged as an integral part of learning. An approach to pedagogy related to science education is Sauder's approach. Here, firstly the hands on investigative labs are arranged. Then, there is active cognitive involvement. Further, learners work in small groups. Finally, this provides the students some opportunities to construct the knowledge. Constructive learning is an active process.

MODERN ASPECTS

In the present technological era, pedagogical framework for a conceptual teaching methodology is required rather than a prescriptive one. It encourages students to think about design and developments critically. Similarly, service learning pedagogy is another concept that allows the students to apply academic concepts and theories in real-life environment. Design science research methodology (Peffers et al., 2007) implemented in modern days has five steps: (1) problem identification and motivation (2) define objectives for solution (3) design and development, (4) demonstration and evaluation and (5) communication. All these are possible with service learning concept only. In teaching and learning process, there are several user friendly tools invented in recent times. Freeware and licensed software tools are available for teachers to interactively explain many technical concepts. In areas like computer aided design, artificial intelligence and modern robotics, graphical user interfaces are very helpful to clarify several difficulties of learners. If the learning tools are available, what is the use of teachers? Really, there is a misunderstanding in younger minds. Teachers have to suggest the correct usage of these tools with relevant examples. Most of the software tools have only front end. The background equations and formulations are to be briefly described before using these tools.

Parametric Concept

While it is easy to teach programs in computer, the pedagogical framework allows a structure to make a non-computational connection to parametric thinking in order to understand and evolve the digital processes. In CAD model development, parametric concept allows to begin the model without required dimensions. Soon after the sketch is completed, the dimensions can be altered conveniently. At the beginning the concept is quite not easy to follow, however in course of time, students find it comfortable to follow the approach. The outlined framework requires students to begin with physical modeling that asks them to perform repeated tasks with only slight variations to the model each time. This approach allows them to know the effect of small changes in design parameters on the final output. The parametric design thinking is not a software application, but a way of thinking about the design process as a whole. Like parametric modeling, today feature based approaches are also playing role in teaching computer aided design tools.

Virtual Simulations

Along with computational or programmatic way of thinking, virtual simulations are also needed during modeling. Virtually recreating physical models gives the technicalities of the parametric software. For example, an aircraft model created can be given an input motion to understand its overall animation along with its position kinematics at various points. Thus, with parametric thinking, virtual simulation and automated construction, students can create and simulate complex models. Parametric software requires very explicit instructions to locate an object in space. For example, to rotate an object, the software needs to know a point of rotation, a direction of rotation and amount of rotation. As the students use their physical objects and notational system to describe these parameters, they are able understand and code the software with required instructions. Today, the pedagogical tools that help students to understand the workflow in any application are necessary, rather than simple programming or drawing modules. In modern mechatronics and manufacturing laboratories, simulation of robots in an environment using generated data is quite common. It provokes the students to think about the sequences. When the project is expected to be produced correctly, many things like rules of gravity, other phenomena are to be accounted. By watching the construction sequence on the screen allows in many new ideas. A complex interaction between machine, environment and materials is important consideration for success of automated process.

Architectural Tools

The students of architecture are facing several doubts during the learning. The academic education in architectural schools is very condensed and requires lot of personal work since the tutors just provide advices and guide the students. The architecture is a creative field and therefore, the difficulties faced by teachers regarding the teaching of their expertise within an entirely specific pedagogic structure may block the knowledge transmission. Recently introduced pedagogical methods (example Ergo) enrich (Mavromatidis, 2018) the architectural conception exercises to strengthen the architectural synthesis foundations.

Artificial Intelligence

For teaching modern computational laboratories, soft computing techniques have become important platforms. Teachers involved in such courses need an integrated technological knowledge with managerial skills for effective transfer of the concepts. Both theoretical and laboratory exercises are required in a coordinated way. The programming platform is on one hand, while the background knowledge of artificial intelligence tools on other hand makes complexity in delivering perfect information. Initial sessions on simple programs of Matlab, or MathCAD or Mathematica would help in further programming with approximate vague data employed in artificial intelligence programs. Under the umbrella of artificial intelligence, the concepts of deep learning via convolution neural networks are required with simple examples apart from the basic aspects of neural networks, fuzzy systems and evolutionary programming, The soft computing laboratory has now become common for all branches of engineering, which introduces various artificial intelligence tools. The application oriented programming concepts with presentation of lectures and simplified programming skills are therefore necessary for modern artificial intelligence pedagogy.

Mechatronics and Robotics

Early twenty first century has witnessed several developments in the Mechatronics curriculum. Several articles were published on teaching and experimentations from students' perspective. Simple experiments like graphical simulations via Simulink tool, programmable logic controllers and hydraulic/pneumatic simulators would help the beginners to understand the theoretical concepts thoroughly. In robotics point of view, today, many basic tutors are avoiding the kinematics and dynamics principles. For example, in trajectory tracking applications inverse kinematics are needed, which are rarely considered. Robot programming and control is a key area, where, the teachers should focus on the specific applications so as to give sufficient information on the related syntax to the students. Due to several unrealized targets, teaching and applications in this area are limited.

The concept of inductive teaching has become recently popular, where the students can learn the fundamental concepts and use them for problem solving. Hands-on experiences are also one part. New requirement of engineering pedagogy program is remote lab access, which are either distance labs or virtual labs. In distance labs, real hardware can be controlled with internet, while in virtual labs, it is simulated by a computer. Management system and supportive study materials help in real time study process. Thus, remote labs are ideal tools helps in teaching collaborative learning skills. One example of collaborative learning is team based teaching, which is preferred in recent times. In the concept of pedagogical robots (introduced by Felah and Noreddine 2017), robots receive the signals from computers enabling the students to carryout simulations like pick and place and assembling operations. Classification and installation, programming skills can transfer the students from theoretical to practical orientation. It refines the student personality in line with requirements of twenty first century. Here, Arduino kits, DC servo motors, sensors (encoders, ultrasonic), wheels, etc., can be used as components. If students use systematic way to solve a given problem in real time, it has no significance. Virtual reality simulators for mobile robot controls, sensor-actuator modules, robot competitions, etc., would help to augment the interest in learning. In total, there are many open application areas of pedagogical techniques in robotics and mechatronics area.

CONCLUSION

Some of the developments in modern pedagogical concepts in technology have been briefed out. Starting from general concept of pedagogy, some issues of pedagogical knowledge, constructive pedagogy were explained. Concept of integrated content-pedagogical-technological knowledge framework was summarized. Few applications of pedagogical developments in computer aided design, robotics, architectural engineering and artificial neural networks have been explained briefly. With several developments in technology, the modern pedagogical concepts are required everywhere. Students' projects should be oriented towards a product development and some practical experiments with an independent study is required from students. Even, technologically the concepts are different in various engineering streams the common pedagogical learning tools should reach every teacher. As a mandatory requirement in teaching, frequent pedagogical training programmes are required in the future.

REFERENCES

Akin, O. (1989). *Computational design instruction: towards a pedagogy CAAD Futures, Digital Proceedings*. Boston: MIT Press.

Bezjak, J. (2010). Contemporary engineer pedagogic's project research - using multimedia at technology classes in technical and vocational schools. *Procedia: Social and Behavioral Sciences*, *2*(2), 407–411. doi:10.1016/j.sbspro.2010.03.035

Chalmers, C. (2018). Robotics and Computational thinking in primary school, Int. *J. Child-Computer Interaction*, *17*, 93–100. doi:10.1016/j.ijcci.2018.06.005

Felah, B., & Noreddine, H. (2017). Pedagogical robotics-A way to experiment and innovate in education teaching in Morocco, *Int. J. Education and Learning Systems*, *2*, 71–75.

Graham, C. R. (2011). Theoretical considerations for understanding technological pedagogical content knowledge (TPACK). *Computers & Education*, *57*(3), 1953–1960. doi:10.1016/j.compedu.2011.04.010

Koehier, M., Mishra, P., & Cain, W. (2013). What is Technological Pedagogical Content Knowledge (TPACK)? *Journal of Education, 193*(3), 13–19. doi:10.1177/002205741319300303

Lanz, M., Pieters, R., & Ghabcheloo, R. (2019). Learning environment for robotics education and industry-academia collaboration. *Procedia Manufacturing, 31*, 79–94. doi:10.1016/j.promfg.2019.03.013

Mavromatidis, L. (2018), Coupling architectural synthesis to applied thermal engineering, constructed thermodynamics and fractal analysis: An original pedagogic method to incorporate sustainability into architectural education during the initial conceptual stages. *Sustainable Cities and Society, 39*, 689-707.

Melonçon, L., Rosselot-Merritt, J., & St.Amant, K. (2019). A Field-Wide Metasynthesis of Pedagogical Research in Technical and Professional Communication. *Journal of Technical Writing and Communication*. doi:10.1177/0047281619853258

Mishra, P., & Koehler, M. J. (2006). Technological Pedagogical Content Knowledge: A Framework for teacher knowledge. *Teachers College Record, 108*(6), 1017–1054. doi:10.1111/j.1467-9620.2006.00684.x

Peffers, K., Tuunanen, T., Rothenberger, M. A., & Chatterjee, S. (2007). A design science research methodology for information systems research. *Journal of Management Information Systems, 24*(3), 45–77. doi:10.2753/MIS0742-1222240302

Royalty, A. (2018). Design-based Pedagogy: Investigating an emerging approach to teaching design to non-designers. *Mechanism and Machine Theory, 125*, 137–145. doi:10.1016/j.mechmachtheory.2017.12.014

Saengbanchong, V., Wiratchai, N., & Bowarnkitiwong, S. (2014). Validating the Technological Pedagogical Content Knowledge appropriate for instructing Students (TPACK-S) of pre-service teachers. *Procedia: Social and Behavioral Sciences, 116*, 524–530. doi:10.1016/j.sbspro.2014.01.252

Scherer, R., Tondeur, J., Siddiq, F., & Baran, E. (2018). The importance of attitudes toward technology for pre-service teachers' technological, pedagogical, and content knowledge: Comparing structural equation modeling approaches. *Computers in Human Behavior, 80*, 67–80. doi:10.1016/j.chb.2017.11.003

Sweet, K. (2016). Resurrecting the master builder: A pedagogical strategy for robotic construction. *Automation in Construction, 72*, 33–38. doi:10.1016/j.autcon.2016.07.001

Wilson, G. (2001). Technical Communication and Late Capitalism: Considering a Postmodern Technical Communication Pedagogy. *Journal of Business and Technical Communication, 15*(1), 72–99. doi:10.1177/105065190101500104

Section 2

Education in the Industry 4.0 Era

Chapter 7
Education in the Era of Industry 4.0:
Qualifications, Challenges, and Opportunities

Dharmendra Trikamlal Patel
https://orcid.org/0000-0002-4769-1289
Charotar University of Science and Technology, India

ABSTRACT

Industry 4.0 has changed the thinking of industry owners in terms of technological usage. With the help of modern digital technology, industry can fulfill the requirements of customers easily and compete strongly against their competitors. In order to achieve good quality of products at an affordable price, industry needs skilled people who are aware of autonomous and intelligent components. To prepare skilled people compatible with Industry 4.0, education plays a very important role. The chapter starts with which kind of qualifications are needed to fit in the smart factory era. In next section, the chapter deals with challenges that emerge in education in order to implement skills suitable for Industry 4.0. Lastly, the chapter describes opportunities for the education sector as far as the smart factory is concerned.

DOI: 10.4018/978-1-7998-2245-5.ch007

THE WAY OF SMART FACTORY: AN EDUCATIONAL PERSPECTIVE

In 1780, the first revolution of industrial manufacturing (Nic Von,1996) had started. No technology was used in that era and manufacturer heavily depended on laborers for any kind of productions. Laborers had to do the mechanical kind of work so not a specific kind of qualifications were expected from them.

The second industrial revolution (Joel Mokyr,1998) had started in 1870 which is considered as the technical revolution as the manufacturer had started using numerous technologies. This revolution comprised of heavy usage of manufacturing machineries, communication via telegraph, electrification, use of petroleum and transportation by means of railroads. This revolution had changed the thinking of the education sector and they started more emphasizing on atomization, telecommunications and electrification concepts.

The era of 1970's is considered as the third revolution of industry (Xiaowen, 2016). The ways of communication in the form of Internet and mobile devices have changed the entire thought process of industry personnel. Artificial Intelligence has replaced laborers with automatic entities like robots. The use of 3D printing is unbelievable. The industry has started by using renewable energy in their production. Agriculture sector became dominant as genetically modified crops, farming started. Due to nanotechnology, new materials became lighter and more durable. The third revolution is the most rapid revolution due to information and electronics technologies. The education sector has boosted up the speed of the third revolution by producing right level of skills in diversifying areas like information technology, electronics technology, nanotechnology and synthetic biology. The third revolution has changed choices of Indian students and educational institutions and universities as well.

Cyber-physical systems (Antsaklis, 2014, Shi J, et al., 2011) and interoperability among machines have given the birth of the next revolution, i.e. Industry 4.0(Aehnelt et al., 2014, Bauernhansl et al.,2014, Brettel, et.al,2014, Kolberg et al.,2015, Weyer et al.,2015) or Smart Factory(Aehnelt et al.,2014). Smart factory revolution facilitates any organization to digitally manage the entire life cycle of the project starting from planning to the testing phase. Smart Factory(Groover,2007, Kane et al.,2015) emphasizes on two things: **(a) Information Technology** that is responsible for business process automation and **(b) Operational Technology** that is responsible for industrial process and factory automation. The Machine to Machine communications and Human Machines Interface permits machines with intelligent sensors to converse as human language to ERP system(Lazovic et al.,2014, Scheifele et al.,2014). Internet of Things (IoTs) based technology (Kovatsch et al., 2012) plays a crucial role in the integration of IT and OT. The main challenge for educational institutions is to produce skills that fulfill demand of IT and OT. The People having fusion

Figure 1.

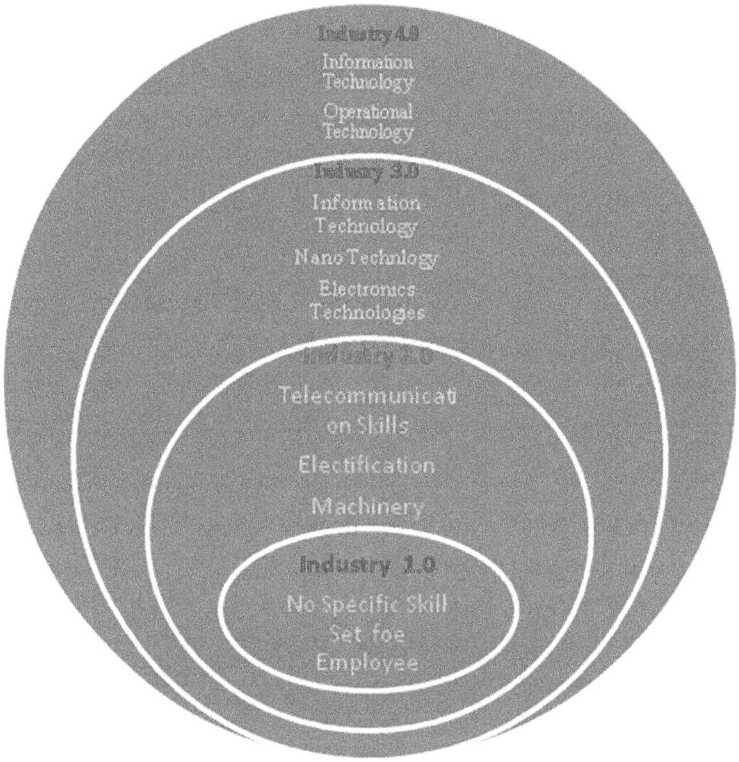

knowledge of the domain and Information technology will survive in today's era. The knowledge in the field of cyber-physical systems, hand held robotics, RFID (Priego et al.2014), NFC, Intelligent networking, etc. is expected from people to survive in this highly digital era.

The following **Figure 1** describes educational evolution in terms of the industrial revolution. From the figure it is determined that Industry 3.0 and 4.0 has completely changed the scenario of educational systems. A person having only domain knowledge will not survive in a smart factory era due to extensive use of digitization.

The Cyber-Physical system has changed the direction of an educational organization. Educational Cyber Physical System is the need of the modern education. Educational Cyber Physical System needs many components to work it efficiently.

1. **Collaborative Learning Tools**: It provides the communication and interaction between teacher and learner.
2. **Learning Management Tools**: Administrative functionalities to manage learning processes and data.
3. **Assessment Tools:** It is needed to assess the learning progress of the students.
4. **Educational Guidelines:** It takes care of the student development from different angles.
5. **Intelligence Tools:** Covers some intelligent aspects for better teaching-learning.
6. **Engagement Tools:** It provides high level participation experience to the students.
7. **Integration Tools**: It provides integration of sound, video,3D animations, text to the system for better experience of teaching-learning.

Expected Qualifications For Smart Factory

Smart factory is a fusion of real world manufacturing with the extensive use of Information and Communication Technology (ICT). Conventional expertise of employee must be tailored to meet up the requirements of this fusion approach(Brauner, P et.al.2015) to improve the quality of the production. Following listed qualifications are expected other than domain knowledge from people to fit in the Industry 4.0 standard:

- **Industrial ICT Specialist**: Fusion expertise of electronics and ICT based hardware and software.
- **Intelligent Network Analyst**: Expertise in distributed sensor/actuator networks.
- **Robotics**: Expertise in autonomous robots.
- **3D Printing**: Expertise in planning of business and factory process.
- **Cloud Computing**: Expertise in store, manage and process the data from remote servers.
- **Big Data Analytics**: Expertise to uncover hidden, valid and useful patterns from large data sets.
- **Augmented Reality Expert**: Expertise to give support to workers by means of real like software.
- **Information Security Expert:** Expertise to solve security issues.
- **ERP Expert**: Expertise in facilitating integration of ERP system and generate desired output.
- **Quality Assurance Expert:** Expertise in assuring quality of product.
- **Fog Computing Specialist:** Expertise in computation among devices.
- **RFID Expert:** Expertise in tracking smart devices.

Interdisciplinary Education For Industry 4.0

Industry 4.0 builds from four main dimensions such as automation, cyber physical system, Internet of things and cloud computing paradigms (Schlechtend et al.,2014). Substantial outcomes could be achieved from this new paradigm of manufacturing system if an organization have a right mix of skills from above mentioned dimensions. A person is considered as an expert in a specific dimension if he has knowledge of all sub dimensions and educational aspects mentioned in **Table 1**.

CHALLENGES OF EDUCATION SECTOR FOR INDUSTRY 4.0

The success of Industry 4.0 is profoundly depending on the skill sets of employees of an organization. Employees can achieve right skill sets, compatible with Industry 4.0, through proper education. Industry 4.0 requires interdisciplinary skill sets and that creates plenty of challenges for the education sector.

Educational Challenges of Smart Robot Based Manufacturing

Traditional Manufacturing was required only core engineering knowledge to absorb in any industry. Historical manufacturing role were: Machine Operator, Mechanist, Technician, Coating Workers, etc. The fourth revolution of industry has changed the entire manufacturing (Schuh et al.,2014) thought process and robots become an integral part of manufacturing. Involvement of robots in manufacturing has created many challenges for education sector to produce right skills of employment. The challenges are:

- **To design the curriculum that comprises of concepts of smart robot based automation (Dawande et al.,2009)**: Robots are one of the key components in automation. If the curriculum of university education lacks of the concepts and methodologies of robotics, it is quite impossible to prepare the students for the future Industrial based education. Now a day's robotics is used in every aspect in industry so inclusion of robotics in effective way is very vital in the curriculum of university education. The organization has to set up appropriate robotics labs to map the theoretical content with the practical aspects.
- **To prepare teachers who can give real industry based training to students or employees of an organization:** The knowledge is transforming from teacher to students in any education system. If teachers do not have appropriate knowledge that cannot be transformed to students so teacher

Table 1. Educational aspects of industry 4.0

Dimension	Sub Dimension	Educational Aspects
Automation	Control System	• Methodology to improve the accessibility of the control system • Behavior of System • Sensors working knowledge • Controller knowledge • Feedback System • Error Signals • Tracking of devices • Stability Concepts • Linear and Non Linear Models • Transmission delay • Soft and Hard Automation • Computer Aided Automation • Robots • Adaptive Control • Tool Interpolation
Cyber Physical System	Architecture	• Seamless Integration of Control, Communication and Computation • Rapid Design and Deployment • Modular Approach of designing • Standards
	Computations	• Distributed Algorithms • Software tools and techniques • Computational Challenges
	Network Control	• Time Delay • Failures • Wireless Networks Concepts • Robustness • Heterogeneous Cooperation • Reliability, Availability and Security issues
	Verification	• Trustworthiness Issues • Interoperability
	Connection	• Sensor Network • Plug and play mechanism • Security based communication
	Conversation	• Analytics for Performance Prediction • Analytics for Data Correlation
	Cyber	• Discrete Mathematics • Data Mining Techniques
	Cognition	• Visualization Techniques • Simulation Techniques • Decision Making
Internet of Things	RFID	• RFID in comparison of Barcode • RFID Tags • Passive and Active Transponders • Middleware • RFID Enabled Applications • RFID Standards • Technical issues of RFID • Implementation of RFID

continued on following page

Table 1. Continued

Dimension	Sub Dimension	Educational Aspects
	Sensors	• Working Principle • Wireless Sensor Network • Sensing Devices and types • Characteristics of Sensors • Digital Sensor Processing • Architecture and adoption of Smart Sensors • Challenges of Sensors
	Actuators	• Working Principle • Types of Actuators • Characteristics of Actuators
	Virtual Objects	• Structure of Virtual Objects • Interactions of Virtual Objects • Integration and Communications issues • Layers of Virtual Objects • Interpretation of Virtual Objects • Behavior of Virtual Objects
	Analytics	• IoT ecosystem and role of analytics • Predictive Modeling • IoTs data flow • Machine Learning Techniques
Cloud Computing	Fog Computing	• Fog Computing Characteristics • Fog Computing Components • Fog Computing Software platform • Fog Computing Applications • Modeling and Simulations

training is very essential in education. Universities have to prepare the teacher in such a way that they can give effective knowledge to their students. Training identification in reputed organization and to adjust the timing of teachers is very challenging for any educational organization.

- **To design simulation that includes proper skill sets of robot:** As discussed in first topic, robotics is an essential in any industry and concepts of robotics can be used in designing of the training module. Industry based simulation needs appropriate skill sets of robotics. For any educational organization, to establish such simulation needs huge funding.
- **To determine how robot ought to work and how changes in circumstances and environment will affect the final outcome**: Testing of all the aspects of robot's behavior is very essential to determine the final outcome of the process. The challenge is to understand all kinds of behavior of robots with different environmental situation.
- **To determine causes of operating errors:** When we deal with robots, identification of causes of operating errors is very crucial. Some proactive mechanisms should be implemented to determine the causes of operational errors.

- **To determine which kind of tests are needed for effective quality control analysis:** Prepare effective test cases to evaluate the quality of the process is very essential.
- **To adopt active learning effectively**: Using robots, students should be able to do analysis, amalgamation and an assessment of the curriculum content.
- **To identify complex problem solving to develop desired solutions:** Collaborative robots should be designed to solve complex problem of industry and it should be the part of the university curriculum. In university curriculum case studies or projects should be designed that meets the current requirements of an industry.
- **To define appropriate learning strategies to select appropriate training methods and procedures:** Training of robotics is very vital and for that teachers should decide their learning strategies.
- **To use technical rules and routines to resolve any problem:** Technical knowledge regarding robotics is very important to resolve any technical problem.
- **To decide the appropriate programming language to write code for various purposes:** Programming language selection is very vital to take the work from robots. Industrial robot languages are very important in this context. Few examples are: RAPID programming language, Kuka Robot Lanaguage,PDL2,INFORM, AS,Karel,VAL3, UR Scripts,ROS Industrial
- **To install programs, machines, etc. to meet desired specifications:** Prepare appropriate technical infrastructure is very important to achieve the desired specifications of the robot.
- **To gain practical exposure of circuit boards, processors, chips, computer hardware and software:** There are a variety of equipments are available to design a robot but selection of an appropriate one is very vital.
- **To gain knowledge of mathematics such as algebra, geometry, calculus, statistics, etc**.: Robot controlling is based on mathematical concepts so deep knowledge of mathematics is very important.
- **To understand knowledge of physics, such as physical principles, laws, their interrelationship etc.:** Physics knowledge is very essential in robot designing such as physical principles, laws, applications, material, mechanical structures etc.
- **To be familiar with knowledge of telecommunications**: Communications through robot is carried out using telecommunications principles so knowledge of it is very essential.
- **To establish e-learning modules to learn from anywhere:** Designing effective e-learning modules about robotics that comprises of basic and advanced knowledge of robotics.

Educational Challenges for Production Line Simulation

Industry 4.0 emphasizes on digital manufacturing so production line simulation is extensively used by industry to simulate diverse facets of their production operations. Use of simulation in industry creates abundant challenges; here challenges in education perspective are listed.

- **To design education and training syllabi after clear understanding of industry needs:** Syllabi should cover current requirements of industry. Before designing the syllabi certain trips to well known industry is required to identify their needs. It is also important to take the feedback of industry experts while designing curriculum. For the educational organization, to identify such experts and take their views is the most challenging task.
- **To identify an effective simulator that gives real exposure of industrial processes:** Practical insights of any process of an industry are very important and that can be available to students in form of simulator. Identification of an effective simulation that exhibits current needs of an industry is the most important challenge for the education community.
- **To afford the cost of the simulator:** If an effective simulator is identified, then another challenge is the cost of it.
- **To prepare effective manuals of simulator:** If the simulator is affordable then how to use it is the main challenge.
- To train the trainer about the simulator
- **To determine the causes of operational errors**: When we deal with the simulator, identification of causes of operating errors is very crucial. Some proactive mechanisms should be implemented to determine the causes of operational errors.
- **To optimize code to achieve good performance:** To achieve good performance of the simulator, the code should be optimized. Remove unnecessary code and apply logic with few lines is a very challenging task.
- **To take effective decision based on the simulator's result or errors**: Sometimes simulator generates very inefficient results due to lack of data availability in that case decision based on a simulator is most challenging tasks.
- **To identify measures or indicators to measure system performance**: Identification of important measures and indicators is most challenging tasks to generate efficient results of the simulator.

- **To establish an e-learning module for quick and anytime-anywhere learning:** Designing effective e-learning modules about production line simulation that comprises of the overall concepts of the modern requirements of an industry is most challenging tasks.

Educational Challenges for Big Data Driven Quality Control

Industry 4.0 based digital technologies generates an enormous amount of data that is termed as Big Data. Analysis of Big Data (Lee et al.,2014) generates very efficient insights in manufacturing, but many challenges are associated with that. In this section, only challenges related to educational perspectives in terms of Big Data Quality Control will be discussed.

- **To select set of appropriate big data analytics tools that gives insights to all manufacturing organizations**. Smart industry generates a gigantic amount of data from sensor connections, on demand computing and cyber-physical devices. Analysis of such data is a very challenging task and it needs appropriate analytical tool. There are various categories of analytics tools are available in the market so selection of an appropriate one is very challenging in context to university education.
- **To train the trainer about set of big data analytics tools:** Big Data Analytics needs knowledge of programming, quantitative skill, knowledge of spreadsheets, structured query languages, Linux, Hadoop, statistical packages etc. To train faculty members in all skill is quite an impossible task for education universities.
- **To identify useful metrics and key performance indicators for better quality control:** To achieve good results of analytics, it is important to identify quality measures and indicators.
- **To establish technological architecture to handle exponential data growth:** To establish an infrastructure that seems to real industry is a major challenge for any education university. It is very costly, so the question is about affordability of it for education institutions.
- **To decide acceptance sampling for better judge the quality of product:** For better analytics, adequate data are very important. The main challenge for educational institutions is to generate such amount of sample data that is needed for better and effective results.
- **To gain knowledge about statistics to decide quality measures:** The main base of analytics is the adequate knowledge of statistics. To achieve statistical knowledge in the context of application is a very difficult thing for faculty members and students of different streams.

- **To awareness about new tools and techniques of big data in terms of quality control:** Regular updating knowledge about the tools and techniques of big data is very essential to provide better results from available data. It is very challenging in perspective of teachers of different streams to gain such knowledge and apply new techniques on big data.
- **To gain knowledge about quality standards and quality assessment methods for big data:** Big Data is different than usual data in many perspectives such as volume, variety, velocity, veracity and value. The traditional methods of quality standards and assessment are not enough for big data and to identify an appropriate one is a very challenging task.
- **To adopt suitable and efficient data integration techniques as diversity of data sources:** Data is of a variety of big data. Integration of a variety of data from heterogeneous systems is most difficult exercises in context to big data.
- **To judge data quality in a specified time frame:** Quality of big data depends from multiple dimensionalities so judging it in a specific term is the most difficult task.
- **To gain knowledge about predictive quality analytics:** The main objective of analytics is to predict about situation of an industrial operation. Predictive algorithms are very difficult to understand and implement. Predictive analysis is effective on accurate data.
- **To modularize material of big data driven quality control for easy and efficient access:** Big data involve multiple interdisciplinary fields. In order to gain appropriate knowledge about big data, the entire coverage should be modularized among different modules. To divide the content of big data in different modules exhibits great challenge as one module needs the knowledge about other module.

Educational Challenges For Smart Supply Network

Industry 4.0 leads to a smart supply network that improves order fulfillment, profit, transparency and return of investment. The Smart Supply network depends on a number of key technologies: autonomous and integrated planning, smart execution systems, autonomous logistics, smart procurement, intelligent warehousing and advanced analytics. The main objectives of such smart supply network are to design entire system that is both flexible and openness. The main challenges of such system in educational perspectives are:

- **To make available software and technologies related to smart network:** Network connection is important element in context to smart factory as it needs for communication among smart devices. The network of smart factory is far different than usual computer network as communication is done among devices. The new way of software and technologies are needed to make efficient communication among devices and it creates a great challenge for education community to establish such smart network.
- **To train human resources in a way that they adopt these technologies easily:** To find appropriate trainer in such a new technology is a great hurdle for education community.
- **To design and integrate multi-organizational service network for achieving real exposure:** Smart network involves integration of operations from multiple organizations and to achieve such kind of infrastructure in education institutions create lots of managerial and technical hurdles.
- **To collaborate with benchmark organizations to prepare right content for smart supply network:** To identify and collaborate with benchmark organizations for smart network is very difficult task for education organization as benchmark industries do not have such time and sometimes they do not disclose confidential data and processes for public.
- **To design proper cases of future network to teach employee in right paths:** Without proper basic knowledge of smart network, it is very difficult to design cases of future network for any educational institution.
- **To measure the effectiveness of processes:** Appropriate data and indicators are needed for education organization to measure effectiveness of any smart network based process in premise of institution.
- **To apply proper tools and techniques for data analytics:** There are number of data analytics tools are available in market so selection of an appropriate is very difficult task for education institution due to lack of knowledge about it.

Educational Challenges For Predictive Maintenance

In manufacturing organization, maintenance is a critical and main challenge associated with it to ensure the maximum availability of machines with minimum cost. Industry 4.0 brings new concept of maintenance that is predictive maintenance (Lee et al., 2013) that covers all maintenance related process steps ranges from automated detection to final recovery of the machine. Education sector plays very important role in the implementation of this new way of maintenance, however, there are a number of challenges associated with that.

Figure 2.

- **To understand an appropriate predictive maintenance tool before using it in training purpose:** Predictive maintenance tools involves complex prediction based algorithms that are very difficult to understand for stakeholders for education organization.
- **To provide the correct type of training encompasses of technology and managerial skills:** To find appropriate trainer in this modern technology is very difficult for education organization.
- **To ensure that employee using the tool correctly:** In order to use tool correctly, efficient training is required for teachers. The tool should have appropriate manuals to operate and efficient and affordable trainer should be available in market.
- **To use proper methodology for efficient predictive maintenance:** There are many prediction algorithms are available and use any one in context to current situation is very difficult task for education
- **To determine technique that deals with large amounts of data with heterogeneous properties:** Use of Big Data Analytics for education community is very difficult task.

Figure 3.

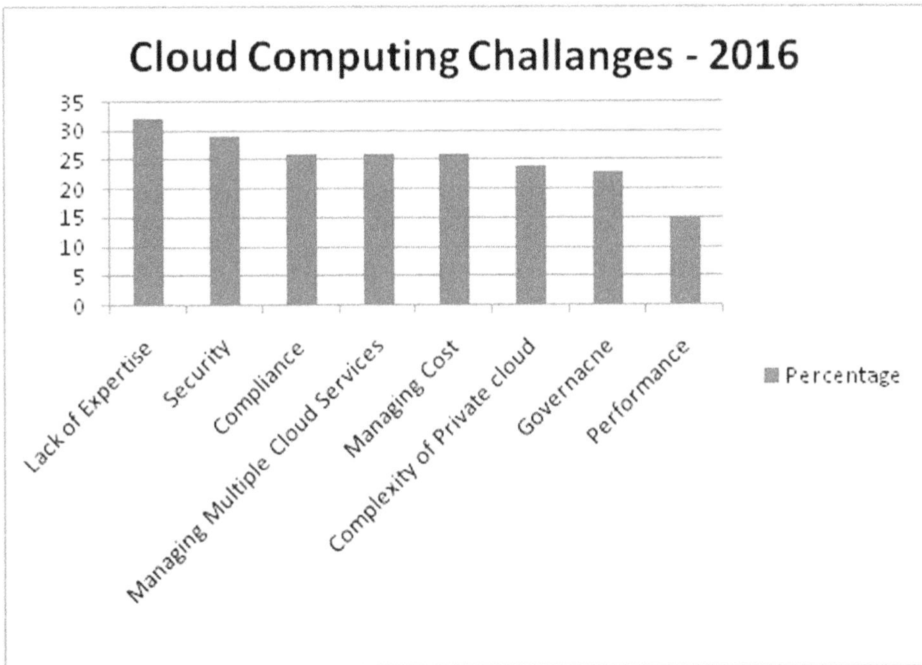

- **To train the people regarding error awareness and derives methods to deal with them:** Exhaustive training is required to solve error in smart network and there is a scarcity of knowledgeable and affordable trainers.
- **To identify meaningful features to indicate upcoming failure:** Prediction about future error depends upon the use of right predictive algorithm. To identify right predictive algorithm needs extensive knowledge about all predictive algorithms that creates huge hurdle for education institutions.
- **To setup experimental environment that continuously monitor the machine and suggest corrective actions:** To establish real industry exposure in educational institution creates great challenge in terms of affordability and technicality.
- **To determine the evaluation techniques for optimum prediction:** Optimum prediction needs exhaustive knowledge about smart network but it is quite difficult for education institutions to get it without qualified trainer.
- **To design training modules effectively:** Designing a training module of smart network needs basic and advanced knowledge of tools and methodologies. These kind of knowledge cannot achieved without proper trainer and to get trainer in this modern topic is quite impossible for education organization.

Educational Challenges in Implementation of Industry 4.0 Based Digital Technologies

Industry 4.0 is all about digital technologies. The following figure 2 describes the major digital technologies that smart factory needs for effective implementation.

- **Cloud Computing:** According to RightScale2016 (RightScale,2016) state of the cloud report, the main challenge in implementation of cloud computing is the lack of expertise in that field. Universities or education sector is not capable to produce skilled people in this area. The figure 3 describes of main cloud based challenges in percentage. Lack of Expertise is highest among them with 32%. Other challenges such as security, compliance, managing multiple cloud services, governance, and performance exist in higher proportion due to lack of expertise.
- **Internet of Things(IoTS):** International Data Cooperation(IDC) did one survey of 2350 large and medium size industries worldwide about the Internet of Things. Near about 50% companies either deployed IoTs or plan to implement in one year. Most survey participants are interested in generating computation at the "edge" rather than at the data center. The main challenge with IoT is security. Security challenges are the main due to lack of expertise in mitigates risks of "edge" computing. Educational universities do not have the right skills to train students in this area. Another issue with the educational sector is that they have a scarcity of professionals who have cross-industry insights that is very essential in this booming technology of the Internet of Things (IoTs).
- **Location Detection Techniques:** Accurate indoor localization is more vital in the era of Industry 4.0. Newly invented ubiquitous indoor positioning system is an ideal for smart factory as it works with heterogeneous devices. People do not have much knowledge about security, applications, spatial algorithms, performance issues etc. about this new technology. The challenge for educational universities is to gain knowledge about all above discussed topics to train people effectively.
- **Human Machine Interface:** Smart human machine interface based requirement is drastically increased due to Industry 4.0 based technologies. Smart HMI enhances the productivity of operators and able to control or maintain machine effectively. The Manufacturer wants to operate machines like smart phone. It must have intuitive visualizations (Chen,2004) options that can integrate with other components of industry. The main challenge for the education sector is to devise the requirements of the interface based on specific industry and generate standards that can be applied to any industry.

- **Fraud Detection:** In the digital revolution of industry, fraud detection is an important aspect to achieve high reliability and availability of any information system. According to ThreatMetrix study of 2015, fraudulent attempts reached 25 million per month. The main challenge for educational universities is to prepare efficient cyber security experts who can devise proper techniques for fraud detections.
- **3D Printing:** With the advancement of commercial manufacturing, the role of 3D printing (Bernd Bickel et al.,2013) becomes vital. According to the report of the Consumer Technology Association (CTA) and United Parcel Service (UPS) 98% of hearing aids worldwide is manufactured using 3D printing. The challenge for the education sector in this context is to prepare software professionals who can design the software for this advanced technology.
- **Big Data Analytics:** In the modern industrial scenario, huge data are generated by machines and devices. Analysis of this data effectively gives valuable insights to industry. The main challenge for education is to select appropriate analytics tool that fits in major domains. To gain knowledge about modern techniques of analytics with very less references also a main hurdle for the education sector.
- **Mobile Computing**: Mobile or ubiquitous devices are integral part of Industry 4.0 based manufacturing. The software designing of mobile applications is totally different than computer based traditional software. The role of educational institutions is to prepared software designers in the context of mobile devices. Another issue is the security of mobile based transactions. The challenge for the education sector is to gain knowledge about all aspects of threats and train employees in such a way that they are in a position to mitigate or monitor all kinds of probable risks.
- **Augmented Reality**: Augmented reality becomes an essential part of manufacturing due to the quick expansion in mobile devices, telecommunications technologies, data storage, and wireless data transfer. Augmented reality can be used for many purposes in industries, but training is the greatest purpose among all. Designing a training module for specific section using concepts of augmented reality is the main challenge for the industry. Educational sectors play a very important role in this matter, but the main hurdle for them to prepared, skilled people of this technology and to gain knowledge about this. A report from Juniper Research shows that the use of Augmented reality apps in the enterprise will grow to $2.4 billion in 2019, up from $247 million in 2014 means tenfold increase over five years. The main challenge for educational sector is how to create skilled manpower to cope up with this situation.

- **Wearable Devices**: Wearable devices can be used in manufacturing for several tasks such as employee and plant monitoring, problem diagnosis, improving employee safety, etc. Gartner research director predicts that wearable could help employees diagnose and revamp problems more quickly, saving up to $1 billion annually in three to five years. The main challenges of wearable technologies in manufacturing industries are design of device and data privacy. The success of any wearable device is how it is able to replace the traditional device. The biggest challenge of wearables is to ensure the privacy and security of client data. The challenge for educational sector is to understand a scenario of wearable devices, in context to manufacturing industries and design effective curriculum according to it so skilled peopled can be produced.
- **Smart Sensors:** Smart factory entrench smart sensors for various tasks such as monitoring of machine performance. Understand and implementation of intelligent, smart sensor standards is crucial for any educational institution to train industry personnel in the right direction. To establish a proper research environment and funding is also a crucial for the education sector.

EDUCATIONAL OPPORTUNITIES FOR INDUSTRY 4.0

Smart factory has created enormous opportunities for the education sector to provide an efficient environment for these new technological stuff. Educational sector plays an important role in the effective implementation of Industry 4.0 based technologies. Following are the main educational opportunities due to Industry 4.0 based technologies.

(1) **An Industry coalition for curriculum designing:** Educational sector has a huge opportunity in designing industry specific curriculum with the coalition with Industry people. Opportunities for educational sector in curriculum designing context are:
   (a) **Industry based Educational Criteria:** Curriculum designing can consist of action oriented, problem solving, and task and issued based curriculum criteria rather than traditional criteria.
   (b) **Integration of more Industry based Internships:** Curriculum can enhance by integrating more internship opportunities with coalition with industry people.
   (c) **Learning outcome based Curriculum:** Curriculum can emphasize learning outcomes based on workforce needs of an industry. Employability can be the main outcome of curriculum designing.

(d) **Fusion Courses:** Curriculum can consist of fusion courses of domain area and emerging technologies related to Industry 4.0 based on the current needs of industry.

(e) **Project Oriented Teamwork**: Curriculum can involve project oriented pedagogy so tacit knowledge of the person can be accumulated. The Project's definition can be decided by consultation of industry experts.

(2) **Digital Chain of Manufacturing:** The opportunity for educational organization is to design entire educational program in digital form rather than the classical way. Educational institutes can strengthen the way of teaching-learning by simulations, virtual figures of machine and online e-contents.

(3) **Set up Labs with Intelligent Devices:** It is a great opportunity for educational institutions to set up a laboratory with intelligent devices and attract students to work with those devices. Educational Institutions can tie up with industry for setting up laboratories and get exposure of technical knowledge of industry experts.

(4) **Develop right skills for Industry 4.0:** Educational institutions have immense opportunities to develop ready-to-absorb people by setting up the right environment for industry 4.0. It consists of right curriculum, trained faculty members, effective learning environment and suitable infrastructure for industry 4.0.

(5) **Establish Research cell for Industry 4.0:** Educational institutions can establish research cell for technologies which involved in Industry 4.0. This is a vital opportunity for educational sector to look differ than other institutions by establishing research initiatives in this modern area.

(6) **Increased Job Opportunities:** Educational sector can increase job opportunities for students by effective implementation of smart factory based technologies in the curriculum.

(7) **Solution to Industrial Problems:** Educational universities can solve problems of Industry by preparing students and faculty members successfully for Industry 4.0 based technologies. The education sector can give right practical exposure to students and faculty members to solve any critical problem of industry.

FUTURE RESEARCH DIRECTIONS OF INDUSTRY 4.0 IN CONTEXT TO EDUCATION

Industry 4.0 merges the virtual world with real productions. It covers several areas such as information technology, logistics, electronics and mechanical. There is a huge scope of the research in this area, particularly in context for education.

There is a huge scope of the research in designing an automatic tool that exhibits, digital chain of manufacturing; based on Industry 4.0. This research involves implementation of Industry 4.0 concepts in educational simulation. The research challenge is to provide an efficient virtual world of production that feels like a real world in simulation software. It is one kind of collaborative research where knowledge of information technology, electronics and mechanical engineering are essential.

Designing a virtual reality based training module apps for students and teachers exhibits good research scope. The research involves the optimization of handheld devices resources without compromising the look and feel of training module. The challenge is to implement concepts of virtual reality in devices with very limited resources.

Designing an analytical module that covers problems of industry, based on appropriate visualization techniques have enormous research scope. This kind of module is very helpful for students to understand industry problems. The module requires a gigantic amount of data that are very difficult to analyze. Deciding appropriate visualization technique in context of an industry problem creates a huge challenge.

The research directions in Industry 4.0 in context to education are to be designed automatic software tools that give exposure about Industry 4.0 to all stakeholders. Designing software is required in any kind of devices such as personal computer, mobile, wearable devices, etc. and extensively use the methodologies of augmented reality. The research direction of Industry 4.0 in the context of education is very limited and constrained to only designing of automated tools. However, there is huge research scope in Industry 4.0 based technologies. Following are several glimpses of it:

- **Automation of industrial machines and plants using distributed embedded systems**. This kind of research work needs expertise of information technology, electronics, robotics and ERP. A special focus is required in designing of the system that deals with transparency and handles the complexity of entire system workflow.
- **Supporting industrial operations by designing and evaluation of Human-Machine-Interfaces**. This research extensively involves visualization techniques and augmented reality concepts.
- **Innovative usage scenarios inside the industrial plan**. This kind of research uses big data analytics along with visualization techniques. This research enhances diagnosis during plant operation.
- **Automated methods and tools for predictive maintenance.** This kind of research involves analytics and visualization techniques. The system is capable of predicting maintenance schedule automatically based on data received from the sensors.

- **Demand oriented adaption of man power**. This kind of research involves analytics based on demand-supply philosophy of production.
- **Modeling for an integrated and comprehensive security in devices.** This research emphasizes on security aspects of cyber physical devices. It involves vulnerabilities and industrial attacks.
- **Quality assurance Analytics.** For industry perspective, quality is the main concern. This research involves analytics from multiple dimensions.
- **Cloud based manufacturing control**. This research involves the cloud computing concept with big data.
- **Sensor based software tools for quality assessment.** This research deals with sensor data for analysis of quality assessment in context to product or process.
- **Human robot collaboration**. The research heavily uses of electronics concepts along with big data analytics.
- **Mobile assisted systems for Industry operations**. This involves mobile computing with visualization techniques.
- **Manufacturing Data Analytics.** This involves predicting operations of manufacturing using data analytics.

CONCLUSION

Industry 4.0 revolution has changed the thinking of the industry as well as educational sector. The success of Industry 4.0 is based on right skilled people. It is the responsibility of the educational sector to prepare human resources in such a way that they can easily absorb in the modern manufacturing paradigm of the industry. For preparing right skills among people, educational sector must know which kinds of skills are vital for them. This chapter had started with educational evolution based on the industrial revolution, so education sector can aware about their objectives of right skills. The second part of chapter dealt with the expected qualifications for the smart factory so educational sector can incorporate the right mix of educational aspects in the teaching-learning process. The chapter also described interdisciplinary educational aspects of students' fraternity. The chapter discussed technological challenges for the educational sector in the implementation of smart factory based curriculum. The chapter briefly described the educational opportunities for industry 4.0. At last chapter gave certain research direction of Industry 4.0 in context for education. Research direction of Industry 4.0 in the context of education is very limited and it is constrained to only designing an automated tool to create a virtual world of real production that is benefitted in the teaching-learning process.

REFERENCES

Aehnelt, M., & Bader, S. (2014). Tracking assembly processes and providing assistance in smart factories. *Proceedings of the 6th International Conference on Agents and Artificial Intelligence (ICAART).*

Antsaklis, P. (2014). Goals and Challenges in Cyber-Physical Systems Research Editorial of the Editor in Chief. *IEEE Transactions on Automatic Control*, *59*(12), 3117–3119. doi:10.1109/TAC.2014.2363897

Bauernhansl, T., ten Hompel, M., & Vogel-Heuser, B. (2014). *Industrie 4.0 in Produktion, Automatisierung undLogistik—Anwendung, Technologien, Migration.* Wiesbaden: Springer. doi:10.1007/978-3-658-04682-8

Bickel, V. B., & Alexa, M. (2013). Computational Aspects of Fabrication: Modeling, Design, and 3D Printing. *IEEE Computer Graphics and Applications*, *33*(6), 24–25. doi:10.1109/MCG.2013.89 PMID:24921096

Brauner, P., & Ziefle, M. (2015). Human Factors in Production Systems. Springer International Publishing. doi:10.1007/978-3-319-12304-2_14

Brettel, M., Friederichsen, N., Keller, M., & Rosenberg, M. (2014). How virtualization, decentralization and network building change the manufacturing landscape: An Industry 4.0 Perspective, International Journal of Science. *Engineering and Technology*, *8*(1), 37–44.

Chen, C. (2004). Searching for intellectual turning points: Progressive knowledge domain visualization. *Proceedings of the National Academy of Sciences of the USA*, *101*(1), 5303-53.

Dawande, M., Pinedo, M., & Sriskandarajah, C. (2009). Multiple Part-Type Production in Robotic Cells: Equivalence of Two Real-World Models. *Manufacturing & Service Operations Management: M & SOM*, *11*(2), 210–228. doi:10.1287/msom.1070.0208

Groover, M. P. (2007). *Automation, Production Systems, and Computer-Integrated Manufacturing*. Prentice Hall Press.

Kane, G. C., Palmer, D., Phillips, A. N., & Kiron, D. (2015). Is your business ready for a digital future? *MIT Sloan Management Review*, *56*(4), 37–44.

Kolberg, D., & Zühlke, D. (2015). Lean automation enabled by industry 4.0 technologies. *IFAC-PapersOnLine*, *48*(3), 1870–1875. doi:10.1016/j.ifacol.2015.06.359

Kovatsch, M., Mayer, S., & Ostermaier, B. (2012). Moving application logic from the firmware to the cloud: Towards the thin server architecture for the internet of things. *Proc. Innovative Mobile and Internet Services and Ubiquitous Computing (IMIS), 2012 Sixth International Conference*, *IEEE*, 751-756.

Lazovic, V., Montenegro, J., & Durickovic, T. (2014). The digital economy in developing Countries -challenges and opportunities. *37th International Convention on Information and Communication Technology, Electronics and Microelectronics (MIPRO)*, 580-1585. 10.1109/MIPRO.2014.6859817

Lee, J., Kao, H.-A., & Yang, S. (2014). Service innovation and smart analytics for industry 4.0 and big data environment. *Procedia CIRP*, *16*, 3–8. doi:10.1016/j.procir.2014.02.001

Lee, J., Lapira, E., Yang, S., & Kao, H. A. (2013). Predictive manufacturing system trends of next generation production systems. *Proceedings of the 11th IFAC workshop on intelligent manufacturing systems*, 150–6. 10.3182/20130522-3-BR-4036.00107

Mokyr. (1998). *The Second Industrial Revolution, 1870-1914, The Lever of Riches*. Academic Press.

Posada, J., Toro, C., Barandiaran, I., Oyarzun, D., Stricker, D., De Amicis, R., ... Vallarino, I. (2015). Visual computing as a key enabling technology for industrie 4.0 and industrial internet. *Computer Graphics and Applications, IEEE*, *35*(2), 26–40. doi:10.1109/MCG.2015.45 PMID:25807506

Priego, R., Orive, D., & Marcos, M. (2014). Maintaining the availability of the Control System in Industrial Automation. In *Agenten im Umfeld von Industrie 4.0* (pp. 15-19). Sierke Verlag.

RightScale. (2016). *State of the cloud report, Hybrid Cloud Adoption Ramps as cloud users and cloud providers mature*. Retrieved from www.rightscale.com/lp/2016-state-of-the-cloud-report

Scheifele, S., Friedrich, J., Lechler, A., & Verl, A. (2014). Flexible, Self -configuring Control System for a Modular Production System. *2nd International Conference on System -Integrated Intelligence, 15*, 398–405.

Schlechtendahl, J., Kretschmer, F., Lechler, A., & Verl, A. (2014). Communication Mechanisms for Cloud based Machine Controls. *CIRP Conference on Manufacturing Systems, 17*, 830–834

Schuh, G., Potente, T., Wesch-Potente, C., Weber, A.R., & Prote, J.-P. (2014). Collaboration Mechanism s to Increase Productivity in the Context of Industrie 4.0. *CIRP Robust Manufacturing Conference, 19*, 51–56.

Shi, J., Wan, J., Yan, H., & Suo, H. (2011), A survey of cyber-physical systems. *International conference on wireless communications and signal processing (WCP)*, 1–6.

von Tunzelmann. (1996). *Engineering and Innovation in the Industrial Revolutions*. STEEP Discussion Paper No 30, Science Policy Research Unit.

Weyer, S., Schmitt, M., Ohmer, M., & Gorecky, D. (2015). Towards Industry 4.0-Standardization as the crucial challenge for highly modular, multi-vendor production systems. *IFAC-PapersOnLine, 48*(3), 579–584. doi:10.1016/j.ifacol.2015.06.143

Zeng. (2016). Study on the Third Industrial Revolution and Paradigm Transformation of China's Manufacturing Industry—Based on Theoretical Analysis of Scale Economy and Scope Economy. *American Journal of Industrial and Business Management*, 73-82.

KEY TERMS AND DEFINITIONS

3D Printing: It is a process used to produce a three-dimensional object of manufacturing industry to view it in a better manner.

Actuator: It is a component of a device that is used to control or move the device.

Adaptive Control: It is a methodology of designing a control system that deals with uncertainties.

Augmented Reality: It is a technology that creates an artificial environment of the real situation by integrating digital information.

Big Data: Data set that is extremely large in size and mostly unstructured in nature.

Big Data Analytics: It is a process to analyze big data to uncover hidden, valid, and useful patterns and insights.

Cloud Computing: Technology that delivers computing services over the Internet. It includes computing services like storage, networking, databases, servers, etc.

Cyber-Physical System: It is a system that is supervised by computer-based algorithms and heavily integrated with the Internet.

Data Mining: It is a process of analyzing large data to discover hidden and useful patterns.

ERP System: It is software that integrates all aspects of manufacturing operations such as production planning, designing, manufacturing, marketing, sales, etc.

Fog Computing: It is a decentralize computing paradigm that performs on the edge of the device.

Industry4.0/Smart Factory: This is the fourth revolution of manufacturing industries. This revolution emphasizes on cyber-physical systems, internet of things, cloud computing, big data analytics, and 3D printing.

Internet of Things (IoTs): It is a networking of physical devices embedded with internet, electronics, sensors, software, actuators that enable devices to store and send the data.

Location Detection Technique: It's a technique that locates objects or people inside the industry.

Machine Learning Techniques: This technique provides an ability to computer to learn without being explicit human intervention.

Predictive Modeling: It is a process that forecasts certain events in manufacturing using data mining and probability.

RFID: It is a technology that automatically tracks tags attached to any object.

Second Industrial Revolution: It considers as a technological revolution of manufacturing industries. In this revolution, industries have started using existing technologies such as telegraph, rail network, gas, water supply, and sewage system.

Sensor: It is an electronic component that is used to detect the behavior of an object in the environment.

Simulation: It is software that imitates the operation of a real-world manufacturing process.

Smart Supply Network: It is a methodology in which things work together. It collects the data in real time and does automatically alteration in the flow of products as demand changes.

Third Industrial Revolution: It considers as the digital revolution of manufacturing industries. Industries have started using computers for their many processes.

Velocity: It deals with the pace at which data flows from heterogeneous sources.

Veracity: It means abnormalities in available data.

Wearable Device: This is the technology that puts on a human body in form of a gadget. This technology is helpful in industry for plant monitoring.

Chapter 8
Digital Technology Integration in Different Educational Fields:
Design, Architecture, Tourism, and Business Engineering

Hridayjit Kalita
Birla Institute of Technology, Mesra, India

Kaushik Kumar
Birla Institute of Technology, Mesra, India

ABSTRACT

The perception of learning and teaching in the educational universities have been affected by digital technology. With the industrial concern over sustainability of resources and efficiency in operation in a digital environment, the need arises to implement digital technologies in the educational setting so that digital competence of the future workforce can be elevated, and better industrial output-based education is provided. In this chapter, an attempt has been made to describe and discuss the current scenario of digital integration in higher educational disciplines. The issues concerning this integration include teacher inability to incorporate digital thinking into student learning, student non-adaptability to modern technologies, unreliability of digital educational resources, and lack of infrastructure/power supply in most of the educational institutes. An active digital learning approach in students and extensive training sessions for digital utilization excellence in teachers and educators are a few ways to solve issues regarding the above-mentioned integration.

DOI: 10.4018/978-1-7998-2245-5.ch008

INTRODUCTION

Digital technologies have emerged as one of the most versatile achievements of mankind which today finds its use in a number of industrial fields and government departments, requiring more skilled manpower. Digital technology integration into the education system has enhanced the learning process and the skills to solve complex tasks of the students for a better future workforce in the modern smart industries. The industries today are driven by the innovations and skills of work personnel who are expected to think fast, innovate, find solution and build optimized product employing digital devices and tools for a smart information exchange, design and co-ordination of departments. A number of research papers suggest integration of digital technologies in the pedagogical field (Selwyn, 2007) so that the efficiency of learning of the students and their skill developments in adapting to new technologies for fulfilling ever changing demands in market enhances. With the higher mismatch in the skill level of the industry requirement and the workforce currently serving in terms of quick adaptation to technology and innovation at a complex level of understanding of processes and phenomenon in a digitally inspired environment, need arises to incorporate computer technology in the higher educational format.

Students go through a critical transformational phase in their transition to universities including high expectation, excitement and motivation.Policy makers have suggested plans for the higher education sectors to implement competence level of the students towards their own well being, active employment, active citizen and entrepreneurship skills.This transitional period to universities can be hard for the students with disabilities with many of them even quitting to enrol their names for higher studies (Tinto, 1993; Kochhar-Bryant et al., 2009). Even if few survive to pursue their dreams, they got to stick to one place with textbook based learning. Digital technology integrated learning can have major contribution towards facilitating students with disabilities (Pacheco et al., 2018) in learning their tasks not only by textbook learning but also by participating in various events and projects employing computer technology as key tool.

Information and communication technology (ICT) including digital technology helps creating a self sustaining learning environment for students where information on any concerned topic or concept can be easily accessed by students without much effort just by browsing the internet and then visualizing using different digital tools. This incorporation of ICT technology enables the concept of personalized learning or the ability to self apply ICT in their learning activity, to flourish in universities and schools, the effectiveness of which can be felt by the discussion provided by (Schmid & Petko, 2019).

3D printing technology involves fabrication of the product by formation of multiple layers of material one above the other having the area dimension of each layer as defined by the slice area of a computer generated CAD model. This technology has the potential to enhance the imaginative, intuition and creative skills in the minds of the learners as well as the teachers by providing an additional dimension to their thoughts and turning a 3D imagination into real physical object. 3D scanners are important devices for scanning an already available solid artefact or model to plot a collection of data points and transforming it into a surface representation to be suited by the CAD software. 3D printers and scanners have always been a tried and tested method for the development of project models and optimizing designs including teaching assistance for visualizing 3D objects to students in the mechanical engineering field (Novakova- Marcincninova et al., 2012). A detailed discussion in the present scenario of the implementation of 3D printing in educational settings (teaching and learning practices) and their challenges has also been given in (Ford & Minshell, 2019).

Educational digital resources in the age of Internet and communication technology have enabled teachers to explore innovative ways for student learning and prepare their doubts and conclusions on any topic by mere accessing the vast free online content in the network. These wealth of digital knowledge though comes with challenges (Davies & West, 2014; Wiley et al., 2014) such as: First is the lack of expertise and self efficacy of the teachers with the digital tools and devices and their inability to relate student learning to educational digital resource contents (Vongkulluksn et al., 2018). Second is the inability of the teachers to distinguish between the reliable, appropriate digital content to the unreliable and un-appropriate ones considering the heterogeneity in the available content (Xie & Luthy, 2017).

In the current paper, digital technology implementation into various fields in the higher education considering Design education, Architectural education, Tourism education and Business Engineering education has been described in detail in the subsequent sections.

DIGITAL TECHNOLOGY (3D SCANNING AND PRINTING) INTEGRATED DESIGN EDUCATION

Digital manufacturing or 3D printing has got a tremendous potential in its role in providing 3D skills and critical thinking to students in the field of design. With recent developments in low cost 3D printers, a person is able to access the 3D world at a cost of less than EUR 1000 (Wohlers, 2012).These include “built it yourself” kits which can be assembled with not much expertise by individuals and operated to obtain 3D artefacts and models at a cheaper cost. Access of these 3D printed models to design students can improve their imaginative skills and visualized learning. Home 3D printers of a plug and play system are easy to make it ready and use.

Application of 3D scanning and printing have been gaining momentum in educational field with the mention of their integration in a number of lectures, studies and textbooks (Sullivan & McCartney, 2017; Gibson et al., 2010; Berger et al., 2013). These technologies have been utilized by students and researchers in their respective laboratories where less number of these systems limits the full potential of a large number of learners. With introduction of low cost*3D printers*, universities are able to accommodate more number of Digital manufacturing systems for students training facilitating allotment of systems to be accessed by individual student. These systems can be integrated with the CAD software where the students are able to play around with the dimensions, surface features and rotation of the 3D virtual objects and can later be produced by transferring the file to the 3D printer in an affordable cost. The produced sample then can be touched, analysed, evaluated and re-optimized for better functionality which gives an added advantage of elevating visualization and intuitive skills in students. Fused deposition modelling system in 3D printers can be utilized to reduce the need for handmade intricate models for teaching design attributes to students at a lesser effort, design and optimization time (Widden & Gunn, 2010). Work pressure of the students is found to be reduced and better results obtained by implementing prototypes into their learning projects (Kriesi et al., 2014) such as introducing 3D molecular tools for pharmacy students (Hall et al., 2017).

There has been effort made to enhance the number of 3D systems to be used by students in various universities (De Beer, 2013; Ford and Dean, 2013; Mostert-van der Sar et al., 2013) so that each student is able to learn the working of the system and the design it needs to fabricate and their limitations at a greater depth. Alsothe students are able to carry out faster optimization of design, change of design ideas and print experiments. Plug and play 3D printers provide easy to use setup that are reported to be bought by different universities in large quantity. These plug and play systems though are easy to use does not provide a deep insight into the guidelines of designing a part and how the technology and the tools areto beoperated for specified production task. The aim is always to provide the students the necessary design skills, critically evaluating design options and to know the limits of digital manufacturing as in the case of University of Maryland, College park where 3D printing and manufacturing has been integrated with the usual regular CAD courses (assignments) and a scalable instructional method been proposed for connecting these links (Reggia et al., 2015).

3D scanners are possessed with the problem of not being able to be used in large quantity in universities and laboratories due to large acquisition cost of these professional devices and its software which limits the understanding of the students in gaining practice to work on 3D scanner and knowing its advantages in data collection. 3D scanning implementation in education system are discussed in a number of papers including (Fang et al., 2012) who demonstrated the advantages of the process chain from scan to reverse engineering and then to CAD in developing design thinking, measurement acquisition skill and evaluation of drafts and design(as can be seen in complex shapes in modern vehicles)in students (Ertu et al., 2006). Artefacts can be easily captured using 3D scanning and sent to students via web browser (Rodrigues et al., 2011).

After assembling "built it yourself" 3D printer kit or any other plug and play systems, students can be arranged in groups of 3 with each grouptasked with designing the 3D model using CAD systems like CATIA or downloading 3D model from a freely accessible database or scanning an already build in part or model for 3D model generation. After each steps in the designing procedure of the model, the students can note down the design guidelines in their diaries. With the scanning and printing of the model, the students can learn into the insight of reverse engineering of an already build part and then 3D print it.

In the *first option* 3D models can be downloaded from a free database by students with all the provisions for technological compatibility or required installation space and having no support structure (as pre-requisite in most of the 3D printers) to be kept in mind. Colour information transfer into the 3D printer by students can be processed to depict different regions or layers on the physical structure. In the *second option*, the students can learn the utilization of the CAD design software in generating 3D models of their own choice. They can learn whether the surface pattern and characteristics supports the accuracy and precision of the 3D printer. For that, the students can be tasked with selecting the appropriate STL interface characteristics for best representation of the 3D model and that is dependent on the 3D printer used. Triangular facets are often employed to represent the surfaces of the CAD model in STL format. Coarser size of the triangular facets used in an STL format requires less information to be stored which roughly depicts the curved surfaces. Finer triangular facets can converge to actual requirement of the curves but the data volume or the file size is inflated. In the *third option*, the students can be provided with 3D scanners to collect the surface points of any uneven complex artefacts for mask building. These collected surface data points can be manipulated, added and deleted for appropriate representation of the actual part using professional software and the revised data points are then edited for surface reconstruction to be read by the CAD software. After the CAD model is transferred to the 3D printer the students can proceed for the selection of appropriate materials, colour dies, binders for the given type of 3D printers.

DIGITAL TECHNOLOGY INTEGRATED ARCHITECTURAL EDUCATION

Computer technologies have impacted the curriculum of Architectural education to great extent. With the term "computer technology" the author (Qaqish et al., 1997) defines all software and applications whether they are integrated into a design studio or any regular courses in both architectural and urban planning departments. In most of the internationally recognized schools and universities, courses have been introduced which integrates the traditional methods in architecture such as plotting, 3D sketching, or developing any structure or architectural designs to the computer applications and software that are based on numerical computational algorithms in 3D design. The technologies like 3D drafting, modelling, analysis and 3D printing provides a suitable architectural environment for the students to elevate their understanding on optimal structures and buildings with contrast to the traditional implementation of educational tools for 2D sketching, 3D physical models and different view 3D drawings. The traditional tools in architectural teaching have been largely replaced with the computer technologies for at least the last three years of most of the bachelor courses (Angelil et al., 2003). Many architectural schools and universities in developed countries have taken the initiative to introduce computer courses into their curriculum (Zhu et al., 2016) which exhibits the level of interaction of the computer technology with the education. Computer technology have enabled the students and also the practitioners to imagine new dimensions in their structural ideas and producing large number of models at a lesser time. Due to modernization in the architectural practice, the computer technology has become the need in its education curriculum (Schenk, 2005).

With contrast to the architectural laboratories becoming the innovation space for the students to get associated with various design studios and media, (Achten, 2003) warned the consequence of over dependence of the students on the computer technology which might degrade their innovation, intuition and creativity skills (Guney, 2015) and that the curriculum of the architectural universities and schools must be reconsidered. Many educators and practitioners including (Breen, 2004) proposed a combination of both traditional architectural implementation of the tools and the digital technologies in the course curriculum that would give the students a new insight into their field and be prepared for future practice (Al-Qawasmi, 2005).

A thorough study on the application of computer applications in the curriculum of architectural courses have been discussed in (Soliman, 2019) considering 20 top universities and schools globally. 8 categories of courses incorporating computer technology applications and software have been identified such as 2D-3D representation, digital fabrication, environmental technology, Parametric design, communication technology, programming and coding, simulation, building technology, etc in the universities and were studied for the level of computer integration in them. The analysis is carried out in the form of questionnaires to architects and experienced practitioners in few principle areas including the role of computer applications in the architectural professional fields, level of computer software use in the architectural works and the upcoming computer technologies in architectural field. The results of the analysis reflects the level of implementation of the computer technology into the architectural curriculum as discussed below:

The *result* of the survey were analysed and the following *conclusions* were made:

It was observed that 68% of the computer application is directly associated with design studios and other 32% employed in stand-alone courses with more integration of the computer courses in the final year of the architectural curriculum. 50% of the computer applications accounts for courses in 2D and 3D representation, 13% for digital fabrication, 9% for programming and coding, 7% for simulation and 5% for environmental technology in all the surveyed international schools. In the final year, the parametric design and digital fabrication courses were generally adopted along with 2D and 3D representation running all throughout the course years. Implementation of digital tools can be recommended to be used by students only after attaining a strong hold on their hand drawing and construction of physical models and buildings to cultivate the practice of imagining, thinking and innovating for exploiting full potential of the computer technology (Kara, 2015).

Autodesk AutoCAD software and Adobe Photoshop were found to be the top 2 frequently used computer application by the respondents followed by Autodesk 3Ds MAX, Microsoft office, Sketch up, Lumion and Adobe Illustrator down the line with no respondents using Processing Software. In undergraduate programme, AutoCAD was studied by 95%, Adobe Photoshop by 88% and Autodesk 3DsMAX by 69% of the respondents.

The respondents also put forward their desire to learn computer applications with 31% of them interested in Autodesk REVIT and 25% interested in Autodesk 3Ds MAX. The analysis also showed that most number (81%) of the respondents wanted to learn and improve on computer applications due to the requirement in their professional fields, 16% for their own interests and 1.5% for research purposes. This signifies the role of digital technologies in generating employment and its utilization in various jobs.

DIGITAL, INFORMATION AND COMMUNICATION TECHNOLOGY (ICT) INTEGRATED TOURISM EDUCATION

Information and communication technology (ICT) have a widespread impact on the tourism education sector and is an important factor in elevating economic and social status of public sector organisations in countries where maximum revenue is drawn from tourism as in South Africa. Its use and adoption by stakeholders, learners, researchers, educators, managers and administrators have accelerated the implementation of digital tools in the education sector. Cantrell and Visser discussed about the domestication of ICT in provincial schools (Cantrell and Visser, 2011).

Few of the discussions on the significant problems and issues in the integration of the ICT in the pedagogical systems can be found in (Vandeyar, 2013; Vandeyar2015) that closely is associated to tourism education as well. It has been mentioned in (Weston and Bain, 2010) that mere physical interaction between the ICT application and the learners and educators cannot be the only solution to non-effectiveness in computer integration into the tourism education. Few important issues that concern the technological adaptation skills of the educators and the learners in this integration are discussed below:

Technology anxiety and self efficacy: Educators play a significant role in introducing effective utilization of the ICT tools and technologies in the learning process of the student and in their teaching methods as well (Sumak et al., 2011). This integration is sometimes a problem as most of the teachers lack awareness of the available technologies in the modern time or are not comfortable with its use in their work. Computer anxiety can be defined as the fear of implementing any computer technologies in their jobs (Leso & Peck, 1992). It happens to be inversely proportional to the computer experience, skill and interest of the person concerned (Shah et al., 2012; Ekizoglu & Ozcinar, 2010). These behavioural traits are affected by their attitude, experience, skills, culture and social interactions towards frequent use of technology (Sumak et al., 2011; Teo, 2011; Van Zyl, 2013).The self efficacy is basically a belief in oneself to be able to perform a certain behavioural need without any hesitation (Bandura, 1977). It is the confidence to be able to use a digital device under different behavioural, environmental and personal impacts. A person with high self efficacy tends to overcome any challenging situation in teaching and learning and can recover at a greater rate while the one who lack self efficacy tend to be reluctant to go for any challenging and complex tasks. Teacher and educators are affected by their lack of computer self-efficacy in accepting integration in the learning and teaching tasks with the computer applications.

Technology domestication: Technological assistance is needed to encourage the teachers and the educators in adapting to the use of computer technology in their teaching habits. The issue of technology domestication in schools and universities is the socio-economic condition of the teachers and learners. It is seen that teachers and learners who have access to computer and internet services at home can easily adapt to the modern techniques of incorporating technology into their teaching habits as contrast to the disadvantaged learners and educator who lacks. Technological domestication in education can be accomplished in 3 stages which can be described as:

1. Commodification is the 1st stage where the technologies are bought and installed, functional values determined and how to put these modern technologies to use in learning and teaching purposes are determined. In many cases, due to lack of resources, infrastructure, power supply and internet access in the schools and universities, commodification cannot be implemented in an effective and efficient manner.
2. In the 2nd stage, the technology is actively integrated to the timetables and routines and put to use in the process of learning and teaching. The teachers and the learners make an effort to change their behaviour in adapting to the ICT integrated learning environment.
3. Conversion is the last stage where the technology is fully adopted by teachers and learners to make new developments and organize projects employing ICT.

A thorough case study of the integration of ICT in the tourism education considering 24 participants was conducted by (Adukaite et al. 2016). The effective integration was found to be basically influenced by the access and management of the ICT, the use of ICT in learning and learner's engagement, teacher's interaction experience and their knowledge on ICT, technology anxiety, lack of ICT resources and mobile technology policies in schools.

For ICT integrated education in the learning of tourism, the teachers and educators that are surveyed, mentioned the use of hardware and projectors, internet, computers and televisions in their activity. Apart from these, the most commonly used technologies were the Google search engine, Youtube, Google earth and Maps by the educators and these applications were considered valuable for tourism instruction. Few teachers belonging to the disadvantaged area of poor internet speed and connection, reported on using CD's and DVD's, presentation, videos and images via projectors for tourism instruction. Few of the teachers also reported on bringing their own data projectors, laptops and dongles and did not rely on the school provisions. ICT help teachers in elevating their knowledge and experience of the subject tourism by providing access to Youtube travel videos, online search engines, online travel magazines and e-books without having to travel 'out of budget' places. Sharing information and materials among their colleagues and other teachers via whatsapp, email, dropbox, facebook and google drives gives an added advantage of ICT implementation.

Another implication of ICT is towards facilitating the students to overcome the lack of any tourism exposure by running video clips, short movies, images or presentations in elevating their interest in the subjects and for more authentic learning experience. Also ICT prepares the learners to compete in their future endeavour and be a part of the actual workforce where technology dominance is more pronounced.

Technology anxiety as already described previously proves to be another major hurdle in integration of ICT in tourism education culture as the teachers and educators themselves in many cases are uncomfortable with the use of it due to lack of experience and skill. Moreover ICT training of the teachers also required funding and time which makes it a non feasible approach with no effective result or usage even after gaining access to these resources.

Lack of ICT resources, logistic challenges in transporting the available resources due to large size class rooms and chances of non availability of free computer labs due to simultaneous running of other IT based subjects creates added disadvantages in integrating ICT.

Many institutions and schools disallow the use of mobile phones in their campuses due to lack of knowledge and guidelines towards proper use of this technology by students in the pedagogical field. This creates an issue for the educators not to be able to give access to images and videos of concerned topics and places of importance in their tourism education.

These issues must be sorted out for active integration of ICT and other digital technologies into the tourism education, giving way to fulfilment of the third stage in technological domestication as described above. Teachers and educators were seen to have a major impact in modernizing the learning skills of the students and getting them accustomed to the practice of implementing technology into their work for enhancing their efficiency and potential. Mobile phones and other visual devices need to be allowed at-least in the tourism class for better exploration of historic or geographic places described in the textbooks.

DIGITAL TECHNOLOGY INTEGRATED LEARNING IN BUSINESS ENGINEERING EDUCATION

There is a need to introduce digital technology integrated learning in educational institutes especially in the higher educational universities including Business and engineering educational field. European enterprises are dealing with the shortage of skilled workforces who are able to perform and adopt well with digital technologies in a collaborative environment and economy (European Commission, 2016).Europe has been witnessing 44% of its citizens unable to adopt to digital technologies in their work environment while on the other hand, in Romania only 26% of the citizens are above the basic levels of understanding of digital technologies (Commission staff working document, 2017, 2017).

With the lack of skilled workforce, the enterprises are unable to perceive the opportunities provided by the modern digital technology integrated businesses. In eliminating the difference in their skill level and the required level in modern industries, the higher education plays a critical role to mould the future workforce towards better adoption of digital technologies and learning to efficiently and effectively use them for optimal outcomes. As the integration of digital technologies in business engineering education also reveals in a number of ways how learners and educators interact in building an environment which is smart, visionary, strategic and skilful. Learners and educators together can work on projects employing digital tools and devices and exploit the potential of far newer technologies (European Commission, 2017). With the ever changing demands in the market, the teachers and learners can contribute immensely to the complete integration of the businesses, research and academics with various organizational collaborations for a complete sustainable growth and development of the society and also of the educational field (Fleaca et al., 2017, 2018).

Analysis has been carried out on the digital competence and its effect on the labour market to identify the mismatch between digital technology skills and the required demand in the market (Titan et al., 2014). Newer models and guidelines to implement digital technologies in educational learning and teaching have also been proposed (Wu et al., 2018).

Digital competence has been widely discussed as the set of skills, practices, awareness, knowledge and attitude that are needed in the society today for a better grip of the potential of digital technology integrated work environment. In a study conducted in (Fleaca and Stanciu, 2019) of the business engineering education in Romania, the authors attempted to present a methodology to assess the student's self perception of the developments in their skills with implementation of digital tools and information into their tasks. The assessment was carried out as the measure of digital competence in students of business engineering discipline which can be described as their skills in information collecting and data processing, problem solving, digital content creation and communication and collaborations. The assessment also intended to synchronise the learning process of the students in the business engineering discipline with the teaching methods of the teachers and educators in the perspective of the student's view of learning a concept or topic employing digital tools and technology.

Questionnaires were provided to around 150 students of the bachelor degree in the business engineering departments through online media, mail, etc, 78 of them of which responded. The questions were related to the digital competence skills i.e. on information and data processing, digital communication, digital content creation and digital problem solving of the students which were ranked in a scale of 1-for strong disagreement and 4-for strong agreement to the questions. A weighted score is thus given to each statement and compared for effective perception of the students on digital technology use in their work practice.

Looking into the skills of *information gathering and data processing*, the respondents have agreed to their ability to integrate information from different sources employing digital technology tools, thus assigning a weighted score of 3.52. Following their ability to adapt and make effective search strategies in a specified search engine is assigned with a weighted score of 3.38. The statement concerning reliability, usefulness, integrity and accuracy of the information gathered through the ICT and digital devices was the 2nd least weighted scorer with only 3.25 score while the issues concerned with distinguishing the reliability and non reliability of the information gathered was ranked the lowest with score of only 2.96.

In the *digital communication* section, the respondents agreed more strongly to the statement concerned with the editing provision of the information for sharing through mails, posts in social networks and for putting up in presentation with a weighted score of 3.48 followed by the statement concerning sharing content through various collaborative platforms, social networks, google drives and dropbox with a score of 3.20. The statements about creating professional e-profiles among friends and selecting information according to one's own need were the lowest scorer with 3.03 and 3.07 weighted scores respectively.

In the *digital content creation* section, almost all the questionnaires shared a similar high weighted score, representing a strong grip of the students in creating visual representations, selecting appropriate application and software for the desired content, knowing license and copyright rules for intellectual property products (highest weighted score of 3.53) and exploring and original creation of the content.

In the 4th section of the *digital problem solving*, the statement concerning smooth adaptability to new technology and integrating it into their learning practice (like pdf reading, online content exploring, etc) has been ranked the highest with a weighted score of 3.99 among all other statements. Other statements in the section also shared a similar high score indicating strong hold of the students in exploiting the potential of the digital technologies, extracting knowledge through interaction with digital resources and awareness on the development of newer technology and their potentials.

Apart from the digital competence parameters, in the *educational resource* section the data analysis also revealed a strong appreciation for the project based learning, teacher's courses and video lessons, from the students in the business engineering education.

Considering all the above four skills in digital competence, it was observed that students had a strong hold on the internet content search, editing the content, applying the content to their work or presentation, sending the content trough mails or social networking platforms, creating original contents, knowing the copyright and licence rules for intellectual property products and integrating the searched content to their learning practices.

The future market and business relies on the young graduates in the business engineering disciplines who are expected to be able to work with different fields, integrating their work in a collaborative environment and learning in a digital environment. Digital technology into these fields therefore provides a skilled modern workforce who can perform critical thinking, creative learning and acquire adaptability with the newer technologies in their endeavour to power the future economy of the country.

CONCLUSION

From the above discussion, it is clear that digital technologies have impacted greatly to learning and teaching practices of students and teachers respectively. The chapter discusses few studies conducted on the digital integration in the educational field of Design, Architecture, Tourism and Business engineering.

In the design educational curriculum, digital technologies in the form of 3D printers and scanners can be implemented effectively in providing the students the 3D experience of the world around them, design and fabricate their own ideas and at the same time, learning the adjustments, calibration, optimization and various other complexities faced during conversion of the CAD design into a physical model. The design students can analyse, optimize, re-design and re-fabricate if errors persist after development which gives the students a hand on experience in designing a product by error an trial method. Design students can select 3 different approaches of obtaining the CAD model such as:searching the network for any relevant 3D model data, preparing the model using CAD modelling and analysis software and obtaining the dimensions of an already produced physical model by 3D scanning.

In the field of architectural education, computer technology can be integrated into the design studios and also into the regular stand alone courses. The students can learn newer approaches in plotting and projecting their 3D structures using modern 3D digital tools while broadening their imagination and visualization of the outside physical world. It must be noted that architectural students need to catch up with the modern 3D applications and software only after gaining full expertise in their hand skills of 2D drafting and building physical model.

Application of ICT devices in the educational setting builds a favourable learning environment for the students in the tourism discipline as it provides an enlarged dimension of information and resource gathering. The teacher plays a critical role in adapting technological devices into the settings of learning and teaching so that the students can experience places around the world visually instead of just going through the textbooks, with having to travel those places. The major issues in integrating ICT to tourism education, regarding the technological anxiety and domestication has been discussed with the suggestion for better teacher training so that his/her involvement in adopting technology to student's learning can be enhanced.

With the growing lack of skilled work force in the modern automated industries, the need for implementation of technologies into the education setting of business engineering discipline proves to be an effective and efficient way of enhancing skills in management, collaboration, designing and developing. Digital competence can be described as the student's readiness to adopt technology into their learning habits and is a measure of their skills in information gathering and data processing, digital communication, digital content creation and digital problem solving.

FUTURE SCOPE

In spite of a number of hurdles in implementing digital technology in the state of present political and technological scenario into the educational institutions, the importance of it can be felt in the coming generation due to complete digitization of industries and small enterprises. The need for skilled workforce with adaptive, imaginative and intuitive skills appeals for a quality higher education and enhanced learning rate of the student. Digital technology will play a major role in enhancing these skills by providing better visualization of 3D objects, multi dimensional problems, textbook artefacts and self assessment of progress. Programming and coding are few other skills which will tremendously benefit the students of the future in shaping the future industrial activities and enhancing effectiveness and efficiency in the value chain.

With the help of 3D printers and scanners, students will be able to design their own idea, optimize their design in the CAD software, solve errors in transferring their designs to the printer, calibrate the 3D printer, select the appropriate material for fabricating and finally printing their product. With the help of 3D scanners, students will be able to directly convert the physical object into digital format with the knowledge of the limitations and possibilities of it. Also in future the scanner might also be able to capture the texture of the physical object which will greatly simplify the steps in preparing the CAD model.

Computer applications in the architectural field can be made more efficient and effective if the entire system gets integrated in all phases across different disciplines. In the tourism education, ICT can transform the way of learning of the students by giving them access to gather information via. internet and build imagination in them to experience potential of the places around the globe without actually travelling. Considering this, the researchers in the future has the scope of involving technology like augmented reality into the tourism education system. It has also been observed that mere involvement of ICT in the education system is not sufficient and that proper training sessions must be provided to the teachers in making them get rid of technological anxiety and fear, thus making them self-sustaining and skillful.

REFERENCES

Achten, H. (2003). New design methods for computer aided architectural design methodology teaching. *International Journal of Architectural Computing*, *1*(1).

Adukaite, A., Van Zyl, I., & Cantoni, L. (2016). The role of digital technology in tourism education: A case study of South African secondary schools. *Journal of Hospitality, Leisure, Sport and Tourism Education*, *19*, 54–65. doi:10.1016/j.jhlste.2016.08.003

Al-Qawasmi, J. (2005). Digital media in architectural design education: reflections on the estudiopedagog. *Art, Des. Commun. Higher Educ.*, *4*(3).

Angelil, M., & Uziyel, L. (2003). Inchoate: An Experiment in Architectural Education. Swiss Federal Institute of Technology (ETHZ).

Bandura, A. (1977). Self-efficacy: Towards a unifying theory of behavioural change. *Psychological Review*, *84*(2), 191–215. doi:10.1037/0033-295X.84.2.191 PMID:847061

Berger, U., Hartmann, A., & Schmid, D. (2013). *Additive manufacturing Fertigungsverfahren: Rapid prototyping, Rapid tooling, Rapid manufacturing*. Europe Lehrmittel.

Breen, J. (2004). Changing roles for (multi) media tools in design: assessing developments and applications of (multi)media techniques in design education, practice and research. *22nd eCAADe Conference Proceedings: Architecture in the Network Society,* 530–539.

Cantrell, S., & Visser, L. (2011). Factors influencing the integration of technology to facilitate transfer of learning processes in South African, Western Cape province schools. *Quarterly Review of Distance Education*, *12*(4), 275–285.

Commission Staff Working Document. (2017). *Europe's Digital Progress Report (EDPR) 2017 – Country Profile Romania.* SWD (2017) 160 final. Part 29/62. Available from https://ec.europa.eu/digital-single-market/en/news/europes-digital-progress-report-2017

Commission Staff Working Document. (2017). *Europe's Digital Progress Report – 2017*. SWD (2017) 160 final. Available from https://ec.europa.eu/digital-single-market/en/news/europes-digital-progress-report-2017

Davies, R. S., & West, R. E. (2014). Technology Integration in Schools. In *Handbook of Research on Educational Communications and Technology* (pp. 841–853). New York, NY: Springer New York. doi:10.1007/978-1-4614-3185-5_68

De Beer, N., (2013). *Additive manufacturing – Turning Mind into Mater.* Sierra College for Applied Competitive Technologies.

Ekizoglu, N., & Ozcinar, Z. (2010). The relationship between the teacher candidates' computer and internet based anxiety and perceived self-efficacy. *Procedia: Social and Behavioral Sciences*, *2*(2), 5881–5890. doi:10.1016/j.sbspro.2010.03.962

Ertu, U., Atkinson, P., & Tancock, D. (2006). Applying 3D scanning and modelling in transport design education. *Computer-Aided Design and Applications*, *3*(1-4), 41–48. doi:10.1080/16864360.2006.10738440

European Commission. (2016). *Communication from the European Commission to the European Parliament, the Council, the European Economic and Social Committee and the Committee of the Regions. A new skills agenda for Europe – Working together to strengthen human capital, employability and competitiveness.* COM (2016) 381 final. Available from https://eur-lex.europa.eu/legal-content/EN/TXT/?uri=CELEX%3A52016DC0381

European Commission. (2017). *Communication from the European Commission to the European Parliament, the Council, the European Economic and Social Committee and the Committee of the Regions. A renewed agenda for higher education.* COM (2017) 247 final. Available from https://eur-lex.europa.eu/legal-content/EN/TXT/?uri=COM%3A2017%3A247%3AFIN

Fang, D. Y., Zhang, A. M., & Qiu, Y. (2012). Research and practice on new measuring and drafting mode of mechanical drawing based on reverse engineering. *International Conference on Measurement, Instrumentation and Automation. Applied Mechanics and Materials, 239*(240), 645–648. doi:10.4028/www.scientific.net/AMM.239-240.645

Fleacă, E., Fleacă, B., & Maiduc, S. (2017). Modeling Stakeholders Relationships to Strengthen the Entrepreneurial Behavior of Higher Education Institutions. *Proc. Eng, 181*, 935–942. doi:10.1016/j.proeng.2017.02.490

Fleacă, E., Fleacă, B., & Maiduc, S. (2018). Aligning Strategy with Sustainable Development Goals (SDGs): Process Scoping Diagram for Entrepreneurial Higher Education Institutions (HEIs). *Sustainability, 10*(4), 1032. doi:10.3390u10041032

Fleaca, E., & Stanciu, R. D. (2019). Digital age learning and Business Engineering Education- A pilot study on Student's E-skills. *Procedia Manufacturing, 32*, 1051–1105. doi:10.1016/j.promfg.2019.02.320

Ford, P., & Dean, L. (2013) Additive manufacturing in product design education: out with the old and in with the new? *Proceedings of E and PDE 2013, the 15th International conference on engineering and product Design education,* 611-616.

Ford, S., & Minshell, T. (2019). Invited review article: Where and how 3D printing is used in teaching and education. *Additive Manufacturing, 25*, 131–150. doi:10.1016/j.addma.2018.10.028

Gibson, I., Rosen, D. W., & Stucker, B. (2010). *Additive manufacturing technologies: Rapid prototyping to direct digital Manufacturing*. Springer. doi:10.1007/978-1-4419-1120-9

Guney, D. (2015). The importance of computer-aided courses in architectural education. *Procedia: Social and Behavioral Sciences, 176*, 757–765. doi:10.1016/j.sbspro.2015.01.537

Hall, S., Grant, G., Arora, D., Karaksha, A., McFarland, A., Lohning, A., Anoopkumar-Dukie, S., (2017). A pilot study assessing the value of 3D printed molecular modelling tools for pharmacy student education. *Currents in Pharmacy Teaching and Learning*.

Kara, L. (2015). A critical look at the digital technologies in Architectural education: When, where, and how? *Procedia: Social and Behavioral Sciences, 176*, 526–530. doi:10.1016/j.sbspro.2015.01.506

Kochhar-Bryant, C., Bassett, D. S., & Webb, K. W. (2009). *Transition to post secondary education for students with disabilities*. Thousand Oaks, CA: Corwin Press.

Kriesi, C., Steinert, M., Meboldt, M., & Balters, S. (2014) Physiological Data acquisition for deeper insights into Prototyping, In *Proceedings of NordDesign 2014 Conference*. Aalto Design Factory and Design Society.

Leso, T., & Peck, K. L. (1992). Computer anxiety and different types of computer courses. *Journal of Educational Computing Research*, *8*(4), 469–478. doi:10.2190/Q1TJ-8JCU-LDAP-84H8

Mostert-van der Sar, M., Mulder, I., Remijn, L., & Troxler, P. (2013). Fablabs in design education. *Proceedings of E and PDE 2013, the 15th international conference on engineering and product design education*, 629-634.

Novakova-Marcincinova, L., Fecova, V., Novak-Marcincin, J., Janak, M., & Barna, J. (2012). Effektive Utilization of Rapid prototyping technology. *Materials Science Forum*, *713*, 61–66. doi:10.4028/www.scientific.net/MSF.713.61

Pacheco, E., Lips, M., & Yoong, P. (2018). Transition 2.0: Digital technologies, higher education, and vision impairment. *The Internet and Higher Education*, *37*, 1–10. doi:10.1016/j.iheduc.2017.11.001

Qaqish, R., & Hanna, R. (Eds.). (1997). The Impact of CAL strategies on CAD: A case study of the effective use of computers in the teaching of architectural design. CAADRIA '97 Conference.

Reggia, E., Calabro, K. M., & Albrecht, J. (2015). *A scalable instructional method to introduce first year engineering students to design and manufacturing processes by coupling 3D printing with CAD assignments.* 122nd ASEE Annual conference and exposition, Seattle, WA.

Rodrigues, M. A., Kormann, M., & Davison, L. (2011). A case study of 3D technologies in Higher education: Scanning the metal work collection of museums Sheffield and its implication to teaching and learning. *ITHET 2011: IEEE 10th International conference on Information Technology based Higher education and Training,* 1-6.

Schenk, P. (2005). Reflections on the teaching of drawing in the digital age: Attitudes of senior academics in the United Kingdom to the place of drawing tuition on the design curriculum in higher education. *Art. Design & Communication in Higher Education*, *4*(3), 189–203. doi:10.1386/adch.4.3.189/1

Schmid, R., & Petko, D. (2019). Does the use of educational technology in personalized learning environments correlate with self reported digital skills and beliefs of secondary-school students? *Computers & Education*, *136*, 75–86. doi:10.1016/j.compedu.2019.03.006

Selwyn, N. (2007). The use of computer technology in university teaching and learning: A critical perspective. *Journal of Computer Assisted Learning*, *23*(2), 83–94. doi:10.1111/j.1365-2729.2006.00204.x

Shah, M. M., Hassan, R., & Embi, R. (2012). Technology acceptance and computer anxiety. *Proceedings of the 2012 international conference on innovation management and technology research (ICIMTR),* 306–309. 10.1109/ICIMTR.2012.6236408

Soliman, S., Taha, D., & El Sayad, Z. (2019). *Architectural education in the digital age Computer applications: Between academia and practice. Alexandria Engineering Journal.*

Sullivan, P., & McCartney, H. (2017). Integrating 3D printing into an early childhood teacher preparation course: Reflections on practice. *Journal of Early Childhood Teacher Education*, *38*(1), 39–51. doi:10.1080/10901027.2016.1274694

Sumak, B., Hericko, M., & Pusnik, M. (2011). A meta-analysis of e-learning technology acceptance: The role of user types and e learning technology types. *Computers in Human Behavior*, *27*(6), 2067–2077. doi:10.1016/j.chb.2011.08.005

Teo, T. (2011). *Technology acceptance in education. Research and issues*. Rotterdam: Sense Publishers. doi:10.1007/978-94-6091-487-4

Tinto, V. (1993). *Leaving college: Rethinking the causes and cures of student attrition* (2nd ed.). Chicago: University of Chicago Press.

Ţiţan, E., Burciu, A., Manea, D., & Ardelean, A. (2014). From traditional to digital: The labour market demands and education expectations in an EU context. *Procedia Economics and Finance*, *10*, 269–274. doi:10.1016/S2212-5671(14)00302-5

Van Zyl, I. J. (2013). *Technology encounters and the symbolic narrative. Localising the 'technology for development' experience in South African education settings* (Unpublished PhD dissertation). Universitàdella Svizzeraitaliana.

Vandeyar, T. (2013). Practice as policy in ICT for education: Catalysing communities of practice in education in South Africa. *Technology in Society*, *35*(4), 248–257. doi:10.1016/j.techsoc.2013.10.002

Vandeyar, T. (2015). Policy intermediaries and the reform of e- Education in South Africa. *British Journal of Educational Technology*, *46*(2), 344–359. doi:10.1111/bjet.12130

Vongkulluksn, V. W., Xie, K., & Bowman, M. A. (2018). The role of value on teachers' internalization of external barriers and externalization of personal beliefs for classroom technology integration. *Computers & Education*, *118*, 70–81. doi:10.1016/j.compedu.2017.11.009

Weston, M. E., & Bain, A. (2010). The end of techno-critique: the naked truth about 1:1laptop initiatives and educational change. *The Journal of Technology, Learning, and Assessment*, *9*(6).

Widden, M., & Gunn, K. (2010). Design build test of model aerofoils for engineering education using FDM. *Virtual and Physical Prototyping*, *5*(4), 189–194. doi:10.1080/17452759.2010.528841

Wiley, D., Bliss, T. J., & McEwen, M. (2014). Open Educational Resources: A Review of the Literature. In Handbook of Research on Educational Communications and Technology (pp. 781–789). New York, NY: Springer. doi:10.1007/978-1-4614-3185-5_63

Wohlers, T. (2012). *Wohlers Report 2012: Additive Manufacturing 3D printing state of the industry. Annual worldwide progress report*. Wohlers Associates Inc.

Wu, Y. J., Yuan, C. H., & Pan, C. I. (2018). Entrepreneurship Education: An Experimental Study with Information and Communication Technology. *Sustainability*, *10*(3), 691. doi:10.3390u10030691

Xie, K., & Luthy, N. (2017). *Textbooks in the digital world*. Retrieved October 24, 2019, from https://theconversation.com/textbooks-in-the-digital-world-78299

Zhu, Z.T., Yu, M.H., & Riezebos, P. (2016). A research framework of smart education. *Smart Learning Environments,* 3-4.

Section 3
Application and Implementation

Chapter 9
Achieving Bloom's Two-Sigma Goal Using Intelligent Tutoring Systems:
Application to Management Education

Owen Hall Jr
Pepperdine University, USA

ABSTRACT

Management education is engaged in significant programmatic reforms in response to the business community's call for web-savvy, problem-solving graduates. Web-based intelligent tutors provide a readily accessible vehicle for enhancing business students' learning performance as well as preparing them for the rigors of the global marketplace. A primary goal of these AI-based systems is to approach Bloom's two-sigma learning performance standard via mastery learning techniques. Furthermore, intelligent tutors can also be used to identify students at risk, to formulate appropriate intervention plans, and to support team learning. Recent evidence suggests that achieving Bloom's goal may be achievable on a routine basis by 2025. The purpose of this chapter is to highlight the growing potential for using intelligent tutors to enhance student and team learning opportunities and outcomes and to outline strategies for implementing this revolutionary process throughout the management education community of practice.

DOI: 10.4018/978-1-7998-2245-5.ch009

INTRODUCTION

Business leaders, accrediting bodies, and management educators alike are calling for a paradigm shift in the content and delivery of management education because of globalization, continuing economic uncertainty, changing demographics, and new learning technologies (Buil, 2016; Dobson, 2019; Doherty, 2015; Philips, 2016). Business schools are now recognizing that digital technology can play an important role in improving both the learning process and learning outcomes (Ghemawat, 2017; Mohapatra, 2015; Damsa, 2019; Stosic, 2015). Specifically, the rapid growth in Web 2.0+ technologies has ushered in a new era for management education including virtual learning environments, intelligent tutors, web-based collaborative learning groups, and crowdsourcing. Additionally, the ubiquitous availability of mobile devices has encouraged educators to utilize this technology as a vehicle for accessing context-based materials, without the constraints or limitations of time, location and pace (Eschenbrenner, 2019). The increased use of learning technologies can help narrow the gap between a graduates' skill set and the ever-changing requirements of business. For example, a survey sponsored by the Association to Advance Collegiate Schools of Business (AACSB) found that many management programs do not emphasize the skills desired by most hiring managers (Lubeck, 2015). These skills include evidence-based decision-making, ethics, and a focus on results and quality.

A number of the concerns such as the lack of integration and too little emphasis on problem finding, international dimensions, entrepreneurship, and practical experience have seemingly not been widely addressed [by the business education community]. Nor have many of the other competencies considered important by employers, e.g., motivation and commitment to the firm, creativity, quality focus, customer focus, ethics–integrity, teamwork, flexibility, and interpersonal skills (Herrington, 2013).

With globalization under increased scrutiny and an unpredictable global economy, the business community is increasingly focusing on adaptive and sustainable positioning strategies (Adbelaal, 2018; Kane, 2017). In that regard, the business community is looking for web-savvy graduates that are both problem-solving and entrepreneurial oriented. This is the ongoing challenge faced by the management education community. Even the business accrediting organizations have gotten into the act of critiquing business education. A report from AACSB (2016) found:

- A larger share of degree-based education will be delivered in flexible formats (modular, part-time) across providers, with students having more control over their curriculum
- Business schools will struggle to align the pace of curriculum development with the pace of evolution in business practice
- A shift toward more experiential learning and business engagement
- Business schools will be increasingly called on to serve the common good

Digital technology provides a vehicle that can be used for addressing these trends. More specifically, providing customized learning opportunities can enhance both student learning outcomes and preparation for the transition to the business community. The mastery of management education material lies at the heart of this proposition. In a 1984 study, Benjamin Bloom found that the average student tutored one-on-one using mastery learning techniques performed two standard deviations better than students who learned via conventional instructional methods (Bloom, 1984). The task at hand is to find technological-based methods of teaching that are scalable, which is not the case with the traditional Bloom model, and that approach the effectiveness of one-on-one human tutoring. Intelligent tutors represent one promising possibility (Alkhatlan, 2018; Harvey, 2017; Sales, 2019). These systems, which consist of a combination of hardware devices and software programs, are designed to provide learners with both customized content and feedback at a performance-driven pace (Tsiakas, 2018). Intelligent tutors can be used to replicate the role of a one-on-one personal tutor on a massive scale and thus move toward the goal of Bloom's Two-Sigma standard. To be effective, an intelligent tutoring system requires a pedagogical framework that is based on the latest learning constructs and that can justify its choice of activities, presentation formats, and decisions (Bagheri, 2015). In recent years, a range of educational technologies has emerged throughout the business school universe including: homework support agents, simulations, massive open online courses (MOOCs), and conditional content release systems (Alexander, 2019; Flavin, 2016).

These same learning support technologies can also be applied to team learning. For example, sensing technologies are now able to assess each team member's behavior and level of engagement. This data along with individual and team outcome metrics provides the agent with the required inputs to enhance the overall learning process. Accurate individual and team performance statistics derived via data mining turn out to be key factors for the agent to identify optimal content delivery and curriculum strategies for enhancing overall learning outcomes. This paper is organized as follows: a review of intelligent tutoring systems (ITS) in management education and an overview of ITS implementation strategies. This article's primary contribution to management education is to identify how intelligent tutors can be used to enhance student learning opportunities and outcomes and thus better prepare students for the growing demands of the business community.

Management education has always striven to inculcate a culture that is both practical and responsive to society's needs. Every tension the industry has experienced in its past continues to affect it today—and will reverberate into its future (Spender, 2016).

INTELLIGENT TUTORS IN MANAGEMENT EDUCATION

The changing demands placed on the management education universe by the business world continues at a torrid pace. Unfortunately, many students now entering a management degree program, especially at the graduate level, often lack the prerequisite skills (e.g., basic statistics) and know-how (e.g., literature search) to successfully compete in both academia and business (Beyrouti, 2017).

The current market trend is that companies look for graduates with diverse skills ranging from critical to soft skills. Due to high demand of skilled graduates by the employers, educators are greatly concerned of the quality of graduates churned out by the university (Mehrotra, 2017).

These competency issues can be addressed using the intelligent tutoring paradigm outline in this paper. An intelligent tutoring system is defined as a learning construct that provides feedback and support during each phase of the learning process (e.g., hints) as well as an overall assessment at the end of the process (Perera, 2015). ITS represent a natural extension to the classical computer-assisted learning (CAL) systems in that they are step-phased oriented allowing the student to advance to the next learning level only after they demonstrate a certain level of proficiency (Farooq, 2017). Additionally, an ITS can further enhance the learning process by making it more personalized based on the learner's characteristics, background, desired learning pace, and business aspirations (Lu, 2018; Dasic, 2016). ITS can acquire real-time information and knowledge via the web to support the learning process. The fundamental challenge is to identify the most effective delivery mode for a given learning application. Again, the goal of incorporating tutors into the curriculum is to improve learning outcomes, which in turn, should lead to greater employment options (Homes, 2019).

If learning goals are not properly aligned with managerial competencies, education and accreditation processes based on those goals are for naught (Costigan, 2015).

Bloom Basics

In the early 1980s, Benjamin Bloom observed that students receiving individualized tutoring in combination with corrective instruction tended to perform two standard deviations better than students receiving conventional classroom instruction. This suggests that the tutored students out-performed 98 percent, or about two-sigma, of the students engaged in a traditional learning environment (Guskey, 2007). Bloom also pointed out that master-based individual and team tutoring do not lend themselves to economies of scale, attesting that one-on-one tutoring would be too expensive for most schools to deploy on a wide scale basis. Accordingly, Bloom called for alternative learning strategies, which when used in combination might approach his two-sigma target. These included, but were not limited to, 1) mastery learning, 2), proactive classroom participation, 3) time on task, and 4) cooperative engagement (e.g., team collaboration). Of these factors, mastery learning was found to yield the largest effect size (i.e., one-sigma). In mastery learning, the class size remains the same as in the traditional classroom setting, but the learning pace for each student can vary. The goal of mastery learning is for each student to achieve general competency for a given topic (e.g., descriptive statistics) before moving on to the next, more advanced subject (e.g., inferential statistics). In mastery learning, the student is given the time needed to achieve the specified level of expertise, even though it may take longer than other students. Two challenges associated with the mastery learning paradigm are: 1) preserving consistency among multiple course sections and between different business disciplines and 2) maintaining the course pace for the entire class. Specifically, what happens to a student who is unable to complete the course requirements by the end of the term? Figure 1 highlights the basic construct behind Bloom's Two-Sigma paradigm. The numbers in parenthesis represent the ratio of instructor to class size. Notice that in both the one-sigma and two-sigma cases, the variability in performance is substantially reduced. Significantly increasing performance while reducing variability lies at the heart of the Bloom two-sigma paradigm and as such, is viewed by many as the Holy Grail of modern education.

Intelligent Tutoring Systems

The proposal to utilize computer-based tutors to enhance the learning process is not new but was first introduced in the late 1990s (Rickel, 1999). Today, the state of the art in intelligent tutoring systems (ITS) has improved dramatically since those early days (Roll, 2016). Specific advancements include the following:

Figure 1. Bloom's two-sigma paradigm

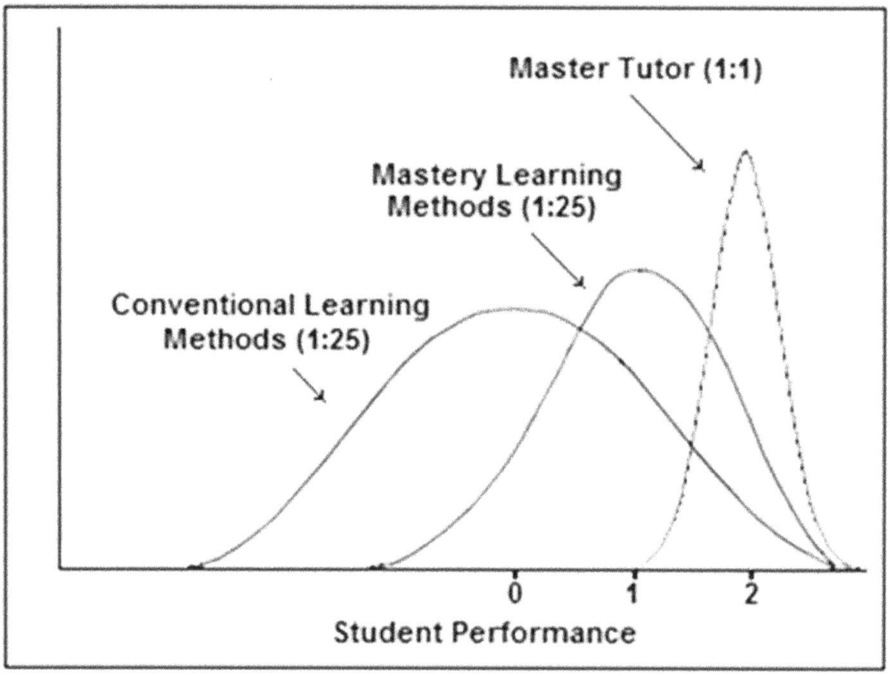

- Inhabits the environment with the student, thus providing increased potential for "physical" collaboration (e.g., telepresence)
- Tracks learners in new ways, for example their visual attention and physical movements
- Maintains learning records over the course of the study program
- Establishes progressive learning goals
- Provides for team collaboration

Typically, ITS embodies a wide range of web-based learning formats including simulations, interactive cases and homework, games, voiceover slide shows, virtual reality exercises, and streaming text. Two recent meta-studies examined the efficacy of ITS (Kulik, 2016; Ma, 2015). The former study found, based on 50 evaluations, that performance scores were raised to 0.66 standard deviations from conventional levels, i.e., 50th to the 75th percentile. This study did observe that performance increases varied based on locally developed or standardized testing methodologies. The latter study, based on a sample size of 107, found a 0.42 sigma increase in student performance across a broad range of subject domains. The first study utilized Glass's effect size estimator while the second study employed Hedges' g. Both these statistics are related to the more common Cohen's d, which translates 0.2 to a small effect, 0.5 to a moderate effect, and 0.8 to a large effect (Ferguson, 2009). This scale has been extended to 2.0, a very large effect, to encompass Bloom's standard (Sawilowsky, 2009). Unfortunately, neither study broke out the results by business category (e.g., finance, marketing). MBA programs, by their very nature, not only embody a broad range of diverse disciplines (e.g., ethics to analytics) but are also highly interdisciplinary. Accordingly, finding the effect per discipline, therefore, is particularly important given the continuing interdisciplinary trends associated with many management education programs (Jacob, 2015). The average of the two meta-studies yielded a 0.5 sigma improvement in student performance. It should be noted that these two meta-studies contain performance data that was based on earlier intelligent tutor technology and usage plans.

A basic tenet of this paper is that intelligent tutors will continue to improve student performance as new learning technologies are introduced that tend to mimic more human behaviors and qualities.

The past fifteen years have seen considerable AI advances in education. Though quality education will always require active engagement by human teachers, AI promises to enhance education at all levels, especially by providing personalization at scale. Similar to healthcare, resolving how to best integrate human interaction and face-to-face learning with promising AI technologies remains a key challenge. Over the next fifteen years the use of intelligent tutors and other AI technologies to assist teachers is likely to expand significantly, as will learning based on virtual reality applications (Stone, 2016).

Recent research suggests that the new generation of management education students face serious concentration, engagement, and socialization issues (Karakas, 2015). This, of course, represents a major issue for business schools in terms of preparing students for the job market. To that end, a well-designed ITS should be able to modify both the lesson plan and content level according to specific student characteristics (Clemente, 2014; Jiménez, 2018; Felijakowski, 2014). Agents can support an adaptive, exploratory learning environment using a multi-layered interface design. Each layer's mechanisms and supportive resources are tailored to facilitate learning for a given situation. Based on a student's background and performance, the agent would select the most appropriate interface layer for that context. For example, if a student is having difficulty mastering a particular subject (e.g., basic statistics) as detected by testing or self-assessment, then the tutor would prescribe specific additional learning content to the student. This content could take the form of interactive tutorials, and simulations. To this end, management education faculty, administrators, and practitioners need to develop coherent and reliable assessments regarding the efficacy of the ITS learning paradigm.

Intelligent tutors can also be used to identify students at risk and to formulate an appropriate intervention plan for students that are challenged (Azcona, 2019). Specifically, early intervention programs can reduce the gap between the lower and higher-performing students (Zhang, 2014). For example, Purdue University's Signals Project has addressed the problem of enhancing student success, which has resulted in improved retention and graduation rates (Pistilli, 2012). This project has led to the development of student success data mining algorithms with intervention messages sent to students based on performance via dashboards. These data mining tools, which often include decision trees, rule induction, and neural networks, can be employed to 1) select student groups with similar characteristics and reactions to learning strategies; 2) detect student misuse and lurking; 3) identify students who, in certain types of test formats, are hint-driven; 4) locate students who exhibit low motivation and find alternate means of reaching them; and 5) predict probable student outcomes (Akcapinar, 2019; Shahiri, 2015). The domain variables for data mining often include previous course grades, class test grades, individual assignment performance, admission exam scores, and team assessments (Day; 2017; Maldonado, 2018).

The emergence of big data and cloud analysis technologies can be used to develop enhanced ITS, which would assess why students are failing or struggling. Typically, students fail or struggle for a variety of reasons including a lack of active participation and an inadequate background in the subject material. Intelligent tutors should assess whether an insufficiency in performance is an isolated incident or a reoccurring phenomenon. Big data driven ITS can be transferred to new student populations, which in turn can be used to provide even more customized learning interventions and educational innovations. Furthermore, ITS can adapt to individual differences in both student know-how and level of experience (Stachowicz-Stanusch, 2018; Suhadi, 2016).

Team Tutoring

Student teamwork has had a long history throughout management education. The empirical evidence suggests that students engaged in team activities (e.g., homework, presentations) exhibit improved learning outcomes (Hsiung, 2012; Mendo, 2018). Many students view their team as a personnel trainer. In this team-oriented learning universe, faculty become designers, coordinators, moderators, mediators, and mentors, rather than pure instructors, whereas students assume a more proactive role including team leadership (Lee, 2015).

Teamwork competencies are becoming a vital part of a student's preparation for the global workplace. Identifying specific challenges and describing how students engage offer useful insights. This in turn provides added perspectives into how teaching might further support student learning (Gonzalez, 2014).

The United States military has seen an increase in the use of intelligent tutors, particularly related to team training (Spain, 2012). The U.S. military's training requirements continue to expand in scope and complexity due to the ongoing introduction of new technology and the changing nature of warfare. In response to these challenges, the military is increasingly relying on an adaptive training paradigm. Presented below are some specific military-based team-oriented organizational principles for enhancing the learning process, which can be applied to management education (Soetanto, 2012):

- Obliged to rely on one another to achieve specific goals
- Tasked with doing their share of the group assignment
- Charged with providing one another with feedback and support
- Encouraged to develop and practice trust-building leadership skills
- Motivated to facilitate changes in the team process as conditions warrant

The overall aim of these principles is to enable group members to monitor their progress and that of their peers in performing specific learning exercises. In addition, practitioners of this strategy can utilize intelligent tutors with information about each trainee's problem-solving behavior and group engagement for assessment and instructional purposes (Freeman, 2018). Currently, most military ITS focus on well-defined domains, where procedures have explicit steps, performance measures, and standards (Sottilare, 2018). Ill-defined domains remain a serious issue in both the military as well as the business universe. For example, hostage negotiations can be considered ill-defined since there could be several successful outcomes based on various strategies. One potential approach to overcome these obstacles is to combine team-based simulations with intelligent agents. Team-based business simulations, which are used extensively throughout the management education community, tend to provide limited external guidance while intelligent tutors usually require a structured knowledge domain (Goi, 2019). However, combining these two pedological platforms where agents provide advice, based on game conditions, to each team over the course of the simulation could prove to be extremely useful from a team learning perspective, particularly in business simulations involving multi-targets (e.g., market share, stock price). In some instances, multi-agents can be used wherein each agent learns from other agents, which will help improve the overall consulting tasks (Jumadinova, 2014). This combination approach is illustrative of the experiential learning process that is receiving increased attention throughout schools of business.

Experiential learning is an interdisciplinary approach based on management,

education and psychology. It is particularly powerful in connection with management education as it is perceived to be effective in support of training and education in fields as diverse as talent management, leadership performance, competence development, change management, community involvement, volunteering, cross-cultural training and entrepreneurship (Bevan, 2012).

MANAGEMENT EDUCATION ITS EXAMPLES

The previously cited two meta-studies contained a paucity of business management courses. Another meta-study on the use of ITS in undergraduate education reported an effect size of a modest 0.26 (Hedges' g) for three business related subjects (Steenbergen-Hu, 2014). To be clear, many of the reports included in the meta-study were over ten years old, which obviously does not reflect the current state-of-the art in ITS technology. Some additional insights into the efficacy of ITS pertaining to specific business management courses are presented in Table 1. Unlike other educational sectors, the number of studies that have assessed the impact of ITS within the management education universe remains somewhat sparse. The application of an ITS in an information technology course revealed an effective size of 0.53 (Wang, 2014). The results indicated even larger performance improvement for those students with low levels of prior subject knowledge. The next example involved an operations management course where the experimental student group was exposed to a series of tutorial videos (Winch, 2015). The resultant effect size was a moderate 0.66. In the supply chain management (SCM) course, the authors noted that the simulation-based, intelligent tutoring system, yielded an effect size of 0.78 (Bhuiyan, 2015). In this ITS application, a survey of both groups was conducted after completing the course. The survey results indicated that students participating in the simulation reported statistically higher levels of confidence and satisfaction than those that did not use the ITS. The use of an ITS in an introductory management course revealed an effect size of 0.63. Perhaps equally significant was the fact that those students exhibiting a minimal understanding of the material covered in the class at the pre-test phase showed the most improvement after utilizing the ITS (Stonebraker, 2016). The next example utilized an ITS that provided an adaptive learning strategy based on the level of assistance the student received with previous problems (Najar, 2016). The results indicated that students using the ITS learned significantly more than their

Table 1. Selected ITS business course examples

Discipline	Effect Size[1]
Information Technology (Wang, 2014)	0.53
Operations Management (Winch, 2015)	0.66
Supply Chain Management (Bhuiyan, 2015)	0.78
Introductory Management (Stonebraker, 2016)	0.63
Database Design (Najar, 2016)	0.75
Project Management (González-Marcos, 2016)	0.67

[1]Cohen's d

peers who were exposed to a fixed sequence of problem sets (Cohen's d = 0.75). The last example featured a course in project management where the experimental student group had access to an ITS-based project management information system (González-Marcos, 2016). The resultant Cohen's d of 0.67 was based on a comparison of numerical course grades between the control and experimental groups.

The overall performance improvements reported in Table 1 are consistent with the outcomes delineated in the previously referenced meta-studies. The general upward trend in effect size over time does suggest that the one-sigma goal of mastery learning via intelligent tutors is within near-term reach. However, the lack of standardized assessment methods and small to modest sample sizes makes assessment comparisons somewhat challenging (Baird, 2019; Cheung, 2016). To this end, one of the significant problems in evaluating ITS performance is establishing and using a consistent definition of content mastery and standardized assessments (Simpson, 2017).

In terms of achieving the prize, that is, Bloom's two-sigma standard, the more recent evidence is encouraging. In a study involving engineering students the results revealed an effect size ranging between 1.7 and 2.0 over several controlled examinations (Grant, 2016). Even more impressive results on the use of intelligent tutoring systems have been reported in-training programs conducted by the United States Navy (Fletcher, 2018). Again, while the sample sizes are modest and the methodologies are different in detail the results are encouraging.

Many contemporary educators understand that for education to persist beyond the boundaries of the classroom students must develop an interactive dialogue with information they are learning in ways that concern them directly, consistent with participatory action research and critical pedagogy (Hills, 2015).

IMPLEMENTATION STRATEGIES

The rate at which technology is adopted in organizations varies considerably. The Rogers' diffusion of technological innovation model, which was first used for predicting the introduction rate of hybrid corn seed, provides a helpful paradigm for better understanding the adoption process (Rogers, 2003). Applying Rogers' model to academia suggests that a significant proportion of faculty tend to react slowly in embracing technological innovation (Beausoleil, 2018; Fathema, 2015). Figure 2 depicts a slight variation from the basic Rogers' adoption distribution model. This format incorporates a fifth category – Luddites. In a commercial setting the ultimate incorporation of new technologies throughout the organization, for the most part, centers on economics. In academia, this is not always the case. There are some faculty members who most likely will not adopt new learning technologies no matter the nature of the incentives and disincentives and, in fact, may erect roadblocks just as their namesakes did in the early 1800s. This reality is the primary reason for adding the Luddite category. The Rogers' model is based on the Empirical Rule gleaned from statistics. The Empirical Rules states that approximately 68 percent of the population resides within ± one standard deviation from the mean, while 95.5 percent are contained within ± two standard deviations. Figure 2 shows that less than 16 percent of the faculty can be regarded as innovators or enthusiasts during the early phase of new technology adoption. Innovators and early adopters tend to embrace new learning modalities based on their own initiative and technical background without the need for significant institutional support. The emergence of ITS may provide enough motivation for these two groups to embrace this technology. Clearly, this percentage does not represent a critical mass. According to Rogers, the majority, which constitutes a little over two-thirds of the faculty, will begin to embrace new technology only after some period of consideration.

In general, adoption will significantly increase if the majority is recognized as a distinct group within the community and are made a part of the planning and policy making process. A technological innovation, such as ITS, must be demonstrated as both an effective and easily applied solution to a specific learning need. The motivation for the majority to adopt new learning technologies may not be the same across disciplines. Innovation by laggards is more likely to occur through ongoing majority involvement and continued administrative support in the form of incentives. Laggards will adopt only if they are certain that the ITS will not fail and that there will be no negative consequences for them (e.g., negative student evaluations). The Luddite segment, as the name implies, can be the source of potentially significant disruptions in the overall adoption process.

Figure 2. Modified rogers' adoption distribution model

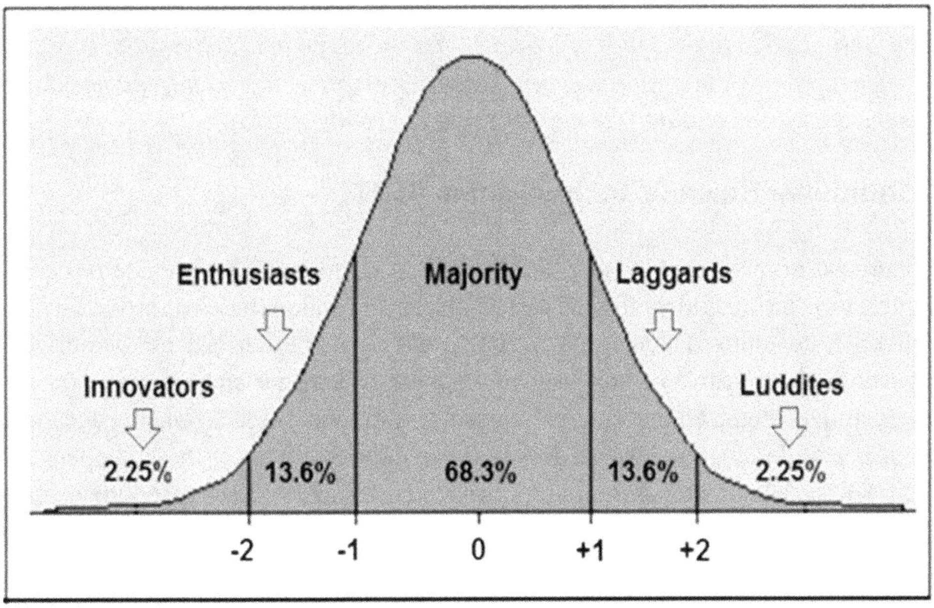

One way to address these adoption concerns in an educational setting is to employ systems that are easy to use, that minimize potential problems, and that yield early success. Specifically, these systems should possess the following four basic characteristics: autonomy, pro-activity, adaptability, and sociability. An additional important feature of an effective tutor is the capability to identify something positive in a student's response (i.e., positive reinforcement). Two promising ITS implementing vehicles are learning management system (LMS) based conditional release technologies and standalone authoring design tools, which can be facilitated via an outside vendor. These two options are not mutually exclusive. Typically, both approaches (LMS and standalone agents) utilize an inner and outer loop format. The function of the inner loop is to execute the steps associated with the task at hand (e.g., taking a quiz). The inner loop can provide feedback and hints at each step. The inner loop also assesses the student's competence in terms of task performance, which is used by the outer loop to select the next task that is appropriate for the student, for example, return to review the current material or advance to new more challenging material. Additional loops can be incorporated into the agent (Niraula, 2015; Chiu, 2016). However, both system complexity and implementation problems will dramatically increase.

User based agents significantly help reduce the administration duties of the course and focus on responding to users' questions or preparing training materials. Agents have been used in many areas of learning. Yet, there remains a myriad of contexts where agents can be incorporated to make learning more efficient and fundamentally change the way education is being delivered (Jawahar, 2015).

Conditional Release Technologies (CRT)

Conditional release technologies (CRT) incorporated within an LMS provide a specific vehicle for offering students a wide variety of learning options based on progressively delivered, customized content (Li, 2015). CRT can be defined as data-driven algorithms that release content based on student performance and characteristics. It is generally recognized that students entering a program of management education possess a wide range of experiences and capabilities. To meet these challenges, the CRT model can be configured to provide customize content based upon such characteristics. Three of the most significant advantages of integrating the ITS based CRT inside the LMS are: 1) ease of implementation, 2) maintenance of student performance records, and 3) flexibility in providing customized content to meet the needs of a wide portfolio of management education programs. CRT offer the capability to assist students in overcoming specific learning challenges by assessing their performance and providing customized content.

The same technology that supports learning activities can also gather data that can be used for assessment. As students work, the system can capture their inputs and collect evidence of their problem-solving sequences (Bienkowski, 2012).

In this regard, students tend to participate more in learning systems that are content rich and that feature a wide variety of delivery modes, which is a hallmark of conditional release technologies. Research on the use of CRTs revealed (Fisher 2014; Shinaberger, 2017):

- Supplementing or replacing face-to-face lectures with active learning exercises enhanced student confidence
- Tutorials that gave students the highest level of bandwidth and engagement without technological interference or latency were the most effective
- Online quizzes were the most challenging activity
- Requiring completion of an assignment before proceeding to the next assignment improved student engagement

Figure 3. CRT design schematic

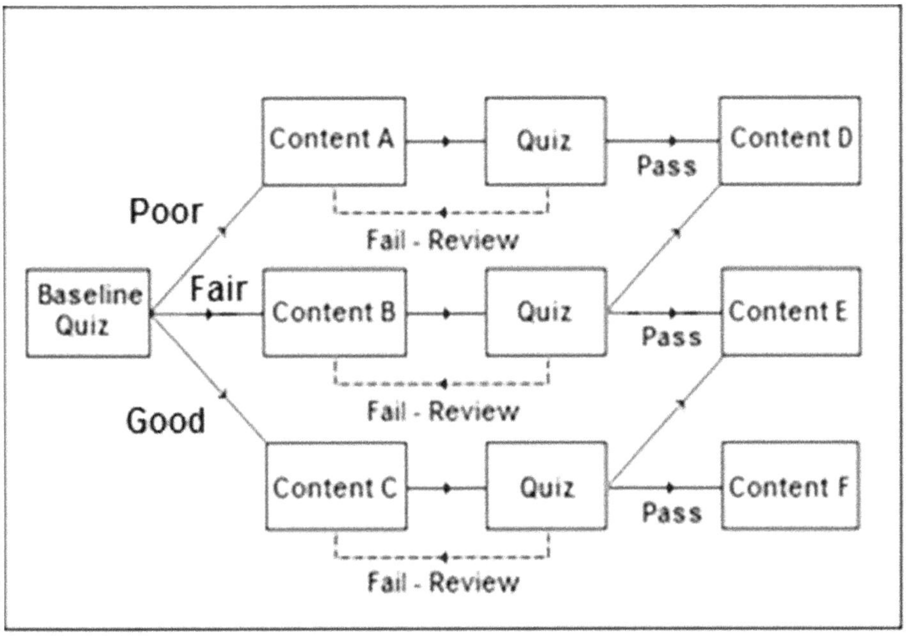

Figure 3 depicts a simplified CRT structure. In this application setting, the student takes a baseline quiz to determine the appropriate content level based on test performance. After reviewing the web-based content the student is presented with a new quiz. If the student passes the quiz a new learning module is displayed, otherwise, the student is directed back to the previous material for further review or given additional supplemental materials based on the quiz results. In some applications, additional review material can be provided via a variety of delivery modes, e.g., interactive exercises and simulations. This process can be further refined by including the student's prior background with the subject material. While quizzes represent an important and easily implemented performance rubric, they can also be augmented by discipline-based simulations as conditions warrant. Figure 3 also illustrates the two-loop configuration. In this example, the quiz represents the inner loop while returning the student to the recently reviewed content or providing the student with new content highlights the outer loop. One approach for incorporating additional information into the ITS is through the use of data mining models (Raju, 2016). For example, Bayesian networks can provide enhanced accuracy in the diagnostic of students' knowledge possession (Ramírez-Noriega, 2017). The Bayesian network is continuously updated as additional information about the student's interaction with the ITS is gleaned. These revised probabilities offer further insights into student learning behavior, which in turn allows for the identification of more effective learning content.

Authoring Tools

Authoring tools represent another approach for developing and deploying ITS. These systems have been shown to decrease the amount of time and resources needed for the development of intelligent tutoring systems (Adetokunbo, 2014; Janghorbani, 2019). Typically, the tutor designer has two classic options: acquire the authoring systems or obtain an already existing tutoring system that can then be modified to meet the specific needs of the learning application. An authoring system, which is usually designed for non-programmers, is a software package that provides the capability to create courseware, multimedia applications, and the corresponding navigational appliances. Authoring tools provide the course designer with more flexibility and thus more creativity than the traditional CRT approach. This capability allows the designer to incorporate more complex and dynamic learning alternatives. Presented in the following are some specific attributes associated with an effective authoring system (Dermeval, 2018; Koedinger, 2012).

- Supports the development of cognitive-based, example-tracing, and collaboration tutors
- Accommodates a wide range of application and technical domains
- Incorporates a dynamic and varied learning environment
- Utilizes the latest empirical evidence and learning techniques
- Skeptical of the student's level of understanding

Rule-based cognitive tutors are designed to capture how students perceive, think, acquire knowledge, and solve problems and are being increasingly used throughout higher education. Example-tracing tutors assess learning performance by comparing student responses against examples of correct and incorrect responses. Typically, this class of tutors provides step-by-step guidance on complex problems and situations, such as business cases, and tends to be less costly to develop than rule-based systems (Aleven, 2016). Collaborative authoring tools represent a natural and important extension to the classical ITS in the sense that they provide for multiple user assessments (Olsen, 2014). To author a collaborative tutor, each of the steps to create an individual tutor are followed for each member of the collaboration team. While the goal of this article is not to evaluate authoring tools, Table 2 lists of some of the better-known development systems. This list is by no means comprehensive. Furthermore, as mentioned earlier, it may be more cost-effective to acquire an

Table 2. Selected ITS authoring tools

System	Source
Situated Pedagogical Authoring (SPA)	University of Southern California
Cognitive Tutor Authoring Tools (CTAT)	Carnegie-Mellon University
Extensible Problem Specific Tutor (PST)	Iowa State University
Authoring Software Platform (ASPIRE)	University of Canterbury (NZ)
AutoTutor Script Authoring Tools (ASAT)	University of Memphis
Generalized Intelligent Framework for Tutoring (GIFT)	U.S. Army Research Lab.

Table 3. Example delivery modes

Topic	Content Delivery Mode
Income Statement	Computing Template
Annual Report	Voiceover PowerPoint Presentation
Net Present Value (NPV)	Intelligent Agent Consultation
Ethics	YouTube Video
Information Technology	Interactive App
Supply Chains	Simulation
Mergers & Acquisitions	Streaming Video

existing ITS application from a third-party vendor than to develop one using one or more of these tools.

Illustrative Business Course ITS Example

The Accounting Education Change Commission (AECC), among other bodies, has been active in calling for the incorporation of both innovative teaching as well as active learning paradigms into accounting courses (Larsen, 2014). Accounting principles courses are well suited to an ITS learning environment, because they tend to be highly rule-based. To illustrate the use of an authoring tool based ITS, consider the following learning objectives for an entry-level accounting principles course:

Figure 4. Illustrative authoring tool format (NPV example)

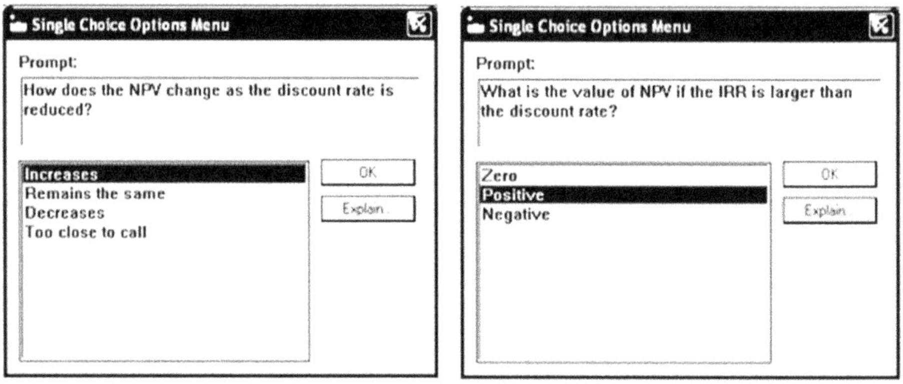

- Prepare and analyze an income statement, balance sheet, and statement of cash flow
- Evaluate the annual report of a public company
- Assess emerging issues such as revenue recognition and restructuring costs
- Appreciate the ethical, global, and regulatory issues of accounting
- Understand and use the information technology of accounting

The ITS can be used to identify and deliver the most effective web-based resources to achieve these objectives as illustrated in Table 3.

Figure 4 shows a sample consultation for a net present value learning application, which was developed using an authoring tool. In this example application, the student is guided through a series of prompts regarding the case and explanations are provided for each prompt. The consultation can be taken more than once since, among other things, many of the prompts are randomized. The use of this expert system is not limited to entering students requiring preparatory work but can also be used as a refresher by continuing students. For example, a student who is struggling with the notion of net present value (NPV) would be directed to a consultative system that outlines the basic NPV process and that provides more specific content.

Implementation Guidelines

Presented in the following are some basic principles for implementing ITS throughout the management education community of practice (Bonner, 2016; Brew, 2014; Fahimrad, 2018; Johnson, 2016):

- **Early Success**: Learning innovations, like intelligent tutors, are most likely to be accepted and used by the majority of management educators if success is experienced early in the adoption process. Early on success extends to the peer network, both within and outside the institution, thereby magnifying the impact on adoption and diffusion.
- **On-Going Peer Support**: Live and virtual peer support not only serves as a repository of encouragement, but also contributes diffusion throughout an educational community.
- **Real Learning Activities**: Initiatives designed to introduce and use ITS should address real activities and requirements. Furthermore, they should support adaptive learning strategies based on students' responses and backgrounds.
- **Design Philosophy**: Intelligent tutors are fit for purpose. The purpose determines how the tutors are structured and used.
- **Multi-User Collaborations**: Intelligent tutors should support a collaborative learning environment that improves student and team opportunities and outcomes.

Agents can also be used support the implementation process. For example, agent driven personal learning environments (PLEs) represent a pedagogical approach for both integrating formal and informal learning and supporting students' self-regulated learning. The evidence is growing that social media can facilitate the creation of PLEs that help learners aggregate and share the results of learning achievements and participate in collective knowledge generation (Bartolomé, 2017). In many circumstances, dynamic information systems like PLEs should be used to support the convergence of learner-centered and knowledge-centered environments. Institutions need to deploy strategies to use dynamic information to support instructional and curriculum reform.

To reap the full potential of integrating ITS throughout the management education universe, the design must consist of more than simply "attaching" a series of standalone tutors to the curriculum. Generally, the program/course structure must be redesigned to provide a seamless transition between face-to-face learning and the ITS. Some specific administrative implementation tasks include: 1) training faculty for successful system deployment and usage, 2) providing high quality and consistent system access, 3) setting specific performance goals and metrics, 4) preparing students for entry and ongoing use, 5) sustaining system operation, and 6) establishing and maintaining the overall culture (Graesser, 2018; Jimenez, 2018). An internalized only approach may not take advantage of ongoing technological advancements. Developing institutional partners represents one possible strategy that can help overcome some of these issues. The current state-of-the-art in ITS technology will continue to improve and learning applications, some of which are unknown at the present time, will continue to unfold. One example is the recent partnership between IBM-Watson and Pearson Publishing, which is designed to apply artificial intelligence to enhancing the education learning process (Luckin, 2016).

The ultimate goal of the field of artificial intelligence in education (AIED) is not to promote artificial intelligence, but to promote education. The leading systems in AIED seem to represent a different paradigm than the classic paradigm of intelligent tutoring systems (Baker, 2016).

CONCLUSION

The management education community is engaged in significant programmatic reforms in light of the business community's call for web-savvy, problem-solving graduates. Today the demands from a globalized economy are causing a sea change in the way business education is being delivered. The traditional one-size-fits-all educational approach of the past is being replaced with a customized and flexible learning paradigm that focuses on student outcomes and performance. Web-based intelligent tutoring systems provide a readily accessible vehicle for enhancing business students' learning performance as well as preparing them for the rigors of the global marketplace. Specifically, ITS can be used to replicate the role of a one-on-one personal tutor on a massive scale and thus move toward the goal of Bloom's Two-Sigma standard. To support these goals, intelligent tutors should be designed to perform three basic functions: 1) perceive the dynamic conditions in the learning environment, 2) formulate problem-solving actions, and 3) implementation of the identified actions. Furthermore, they need to exhibit both a user-friendly and adaptive format as a vehicle for enhancing student learning outcomes.

The empirical evidence to date suggests that ITS performance, in the management education universe, is approaching the one-sigma standard (i.e., mastery learning). While Bloom's two-sigma goal remains elusive, recent evidence suggests that it may be technologically achievable by the mid-2020s. However, technological achievability and adoptability are two quite different issues. The Rogers' diffusion of technological innovation model provides a helpful paradigm for better understanding the adoption process in business management. Applying Rogers' model to academia suggests that a significant proportion of faculty will react slowly in embracing technological innovation. Some specific research areas to help accelerate both achieving Bloom's two-sigma goal and the ITS adoption process include improving the effectiveness of authoring tools to reduce development costs and time, enhancing ITS assessment methodologies to provide improved consistency, reducing students' and student teams' cognitive overload, and enriching adaptation methodologies with a focus on student learning behaviors.

REFERENCES

AACSB. (2016). *A collective vision for business education*. AACSB International.

Adbelaal, M., Ismail, M., Khater, M., & Zaki, M. (2018). *Digital Business Transformation and Strategy: What Do We Know So Far?* Working Paper, Cambridge Service Alliance.

Adetokunbo, A., Adenowo, A., & Patel, A. (2014). A metamodel for designing an intelligent tutoring system authoring tool. *Computer and Information Science*, *7*(2), 82–98.

Akcapinar, G., & Hasnine, M., Nehal, M., Majumdar, R., Flanagan, B., & Ogata, H. (2019). Developing an early-warning system for spotting at-risk students by using eBook interaction logs. *Smart Learning Environments*, *6*(4), 1–18.

Aleven, V., McLaren, B., Sewall, J., van Velsen, M., Popescu, O., Demi, S., ... Koedinger, K. (2016). Example-tracing tutors: Intelligent tutor development for non-programmers. *International Journal of Artificial Intelligence in Education*, *26*(1), 224–269. doi:10.100740593-015-0088-2

Alexander, B., Ashford-Rowe, K., Barajas-Murphy, N., Dobbin, G., Knott, J., McCormack, M., . . . Weber, N. (2019). EDUCAUSE horizon report 2019 higher education edition. *EDUCAUSE*. Retrieved from https://library.educause.edu/resources/2019/4/2019-horizon-report

Alkhatlan, A., & Kalita, J. (2018). *Intelligent tutoring systems: A comprehensive historical survey with recent developments*. Cornell University. Retrieved from: http://arxiv.org/abs/1812.09628

Azcona, D., Hsiao, I., & Smeaton, A. (2019). Detecting students-at risk in computer programming classes with learning analytics from students' digital footprints. *User Modeling and User-Adapted Interaction*, 1–30. doi:10.100711257-019-09234-7

Bagheri, M. (2015). Intelligent and Adaptive Tutoring Systems: How to Integrate Learners. *International Journal of Education*, *7*(2), 1–16. doi:10.5296/ije.v7i2.7079

Baird, D., & Pane, J. (2019). Translating standardized effects of education programs into more interpretable metrics. *Educational Researcher*, *48*(4), 217–229. doi:10.3102/0013189X19848729

Baker, R. (2016). Stupid tutoring systems, intelligent humans. *International Journal of Artificial Intelligence in Education, 26*(2), 600–614. doi:10.100740593-016-0105-0

Bartolomé, A., & Cebrian-de-la-Serna, M. (2017). Personal Learning Environments: A study among Higher Education students' designs. *International Journal of Education and Development Using Information and Communication Technology, 13*(2), 21–41.

Beausoleil, A. (2018). Revisiting Rogers: The diffusion of his innovation development process as a normative framework for innovation managers, students and scholars. *Journal of Innovation Management, 6*(4), 73–97. doi:10.24840/2183-0606_006.004_0006

Bevan, D., & Kipka, C. (2012). Experiential learning and management education. *Journal of Management Development, 31*(3), 193–197. doi:10.1108/02621711211208943

Beyrouti, N. (2017). Digital technology management and educational innovation: The marketabilityand employability of the higher education degrees. *Journal of Developing Areas, 51*(1), 392–400. doi:10.1353/jda.2017.0022

Bhuiyan, T., Peng, W., & Mahmud, I. (2015). Measuring learning motivation of students in supply chain management games setting: A case study of Innov8.0 game. *Problems and Perspectives in Management, 13*(4), 81–90.

Bienkowski, M., Feng, M., & Means, B. (2012). Enhancing teaching and learning through educational data mining & learning analytics: An issue brief. U.S. Dept. of Education, Center for Technology in Learning.

Bloom, B. (1984). The 2-Sigma problem: The search for methods of group instruction as effective as one-on-one tutoring. *Educational Researcher, 13*(6), 4–16. doi:10.3102/0013189X013006004

Bonner, D., Gilbert, S., Dorneich, M., Winer, E., Sinatra, A., Slavina, A., … Holub, J. (2016). The challenges of building intelligent tutoring systems for teams. *Proceeding of the Human Factors and Ergonomics Social Annual Meetings, 60*(1), 1981-1985. 10.1177/1541931213601451

Brew, A., & Cahir, J. (2014). Achieving sustainability in learning and teaching initiatives. *The International Journal for Academic Development, 19*(4), 341–352. doi:10.1080/1360144X.2013.848360

Buil, M., Aznar, P., Galiana, J., & Rocafort-Marco, A. (2016). An explanatory study of MBA students with regards to sustainability and ethics commitment. *Sustainability, 8*(3), 280–293. doi:10.3390u8030280

Cheung, A., & Slavin, R. (2016). How methodological features affect effect sizes in education. *Educational Researcher*, *45*(5), 283–292. doi:10.3102/0013189X16656615

Chiu, P., & Huang, Y. (2016). The development of a decision support system for mobile learning: A case study in Taiwan. *Innovations in Education and Teaching International*, *53*(5), 532–544. doi:10.1080/14703297.2015.1020328

Clemente, J., Ramírezb, J., & Antoniob, A. (2014). Applying a student modeling with non-monotonic diagnosis to intelligent virtual environment for training/instruction. *Expert Systems with Applications*, *41*(2), 508–520. doi:10.1016/j.eswa.2013.07.077

Costigan, R., & Brink, K. (2015). Another perspective on MBA program alignment: An investigation of learning goals. *Academy of Management Learning & Education*, *14*(2), 260–276. doi:10.5465/amle.2013.0315

Damsa, C. (2019). Learning with digital technologies in higher education. *Journal of Educational Sciences & Psychology*, *9*(2), 5–9.

Dasic, P., Dasic, J., Crvenkovic, B., & Serifi, V. (2016). A review of intelligent tutoring systems in e-learning. *Annals of the Oradea University – Fascicle of Management and Technological Engineering, 15*, 85-90.

Day, I., van Blankenstein, F., Westenberg, M., & Admiraal, W. (2017). A review of the characteristics of intermediate assessment and their relationship with student grades. *Assessment & Evaluation in Higher Education*, *43*(6), 908–929. doi:10.10 80/02602938.2017.1417974

Dermeval, D., Paiva, R., Bittencourt, I. I., Vassileva, J., & Borges, D. (2018). Authoring tools for designing intelligent tutoring systems: A systematic review of the literature. *International Journal of Artificial Intelligence in Education*, *28*(3), 336–384. doi:10.100740593-017-0157-9

Dobson, J., & Bauerly, R. (2019). Ten years after: An update on globalization and the future of business education. *Proceedings of the American Society of Business & Behavioral Sciences*, *26*, 182–188.

Doherty, B., Meehan, J., & Richards, A. (2015). The business case and barriers for responsible management education in business schools. *Journal of Management Development*, *34*(1), 34–60. doi:10.1108/JMD-06-2013-0082

Eschenbrenner, B., & Fui-Hoon Nah, F. (2019). Learning through mobile devices: Leveraging affordances as facilitators of engagement. *International Journal of Mobile Learning and Organization*, *13*(2), 152–170. doi:10.1504/IJMLO.2019.098193

Fahimrad, M., & Kotamjani, S. (2018). A review on application of artificial intelligence in teaching and learning in educational contexts. *International Journal of Learning and Development*, *8*(4), 106–118. doi:10.5296/ijld.v8i4.14057

Farooq, M., Kawoosa, H., & Muttoo, M. (2017). CAL: Computer assisted learning. *International Journal of Computer Science and Mobile Computing*, *6*(6), 254–258.

Fathema, N., Shannon, D., & Ross, M. (n.d.). Expanding the technology acceptance model (TAM) to examine faculty use of learning management systems (LMSs) in higher education institutions. *MERLOT Journal of Online Learning and Teaching, 11*(2), 210-232.

Felijakowski, K., & Kosinski, R. (2014). Opinion formation and self-organization in a social network in an intelligent agent system. *Acta Physica Polonica B*, *45*(11), 2123–2134. doi:10.5506/APhysPolB.45.2123

Ferguson, C. (2009). An effect size primer: A guide for clinicians and researchers. *Professional Psychology, Research and Practice*, *40*(5), 532–538. doi:10.1037/a0015808

Fisher, L., Brinthaupt, T., Gardner, J., & Raffo, D. (2014). Conditional release of course materials: Assessing best practice recommendations. *Journal of Online Learning and Teaching / MERLOT*, *10*(2), 228–239.

Flavin, M. (2016). Technology-enhanced learning and higher education. *Oxford Review of Economic Policy*, *3*(4), 632–645. doi:10.1093/oxrep/grw028

Fletcher, J., & Sottilare, R. (2018). *A perspective of intelligent tutoring systems design and use, in Assessment of intelligent tutoring systems technologies and opportunities.* North Atlantic Treaty Organization, STO Technical Report, AC/323(HFM-237) TP/764, 2, 1-11.

Freeman, J., & Zachary, W. (2018). Intelligent Tutoring for Team Training: Lessons Learned from US Military Research. In Building Intelligent Tutoring Systems for Teams. Emerald Publishing Limited.

Ghemawat, P. (2017). Strategies for higher education in the digital age. *California Management Review*, *59*(4), 56–78. doi:10.1177/0008125617717706

Goi, C. (2019). The use of business simulation games in teaching and learning. *Journal of Education for Business*, *94*(5), 342–349. doi:10.1080/08832323.2018.1536028

Gonzalez, M. (2014). Virtual teams and international business teaching and learning: The case of the global enterprise experience. *Journal of Teaching in International Business*, *25*(3), 200–213. doi:10.1080/08975930.2014.925738

González-Marcos, A., Alba-Elíasa, F., Navaridas-Naldab, F., & Ordieres-Meréc, J. (2016). Student evaluation of a virtual experience for project management learning. *Computers & Education*, *102*, 172–187. doi:10.1016/j.compedu.2016.08.005

Graesser, A., Hu, X., Nye, B., VanLehn, K., Kumar, R., Heffernan, C., ... Woolf, B. (2018). Intelligent tutoring systems. In *International Handbook of the Learning Sciences* (pp. 246–255). Routledge. doi:10.4324/9781315617572-24

Guskey, T. (2007). Closing achievement gaps: Revisiting Benjamin S. Bloom's earning for mastery. *Journal of Advanced Academics*, *19*(1), 8–31. doi:10.4219/jaa-2007-704

Harvey, E., & Reitz, E. (2017). Enhanced stem subject outcomes from the use of intelligent tutors. Interservice/Industry Training. *Simulation, and Education Conference.*, *17239*, 1–10.

Herrington, J., & Arnold, D. (2013). Undergraduate business education: It's time to think outside the box. *Journal of Education for Business*, *88*(2), 202–209. doi:10.1080/08832323.2012.672934

Hills, T. (2015). Crowdsourcing content creation in the classroom. *Journal of Computing in Higher Education*, *27*(1), 47–67. doi:10.100712528-015-9089-2

Homes, W., Bialik, M., & Fadel, C. (2019). *Artificial intelligence in education: Promises and implications for teaching and learning*. Boston: Center for Curriculum Redesign.

Hsiung, C. (2012). The effectiveness of cooperative learning. *Journal of Engineering Education*, *101*(1), 119–137. doi:10.1002/j.2168-9830.2012.tb00044.x

Jacob, W. (2015). Interdisciplinary trends in higher education. *Palgrave Communications*, *1*(15001), 1–5.

Janghorbani, S., Modi, A., Buhmann, J., & Kapadia, M. (2019). Domain Authoring Assistant for Intelligent Virtual Agents. *Proceedings of the 18th International Conference on Autonomous Agents and Multi-gent Systems*, 1-9.

Jawahar, J. (2015). A study on role of intelligent agent in education. *International Journal of Applied Engineering Research*, *10*(76), 497–501.

Jiménez, S., Juárez-Ramírez, R., Castillo, V., & Ramírez-Noriega, A. (2018). Integrating affective learning into intelligent tutoring systems. *Universal Access in the Information Society*, *17*(4), 679–692. doi:10.100710209-017-0524-1

Johnson, D., & Samora, D. (2016). The potential transformation of higher education through computer-based adaptive learning systems. *Global Education Journal*, *16*(1), 1–17.

Jumadinova, J., Dasgupta, P., & Soh, L. (2014). Strategic capability learning for improving multi-agent collaboration in ad hoc environments. *IEEE Transactions on Systems, Man, and Cybernetics. Systems*, *44*(8), 1003–1014. doi:10.1109/TSMC.2013.2285527

Kane, G., Palmer, D., Phillips, A., & Kiron, D. (2017). Winning the Digital War for Talent. *MIT Sloan Management Review*, *58*(2), 17–19.

Karakas, F., Manisaligilb, A., & Sarigolluc, E. (2015). Management learning at the speed of life: Designing reflective, creative, and collaborative spaces for millennials. *International Journal of Management Education*, *13*(3), 237–248. doi:10.1016/j.ijme.2015.07.001

Koedinger, K., Corbett, A., & Perfetti, C. (2012). The knowledge-learning-instruction framework: Bridging the science-practice chasm to enhance robust student learning. *Cognitive Science*, *36*(5), 757–798. doi:10.1111/j.1551-6709.2012.01245.x PMID:22486653

Kulik, J., & Fletcher, J. (2016). Effectiveness of intelligent tutoring systems: A meta-analytic review. *Review of Educational Research*, *86*(1), 42–78. doi:10.3102/0034654315581420

Larsen, R., Blocher, E., Brewer, P., Cokins, G., Sorensen, J., Stout, D., ... Wouters, M. (2014). Focusing accounting curricula on students' long-run careers: Recommendations for an integrated competency-based framework for accounting education. *Issues in Accounting Education*, *29*(2), 295–317. doi:10.2308/iace-50673

Lee, H., & Mehta, M. (2015). E-Mentoring through a network of practice on Facebook. *International Journal of Knowledge-Based Organizations*, *5*(1), 34–45. doi:10.4018/ijkbo.2015010103

Li, N., Matsuda, N., Cohen, W., & Koedinger, K. (2015). Integrating representation learning and skill learning in a human-like intelligent agent. *Artificial Intelligence*, *219*, 67–91. doi:10.1016/j.artint.2014.11.002

Lu, O., Huang, A., Huang, J., Lin, A., Ogata, H., & Yang, S. (2018). Applying learning analytics for the early prediction of students' academic performance in blended learning. *Journal of Educational Technology & Society*, *21*(2), 220–232.

Lubeck, J., Cheng, B., Myszkowski, G., Doedijns, G., Drew, D., & Snow, S. (2015). *Future Trends in Business Education.* Association to Advance Collegiate Schools of Business (AACSB), the Executive MBA Council (EMBAC), and the Executive Education Consortium (UNICON), Executive Core.

Luckin, R., Holmes, H., Griffiths, M., & Forcier, L. (2016). *Intelligence unleashed: An argument for AI in education. UCL Knowledge Lab.* London: Pearson.

Ma, W., Adesope, O., Nesbit, J., & Liu, Q. (2014). Intelligent tutoring systems and learning outcomes: A meta-analysis. *Journal of Educational Psychology*, *106*(4), 901–918. doi:10.1037/a0037123

Maldonado, E., & Seehusen, V. (2018). Data mining student choices: A new approach to business curriculum planning. *Journal of Education for Business*, *93*(5), 196–203. doi:10.1080/08832323.2018.1450212

Mehrotra, A., & Elias, H. (2017). Employers' Feedback on Business Graduates. *US-China Education Review*, *7*(4), 190–199.

Mendo, S., León-del-Barco, B., Felipe-Castaño, E., Polo-del-Río, M., & Iglesias-Gallego, D. (2018). Cooperative team learning and the development of social skills in higher education: The variables involved. *Frontiers in Psychology*, *9*, 15–36. PMID:29445346

Mohapatra, S. (2015). Business school education and technology: A case study. *Education and Information Technologies*, *20*(2), 335–346. doi:10.100710639-013-9287-3

Najar, A., Mitrovic, J., & McLaren, B. (2016). Learning with intelligent tutors and worked examples: Selecting learning activities adaptively leads to better learning outcomes than a fixed curriculum. *User Modeling and User-Adapted Interaction*, *26*(5), 459–491. doi:10.100711257-016-9181-y

Niraula, N., Rus, V., & Banjade, R. (2015). DeepTutor: An effective, online intelligent tutoring system that promotes deep learning. *Proceedings of the Twenty-Ninth AAAI Conference on Artificial Intelligence.*

Olsen, J., Belenky, D., Aleven, V., Rummel, N., Sewall, J., & Ringenberg, M. (2014). Authoring tools for collaborative intelligent tutoring system environments. *Lecture Notes in Computer Science*, *847*, 523–528. doi:10.1007/978-3-319-07221-0_66

Perera, L., Nguyen, H., & Watty, K. (2015). Formative feedback through summative tutorial-based assessments: The relationship to student performance. *Accounting Education*, *23*(5), 424–442. doi:10.1080/09639284.2014.947093

Philips, F., Hsieh, C., Ingene, C., & Golden, L. (2016). Business schools in crisis. *Journal of Open Innovation*, *2*(1), 1–21. doi:10.118640852-016-0037-9

Pistilli, M., & Arnold, K. (2012). *Course signals at Purdue: Using learning analytics to increase student success.* 2nd International Conference on Learning Analytics and Knowledge, Vancouver, Canada.

Raju, D., & Schumacker, R. (2016). Comparing data mining models in academic analytics. *International Journal of Knowledge-Based Organizations*, *6*(2), 38–53. doi:10.4018/IJKBO.2016040103

Ramírez-Noriega, A., Juárez-Ramíreza, R., & Martínez-Ramírezb, Y. (2017). Evaluation module based on Bayesian networks to intelligent tutoring systems. *International Journal of Information Management*, *37*(1), 1488–1498. doi:10.1016/j.ijinfomgt.2016.05.007

Rickel, J., & Johnson, W. (1999). Animated agents for procedural training in virtual reality: Perception, cognition, and motor control. *Applied Artificial Intelligence*, *13*(4-5), 343–382. doi:10.1080/088395199117315

Rogers, E. (2003). *Diffusion of innovations* (5th ed.). New York: Free Press.

Roll, I., & Wylie, R. (2016). Evolution and revolution in artificial intelligence in education. *International Journal of Artificial Intelligence in Education*, *26*(2), 582–599. doi:10.100740593-016-0110-3

Sales, A., & Pane, J. (2019). The role of mastery learning in an intelligent tutoring system: Principal stratification on a latent variable. *The Annals of Applied Statistics*, *13*(1), 420–443. doi:10.1214/18-AOAS1196

Sawilowsky, S. (2009). New effect size rules of thumb. *Journal of Modern Applied Statistical Methods; JMASM*, *8*(2), 467–474. doi:10.22237/jmasm/1257035100

Shahiri, A., Husain, W., & Rashid, N. (2015). A review on predicting student's performance using data mining techniques. *Procedia Computer Science*, *72*, 414–422. doi:10.1016/j.procs.2015.12.157

Shinaberger, L. (2017). Components of a flipped classroom influencing student success in an undergraduate business statistics course. *Journal of Statistics Education: An International Journal on the Teaching and Learning of Statistics*, *25*(3), 122–130. doi:10.1080/10691898.2017.1381056

Simpson, A. (2017). The misdirection of public policy: Comparing and combining standardized effect sizes. *Journal of Education Policy*, *32*(4), 450–466. doi:10.1080/02680939.2017.1280183

Soetanto, R., Childs, M., Poh, P., Austin, S., & Hao, J. (2012). Global multidisciplinary learning in construction education: Lessons from virtual collaboration of building design teams. *Civil Engineering Dimension*, *14*(3), 173–181.

Sottilare, R., Burke, S., Salas, E., Sinatra, A., Johnston, J., & Gilbert, S. (2018). Designing adaptive instruction for teams: A meta-analysis. *International Journal of Artificial Intelligence in Education*, *28*(2), 225–264. doi:10.100740593-017-0146-z

Spain, R., Priest, H., & Murphy, J. (2012). Current trends in adaptive training with military applications: An introduction. *Military Psychology*, *24*(2), 87–95. doi:10.1080/08995605.2012.676984

Spender, J. (2016). How management education's past shapes its present. *BizEd*. Retrieved from: http://www.bizedmagazine.com/

Stachowicz-Stanusch, A., & Amann, W. (2018). Artificial intelligence at universities in Poland. *Organization and Management Scientific Quarterly*, *2*(42), 62–82.

Steenbergen-Hu, S., & Cooper, H. (2014). A Meta-Analysis of the effectiveness of intelligent tutoring systems on college students' Academic Learning. *Journal of Educational Psychology*, *106*(2), 331–347. doi:10.1037/a0034752

Stone, R., Brooks, R., Brynjolfsson, E., Calo, R., & Etzioni, O. (2016). *Artificial intelligence and life in 2030*. Stanford University. Retrieved from: https://ai100.stanford.edu/2016-report

Stonebraker, I., Brookem, M., Shaw, R., & Moss, J. (2016). Student see versus student do: A comparative study of two online tutorials. *TechTrends*, *60*(2), 176–182. doi:10.100711528-016-0026-7

Stosic, L. (2015). The importance of educational technology in teaching. International Journal of Cognitive Research in Science. *Engineering and Education*, *3*, 111–114.

Suhadi, S., Bahru, J., Mohamed, H., & Abdullah, Z. (2016). Enhancing student's higher order thinking skills (HOTS) through the Socratic Method approach with technology. *International Journal of Knowledge-Based Organizations*, *6*(4), 14–28. doi:10.4018/IJKBO.2016100102

Tsiakas, K., Abujelala, M., & Makedon, F. (2018). Task engagement as personalization feedback for socially-assistive robots and cognitive training. *Technologies*, *49*, 1–17.

Wang, D., Han, H., Zhan, Z., Xu, J., Liu, Q., & Rena, G. (2014). A problem solving oriented intelligent tutoring system to improve students' acquisition of basic computer skills. *Computers & Education*, *81*, 102–112. doi:10.1016/j.compedu.2014.10.003

Winch, J., & Cahn, E. (2015). Improving student performance in a management science course with supplemental tutorial videos. *Journal of Education for Business*, *90*(7), 402–409. doi:10.1080/08832323.2015.1081865

Zhang, Y., Fei, Q., Quddus, M., & Davis, C. (2014). An examination of the impact of early intervention on learning outcomes of at-risk students. *Research in Higher Education*, *26*, 1–12.

Chapter 10
Training and Evaluation of Mystery Customer and Customer Satisfaction

Loukas K. Tsironis
University of Macedonia, Greece

ABSTRACT

This chapter increases the customer service level by expanding the method of application used by mystery customer (MC) and resolves practical and concrete problems concerning the status of the chain stores. The survival of organisations can often be dependent on their customer service level; therefore, there is an immediate need to form a permanent measurement to act as an indicator of that performance. A use case of MC method was applied in order to observe the level of quality service in a large retail network placed in Greece. The results suggested an explicit determination of the factors of satisfied customers through the application of MC method. Action diagrams were used as a performance-importance map to indicate the strong and weak points in terms of criteria and to define the required improvement efforts. MC data showed the points or issues that service activities performance is likely to cause dissatisfaction and enables proactive measures taken on employee performance and classifies the improvement areas determined by the analysis and used as a benchmarking tool.

DOI: 10.4018/978-1-7998-2245-5.ch010

INTRODUCTION

In an ever-changing environment, companies must create new products, services and processes to compete and survive (Avlonitis et al, 2000).

The element of differentiation is not only the benefit that exceptional quality of products or services brings, but also excellent customer service (Kafchehi et al, 2016).

In this respect, satisfaction of customers is a very important key element for organizations. The expression "*satisfied customer*" in practice means high-exceptional quality, loyal customers and positive reputation due to increased customer satisfaction (Homburg and Rudolf, 2001). *"Customer loyalty can only be earned and cannot be bought"* (Berry and Gresham, 2001). Experts estimate that most North American companies spend about 3% of revenues on customer satisfaction measurement. Customer Satisfaction (CS) surveys addresses the end product of the production line, revealing expectations and perceptions in total. In contrast, Customer Service Measurements (CSM) reveal performance at each identified stage of the production process. Clearly, both methods are important, but different (Buxton, 2000). According to Buxton (2000), both CS and CSM (figure 1) affect each other.

Currently, Mystery Customer (MC) is used by many private and public organizations (Wilson, 1998; Anderson et al., 2001; Douglas & Davies 2007; Douglas 2015; Jankal & Jankalová 2011; Kehagias et al., 2011; Wiele et al., 2005)). The use of MC is dated roughly from the 40s, when, naturally, it served different objectives and needs of organisations. Substantially it covers the need for control of

Figure 1. The cycle of customer satisfaction and customer service measurement (adapted from Buxton, 2000)

integrity and efficiency of personnel (Yanli & Yao, 2015). However, in our times, because of the inexorable competition that prevails in the market, it has taken the form as a tool to measure the quality of customer service and customer satisfaction (Hesselink & Wiele, 2003). Broadly speaking, we could argue that it offers essential information to management about the level of quality its products and services provide as well as on the level of customer satisfaction. Furthermore, if we consider that the meaning of customer includes not only external but also internal ones, then we can understand how important the content of MC chain is, on monitoring and controlling the retail process. It is now a fact that MC survey results can be utilized to lead the marketing strategies (Hesselink & Wiele, 2003; Anderson et al., 2001; Douglas & Davies 2007; Douglas 2015; Jankal & Jankalová 2011; Kehagias et al., 2011; Wiele et al., 2005). It is a tool that supports the managers' decision-making in order to maintain a high level of service.

In this respect, this study attempts to expand the method of application of MC in retail networks, by seeking to improve areas within the network (Singh et al., 2014; Anderson et al., 2001; Douglas & Davies 2007; Douglas 2015; Jankal & Jankalová 2011; Kehagias et al., 2011; Wiele et al., 2005). In particular, it attempts to resolve practical and concrete problems of the network concerning the status of the chain stores in order to increase the customer service level.

The research seeks to utilize recent information concerning the performance in the retail operations to educe specific suggestions for the improvement of quality service. It targets the classification of chain shops according to the performance level of customer service (Korhan, 2015). In addition, this chapter determines both first priority and second priority factors that need improvement. The goals of the determination of factors satisfy the following:

1. To reveal the necessity of combination of two different measurement tools
2. To provide a more flexible and effective managerial tool and
3. To present a visualized managerial technique that can monitor and direct strategic alternatives for a better service quality experience.
4. To link the academic knowledge of MC research with Greek reality and its application in the market.

Through this research, the effectiveness of MC research is evaluated as a critical diagnostic tool for companies, the conclusion of conclusions on the quality of services provided in the services industry, their link with the sales trend, and conclusions about the required frequency such research in order to draw safe conclusions.

It is well known that it is very difficult to evaluate service quality; most of the classical approaches are surveys that use questionnaires to investigate the level of service quality after the service process (Yang, 2002). MC investigation has already been broadly applied, although the information gathered is usually withheld by organizations and therefore the subject is rarely discussed in management literature. According to this method an anonymous expert participates and observes the whole transaction with the customer during the sale process by examining the level quality service from customer's perspective (Chen, 2005; Finn & Kayande, 1999). This method is a diagnostic tool through which the management can understand the frontline staff's performance (Dawson & Hillier, 1995). It is also a measurement tool to reward and encourage employees (Grove & Fisk, 1992).

There is a wide literature on MC. Some parts of it referred to its application in the banking (Holliday 1994; Morral 1994; Dorman 1994; Hotchkiss 1995; Hoffman 1993; Stoval 1993; Leeds 1992 and 1995; Hanke 1993; Tepper 1994; Anderson et al., 2001; Douglas & Davies 2007; Douglas 2015; Jankal & Jankalová 2011; Kehagias et al., 2011; Wiele et al., 2005) or tourism sector (Anderson et al 2001). Some aspects of this literature for example attempt to test if customers are treated equally or are discriminated (Morral 1994; Tepper 1994). Other scholarly works concern the relationship between MC results and specific managerial issues such as rewarding employees, (McNerney, 1996; Eisman, 1993; Boyd, 1995) the prerequisites and conditions and the appropriate rules when implementing MC method (Cramp, 1994; Miles, 1993; Dwek 1996; Burnside, 1994; Cobb, 1995).

Related research emphasises the aspects of reliability and validity of the method (Dawes et al., 2000). Several authors have focused on the role of MC in the measurement and management of the service delivery process (Wilson, 1998; Wilson and Gutmann, 1998)

Finn (2001) describes research on retail chains focusing on the quality of MC data and the number of visits needed. In other cases, the method concerns the control of sales of inadequate products in specific target groups. For example, Gosselt et al., (2007) deal with the sale of alcoholic drinks to underage customers in supermarkets and liquor stores.

Another group of literature studies focuses on the MC method as a useful tool for improving service delivery (Hesselink et al., 2004). Specifically, Hesselink et al., (2004) examined the service delivery with the aim to manage organisational changes through the contribution of MC. Their work provides an assessment of the current level of service provision. Specifically, one goal was to get an organisational dashboard from which management can easily derive the performance of its divisions, regions and branches and it was achieved by visiting branches periodically (Chanal, 2017).

The present chapter is focusing on identifying improvement areas of the organisation and presents the classification of them in comparison with the article presented above in which the focus is to observe the change process and prove that MC gives more stimuli to develop changes in an organization.

Beyond the various aspects of the literature on the application and impact of MC, another part of it emphasizes the actual process of the method attempting to investigate its improvement.

WHAT IS A MC SURVEY?

The MC method is a form of market research through observation and evaluation that companies use researchers as clients to mislead staff so that employees believe they serve existing or potential clients. MC is defined as the process of measuring the quality of service provided by employees at the company's first line of business (Schlossberg, 1991, p.10). Since 1970, it has been widely used, in the banking sector, in the catering industry and recently in the hotel services sector. In the 1970s, 25% -35% of banks with deposits over $ 300 million used MC research (Leeds, 1995). Catering companies such as Kentucky Fried Chicken, Domino's Pizza, Burger King, Taco Bell and Arby's have been carrying out such research since the early 1970s. Its use is widespread in developed economies, with turnover in Great Britain alone of 45 million dollars per year (Wilson, 1998). MC programs have initially been applied as a tool for assessing and improving customer service. It is a diagnostic tool that demonstrates mistakes and weak points in organizing the points both in relation to the desired level of service offered by the company and in relation to competition. In essence, it assesses the competitiveness of the services provided. Another role is to develop teamwork and commitment to achieving first-person staff goals, as well as people's skills through education assessment and rewarding them in the critical areas that they want to develop.

TYPES OF A SECRET VISITS SURVEYS

There are several types for collecting data from a secret visits survey. The most common types are:

Phone Surveys

In such a case, people conducting research should not be recognized and conversations should not be recorded. The average time of the conversation should reflect a real transaction and should not exceed 15 minutes for retailers and construction companies, and 30 minutes for economic sectors and carmakers. Secret visitors should know that their identity is likely to be discovered if the person who receives the call uses call-based calling technology. Among the most important points covered by phone estimates is:

- How many times the phone rings before someone answers. Two are ideal.
- The way of greeting.
- If the person answering the phone is polite and friendly
- If you please the caller for the phone call

Face to Face Surveys

In this case, individuals conducting the survey should not be identified and conversations should not be recorded without the consent of the interlocutor. The average time of the conversation should reflect a real transaction and should not exceed 15 minutes for retailers and construction companies, and 30 minutes for economic sectors and carmakers. The time spent observing or checking is not included in the above limits. Also, secret visitors should make a purchase that reflects the type of business they visit, for example, gas filling the car at a petrol station.

What do MCs observe in face-to-face surveys?

- Is the environment (internal and external) clean?
- Is the dressing of the employees appropriate?
- Is the employee polite and smiling?
- Does the employee drink, smokes or eat during work?

Audio and Visual Surveys

Many businesses that provide tape services are training secret visitors to learn how to use and hide equipment well during an investigation. Secret visitors should place the microchips allocated to them by the company in unseen places as well as where sound is appropriate and clean. Automotive, financial, construction and insurance are some of the many companies that use audio secret visits on a regular basis.

Video shopping may include sound, depending on the wishes of the companies and the company providing the research. Approximately the same procedure as audio shopping follows, except that the secret visitor is equipped with micro-cameras, transmission equipment and video recorder. It is necessary to use "professional visitors", who are trained in the proper use of the equipment.

The advantages of this method are that employees themselves can listen and see themselves on the client's side, which enables them to assess and improve their performance. The company is also in a position to test whether the training programs are adopted and implemented by the employees as well as to assess the value and usefulness of the training programs.

Postal, Fax and E-mail Estimates

In this case, individuals conducting the survey should not be identified and any material copied and used in the report should be anonymous. The nature of the question should be such that it does not take too much time or effort to respond.

Estimates Via the Internet

Persons conducting research should not be identified. But they should be informed that their identity may be revealed if personal credit cards are used in case of purchase.

MYSTERY CUSTOMER SURVEY: A TOOL FOR DIAGNOSING THE SERVICES PROVIDED

Quality as a concept has evolved since its appearance as a quality control of the production line, quality in the development and implementation of processes, and finally, the quality that characterizes the whole organization. Recently, beyond the overall quality of an organization, emphasis is placed on the relationships it has with the internal and external factors of interaction, such as shareholders, employees, local society, environment, etc. This development marks the shift from the tendency to achieve quality in the processes to the achievement of quality in the organization's relations, i.e. from internal to external quality upgrading. At the same time, there is a tendency to improve relationships with an organization's customers. This happens for two reasons:

1. The service sector in developed markets is growing.
2. Establishing good customer relationships takes a leading role in the competitiveness of companies in the modern environment.

Several methods of assessing companies are moving in this direction: A recent ISO (ISO 9000: 2000) certificate includes procedures that approach the customer, although they practically refer to the collection of customer satisfaction data that are rarely evaluated for development and improving the organization. In recent years, through excellence awards and business awards, importance is given to the role of customer satisfaction metrics, how they are used, how they fit into the organization, how they are combined with other functions of the organization, and ultimately how they contribute to improving of. A much bigger challenge than customer satisfaction is to create excellent purchasing experience for customers as described by Pine & Gilmore (1999).

Customer Experience

It is about creating a positive experience for the customer, which in turn leads to customer loyalty and customer oriented sustainability. It is the most recent challenge faced by the business world, (Shepherd, 2004), (Pine & Gilmore 1999). According to Shaw & Ivens (2002), seven pillars characterize the customer's purchasing experience as important:

1. Recognizing the excellent purchasing experience of our customers as a long-term competitive advantage
2. Excellent buying experience is achieved through the constant overcoming of the material and emotional expectations of the client
3. Excellent buying experience differs when we focus on creating stimulating planned emotions that satisfy the customer
4. Creating excellent customer experience is built into an organization where there is inspirational leadership, culture of growth and empowerment, and executives who can take advantage of the opportunities they are given to improve and develop their skills.
5. Excellent buying experience is designed from outside to media rather than the opposite
6. Excellent customer buying experience generates revenue and reduces costs
7. Excellent customer buying experience is the trunk and expression of the corporate image.

The above factors when adopted lead to the continuous improvement of each one individually. Some authors (Smith & Wheeler 2002) emphasize the important role of customer satisfaction in supporting the organization by its own customers. They also claim that customers are now looking for the experience of a service that is consistent and responsive to their lifestyle and, of course, they are willing to pay the cost. The *value* is transferred from the products, initially to the goods and services and then to the experiences. This development implies that companies need to segment the markets more, diversify customer groups to meet the requirements of each individual customer in the best possible way. The only way to create conditions of excellent purchasing experience is to balance and consolidate processes (in order to produce the product / service) of the environment in which the purchasing experience is made, and of the employees (the employees who will establish customer relations.

The completion and application of these factors must be in line with the service standards of the company or higher quality objectives (Disney Institute 2001).

The Chain of Excellence

One of the major factors influencing the provision (Shaw & Ivens, 2002) of excellent buying experience is leadership. This factor is of particular importance in Church (1995) who argues that the style of management affects the performance of the business, and the quality of its services. In particular, the development of human resources has a major impact on the assessment of the organization by customers. This conclusion is also led by House (1992) and Youki (2002). These researchers argue that existing ones respond better to customer needs when inspired and motivated by charismatic leaders who impart their own self-motivation, enthusiasm, and firm commitment to achieving their goals.

Excellent leadership will be achieved through a clear mission statement of the business. Through a clear, distinct behavioral pattern of senior management whose executives feel proud of the organization they work in, they have clear goals and know the work that is expected of them.

Exceptional employees will be developed through the guidance of an excellent leadership, selection and recruitment procedures, excellent training and appropriate training tools, as well as a clear direction of how they will carry out the work they expect.

Satisfied customers will be attracted by excellent service, through excellent employees who treat their clients as individuals whose needs are covered in the slightest detail and whose needs are best recognized and best met and the most appropriate proposal to meet their needs .

Satisfied customers who have become firm in selecting the company will transfer their positive experiences to other, new customers. This will establish long-lasting relationships with the organization that lead to excellent economic returns.

More and more scholars agree that there are specific reasons that define customer satisfaction as one of the most important parameters of success:

- Satisfied customers buy more and are more loyal.
- Satisfied customers are buying complementary products and services.
- Satisfied customers support and disseminate positive feedback to other potential customers.
- Satisfied customers buy more and are more loyal.
- Satisfied customers are buying complementary products and services.
- Satisfied customers support and disseminate positive feedback to other potential customers.

Customer Satisfaction Measurement in the Retail Industry

Usually, MC research is used to complete the overall customer satisfaction measurement framework. But there is a lot to be done to assess the information of unhappy customers and to compare with the levels of service offered by competitors. And here comes the MC survey to fill these business information gaps.

Customers make the decision to purchase a new product or change supplier based on the expected value of the service package offered to them, rather than the satisfaction they receive from existing coverage of their need. Therefore, customer loyalty and customer satisfaction during the purchase process cannot be met by customer satisfaction surveys alone. But, what about the satisfaction of our customers? The evaluation of the product or service in relation to the expected coverage of the customer's needs. Why measure customer satisfaction? In addition to the obvious reasons for improving the services, products and processes of a business, it is also largely related to customer sustainability (Jones & Sasser 1995), while among other researchers there is a small correlation between customer satisfaction and sustainability (Van Looy et al., 1998).

Going further, there is the view that satisfaction is a necessary but not sufficient condition for sustainability of loyalty and therefore repeat purchase. In measuring customer satisfaction, it has been observed that the majority of customers say happy (70-90%), which does not explain the degree of supplier change. This may be due to weaknesses in the measurement-assessment system or simply to the Hawthorne phenomenon, i.e. increasing satisfaction from the fact that the research takes place and not the product or service itself. Therefore, there are opinions (Bateson & Hoffman, 1999) that feedback conclusions should focus on:

1. The needs / concerns of unhappy customers and
2. How the findings are evaluated against competitors.

Bateson and Hoffman (1999), integrated and sometimes costly viewpoint, refers to an organized system that integrates customer complaints, post-sale inquiries, customer focus interviews, MC inquiries, employee surveys, and market research (Ahearne et al., 2007; Evanschitzky et al., 2008). In recent years, customer service has been evaluated, as well as the continuous improvement of processes and people through the exploitation of its findings.

However, in order to achieve a substantial improvement, the company's assessment should also be made in relation to its competitors. Both through questionnaires addressed to their common clients and through a MC survey, which is thus best suited to the fact that in this case the company has control over the scenarios it wishes to evaluate but also concerns an objective and professional look at parameters that potentially unsuspecting customer does not focus in advance. Besides, the limitations of customer satisfaction research have other weaknesses, such as possible limitations to the spoken or written expression of respondents, subjectivity due to the difference between the actual and the declared ex-post experience (Ahearne et al., 2007; Evanschitzky et al., 2008).

ADVANTAGES AND DISADVANTAGES OF MYSTERY CUSTOMER RESEARCH

We all know that customers often do not see any errors or defects in the processes or even when these mistakes are perceived they rarely make complaints and declare their dissatisfaction. According to Hesselink & Wiele (2003), Technical Assistance Research Program (TARP) research shows:

- 26 of the 27 unsatisfied customers with low-priced products do not complain, with 63% of them not buying again
- 45% of dissatisfied consumers about a low level of service will not complain, 45% of them will not come back
- 27% of dissatisfied customers of expensive products will not complain, but 41% of them will not come back
- 37% of unsatisfied consumers of expensive, low-grade products will not show dissatisfaction, with 50% of them not buying again.

Many companies, in order to minimize unsatisfied customer rates and improve their business image both in terms of service and in terms of providing high-quality products and services, proceed with the implementation of secret visits programs.

In particular, the benefits of this research are:

- It is an excellent way to measure and control it efficiency of service and service.
- Helps control the workers' clothing, clean the interior and exterior, if the company's brand name, etc.
- Improves and helps maintain customer consistency
- Informs employees about what is important for customers during the service process
- Encourages positive actions on the part of employees / management, giving incentives to reward employees who are right in their work
- Confirms delivery quality of products and services
- Supports promotional programs

- Controls the price and elasticity of products
- Provides information to help analyze it competition
- Fill in data for product promotion research
- Recognizes any additional training needs employees as well as sales opportunities
- It is a training and development tool
- Ensures positive relationships between employees and customers
- Enhances the integrity of employees
- Program variables are easily measurable. This means that objectives and policy are easy to define and measure. This makes the results of the survey valuable. (Michelson, 2001). To make the secret visits program more objective, reference should also be made at this point to the drawbacks:
- The program is relatively expensive, especially in case a business has many branches to control.
- Most programs rely solely on objective elements, resulting in the loss of subjective awareness of space. There is no measure about the atmosphere of the store or the informal behavior of employees, but they play an important role in generating customer satisfaction from the business.
- The secret visits survey is conducted by professional secret visitors or by people who have made several visits during a period. Although attempts are made to include subjective measures, professional secret visitors are often biased in their judgment. They become either unconscious or overly sensitive to the atmospheric issues mentioned above.
- The secret visits survey does not count the most important variable - the opinion of the consumers themselves. Ultimately, this is what it really is worth, not if the employee follows faithfully every detail of the company policy. (Bryson, 1991);

According to Moore (1999), investigations of secret visits can be used in many ways. And in the more ways you use, the more valuable you invest in the program:

- Preparing for competition.
- Control / Tracking of Competition
- Recognizing the right employees
- Incentive programs
- Measurement of education

Many times, mainly due to misinterpretation of the secret search investigation concept, some businesses misuse and mistakenly

THE PROFILE OF THE MYSTERY CUSTOMER

To ensure an objective investigation of secret visits - and therefore a credible secret visit program - the business should, among other things, carefully select individuals who are eligible to become secret visitors and educate them. The process is extremely important and due care should be taken because secret visitors also have their own share of responsibility in the failure or success of a program.

In the process of selecting secret visitors, the business will have to face the following dilemma: will it use the company's employees (internal partners) or is it preferable to choose external partners?

Internal collaborators are usually quality control analysts (Stefanelli, 1994). The main advantages of choosing this kind of secret visitors are in principle the lowest cost for the company and then the best knowledge of the products and goals of the business. On the other hand, the disadvantages include the reaction of the employees themselves to being "checked" by someone they know, the likelihood of recognizing the secret visitor is greater and there is a bias, either positive or negative (Aftermarket Business, 1994).

On the other hand, external contractors conclude temporary contracts with the company and no one-time salary (Aftermarket Business, 1994) is required, thus reducing overheads. According to Bruno (1988), external secret visitors are becoming necessary, mainly due to the large size of today's businesses, as internal affiliate visits are no longer feasible. The disadvantages of employing external visitors are that it is necessary to prepare the secret visitor for the execution of the program and that the high number of external secret visitors can influence the quality of the evaluation.

Almost everyone can become a secret visitor. The business chooses those people who can do this job faithfully, accurately and objectively. It's harder than it sounds. According to Prince (1996), the qualifications that one must have are:

- Do not have a clear bias in favor of or against the company that is going to visit.
- The ability to personate the role required to make the visit convincing.
- Have normal behavior, i.e. be neutral during the transaction.

Have a good memory to be able to record them as soon as the visit is over.

MYSTERY CUSTOMER APPLICATION INSIGHTS

Step-By-Step MC Investigation

One of the ways to measure this degree of customer satisfaction is also the MC survey. Research can be used to evaluate many factors, although the most common is the quality of service. Other factors that are assessed are:

- Staff training programs (Morrison et al., 1997)
- Equal treatment of customers (in terms of racism or discrimination) (Morral, 1994; Tepper, 1994)

In any case, the key feature of MC research is the credibility and quality of the findings. MCs are trained professionals who are familiar with the internal processes of a business, and in this way they can accurately and clearly identify any weaknesses.

The first step in designing a MC survey is to determine what goals the administration wants to achieve. These goals will determine the structure and structure of the research. They should be objective and clearly defined in order to produce safe, useful conclusions that will help improve the organization. In most cases, the recommended method is to resume visits to a specific store and at different times of the day. In this way, the issue of uneven traffic in a shop is dealt with during the day, when at certain times traffic is particularly high compared to others.

The second step is to gather the findings. The results should not only refer to general points and issues relating to customer service in general (Zeithaml et al., 1990) but also key performance indicators, i.e. those indicators that if achieved will lead the company in fulfilling goals, and will support its mission and vision. Researchers conducting research should be independent, have a critical spirit, be objective and preserve their anonymity.

The main purpose of the MC survey is the reliable presentation of the daily image of the operation of the stores. Although its findings should be presented to employees in order to improve the quality of their work, the MC visit program should be announced in advance in order to avoid a change in employee behavior.

The final stage is the submission of the report and the reference to the findings of the survey. In recent years there has been a change in the type of information requested. Previously, the survey consisted of open questions focusing on general information and categories. Recently, objective measurements based on structured questions were used. The results should be presented to the responsible supervisors as soon as possible.

The bibliography published with regard to the MC survey can be categorized into four groups:

1. banking sector (Leeds, 1992;1995, Hoffman, 1993, Stovall, 1993; Hanke, 1993; Tepper, 1994; Holiday, 1994; Morral; 1994; Doorman, 1994; Hotchkiss, 1995)
2. tourist industry: Where MC research can provide essential services to both travel and hotel industries.
3. (Erstad, 1998; Wilson & Gutmann, 1998; Anderson et al., 2001)
4. The relationship between MC research and pay systems. Examples include linking research to adopting incentive, incentive, and reward systems for employees. (Eisman, 1993; Boyd, 1995; Mc Nerney 1996)
5. Marketing: In this category are included publications referring to the conditions and the necessary infrastructure for the implementation of the MC survey as well as the rules governing its implementation (Mile,s 1993; Burnside, 1994; Cramp, 1994; Cobb 1995; Dwek, 1996).

Academic research has devoted a great deal to the credibility of MC research. Research on this issue has been done by academic researchers such as Wilson (1998a, 1998b) and Wilson & Gutmann (1998), in the United Kingdom. In particular, areas such as measuring and managing service delivery with the help of MC research but and Morrison et al., (1997), while Finn (2001) describes research into retail networks focusing on the quality of the data and the number of visits needed. In general, satisfaction metrics are not the primary tool for exploiting at the moment actions to be taken to achieve the strategic goals of an organization. They are usually done to meet more general goals without exploiting their findings in relation to corporate goals and strategies.

Steps to Develop a MC Research Model

The steps in creating a MC research model concern both the interior (employees, senior management engagement, organization in its implementation, results analysis and decision making mechanisms after exploiting the findings) and the outside (customers, competitors) environment of the company. So we conclude that it concerns the whole of the company, and it affects the entire organization. The following are the steps of creating and conducting this research:

Step 1. Objectives
Knowing what we want to achieve by implementing a MC survey. The end result should be the satisfaction of both satisfied customers and employees. Research should reinforce the positive and improve improper behavior without penalizing employees.

Step 2. The evaluation form

Employee involvement in setting measurable standards. What points the customer considers important, finding them and stating them in the rating form.

Step 3. Mystery customer

Selection and training of the MC in relation to the objectives of the company. The researcher should have the characteristics that will be consistent with the profile required to carry out his / her mission in the best possible way.

Step 4. Evaluation

Create an impartial evaluation system consisting of objective criteria (possibly a small percentage of subjective criteria), an assessment of the customer contact point.

Step 5. Analysis

Identify loopholes in the service chain to the customer, diagnose the causes, and locate their source.

Step 6. Required actions

Develop a remuneration and motivation system related to employee performance in MC investigations. Creation of behavioral and cognitive skills development programs. Improve weaknesses in existing processes and products of the business. Resume the search.

Defining objectives

According to Spooner (1985), there are the following objectives that MC research can achieve:

- Demonstrate the treatment of a company's customers at the forefront.
- It enables the company to investigate and analyze the human factor in detail
- Determine the level and progress of employee training
- Explain ways of contacting employees with the customer.
- Identify areas requiring education or improving existing ones
- To determine the level of service offered by it in relation to competition.

Type of Questions

In generating the specific questions that the survey will contain, Spooner (1985), supports the use of specific but open questions instead of a question like, "*Was the employee willing to help?*". It is not possible to extract clear information. Therefore, the following question is proposed: "*How did the seller / employee describe the service/product?*". Hotchkiss (1995), reiterates the need for more in-depth questions, since knowing whether or not the customer is happy is not enough. Businesses are increasingly using objective questionnaires focusing on objective data that can be quantified and lead to measurable conclusions. The questionnaire should be thoroughly easy to understand and complete (Leeds, 1995).

Form of the Research Form

Evaluation forms can take the form of a list or an open architectural document, but this is more time-consuming to complete. Performance indices, weights, and rating ranges (from 1 to 10 or percent) are widely used to evaluate the service offered. In any case the acceptable limits and the desired level should be communicated to the employees, but also easily imprinted by the researchers.

Selection and Training of MC

When developing the research, the company should choose whether to use its employees or outsourcers. Employees usually belong to the internal audit department, which is headquartered and works closely with the company's management (Stefanelli, 1994). The advantages of this choice are lower costs, knowledge of the processes that are being applied, the company's internal environment, its products and its objectives. The drawbacks are the likelihood of one of colleagues being identified, possible reactions by officials who know that research is being conducted by colleagues, as well as possible bias by both parties. External MCs are outsourced, working on fixed-term contracts, and therefore their remuneration is not an expense that is part of the company's total payroll, with whatever benefits this entails. According to Bruno (1989), external partners are now essential, mainly because of the growing size of companies, making it difficult for staff to make permanent visits because of the volume of items under consideration. Disadvantages of this option are the need for permanent 'new client' training mechanisms and a possible high rate of 'visitor' renewal may lead to a downgrading of the rating quality. Fixed partnerships with companies exclusively conducting such surveys are a solution with fewer weaknesses and are widely used lately. In this case there is know-how, the company, besides case-by-case research, can also provide information to the customer, draw conclusions, quantify and present the findings, and the researchers have as a steady job the MC survey, so have increased opportunities to perform as best they can.

Frequency of Visits

Frequency can be determined by many factors, such as size. For example, in a chain with many points of sale such as KFC, monthly visits are made (Bruno, 1988), which means that over 300 visits per year will be required to cover the 300 points. Another factor influencing is the subject of the industry in which the company operates. Leeds (1998), suggests the frequency of quarterly ratings in the banking sector. However, the main factor determining the frequency of visits is research itself with the purposes and objectives it serves. In order to achieve this, a smaller or a higher frequency may be the one which, in the best and most cost-effective way, will lead the services provided to the desired level as defined by the organization's management

Analysis and Communication of the Results

The findings of the survey should be analyzed in the light of the historical and performance of previous periods rather than as individual events. According to Cramp (1994), a short-term MC investigation should not be used as an employee's performance appraisal. The results of the survey should be taken into account in conjunction with the results of other mechanisms such as customer satisfaction surveys, customer complaint records, or customer service reports. If the findings point to more than one source, there is a strong indication of a problem that we must immediately resolve from the responsible person.

Recipients of the Survey Results

According to Rubel (1995), the conclusions have more than one recipient, and they have to give different gradations of information each time. The results of the visits should not be communicated to the senior management but to the directly involved employees in a way that is related to their daily work and to a positive climate (Burnside 1994).

Customers and vendors should also be informed of some of the results as well as of the changes that have been launched since the survey (Aftermarket Business 1994).

When the round of visits is complete, after a series of visits and conclusions are drawn, they are recorded, the actions required, and a report of all the above, which should be communicated to the senior management. This presentation should also mention the progress and improvement achieved by Morrall's (1994) research program as well as any further changes proposed. While it is widely recognized that the more participants involved in the program, the greater the acceptance of the research results.

Usefulness of MC Research to Measure the Service Provided

Secure customer research in advanced economies such as the UK is extensively used by financial institutions, retailers, automakers, public transport companies, hotel businesses, and utilities. The need to measure and evaluate the service offered to companies becomes imperative for two reasons:

1. The way it is offered in the front line may sometimes be away from the planning of the administration.
2. Purchasing experience may be as important as the benefit resulting from the use of the product / service (Bateson 1992).

Unlike market and consumer surveys, MC research is used to evaluate processes and operations rather than the result of these. It is important to emphasize the widespread view that customer satisfaction surveys do not provide the details that the administration needs in order to identify and correct weaknesses in the procedures for the final consumer. In addition, MC research is also useful in staff assessment by measuring specific indicators and evaluating factors that are important

to achieve the level of service required by the administration to the final customer. So while Customer Satisfaction Surveys do not show the sensitivity required, MC research ensures objective staff and process evaluation by measuring indicators such as wait time, queue size, welcome mode, meet needs.

Areas That Can Be Covered by MC Research

1. Be the diagnostic tool by identifying imperfections and weaknesses in a company's service chain.
2. Strengthen and develop capabilities and promote its executives. This will be achieved through the use of the MC in an integrated mechanism framework and assessment, education and rewarding process.
3. Assess the competitiveness of an organization by comparing it with competing companies, or even the market leader.
4. MC research can also be used to compare the services provided by an organization with its competitors.

This should be done in line with codes of conduct (Dawson & Hillier, 1995). Outside, the Market Research Society Code of Conduct (MRS, 1994) states that MC research in competing companies should not be applied in depth using a straightforward human and material resources, but in a context of research, leaving more in-depth research into sectoral studies involving all the companies in one industry. In our country there is no such clear delimitation.

As a diagnostic tool, it is used to identify and evaluate the most critical factors that determine the required quality in service delivery. Once critical factors have been determined, how the logistical and human resources will be made available. In this way, the operational function as perceived by the final consumer is diagnosed and improved. This point identifies the difference from other methods of assessing the quality of service provided such as customer satisfaction surveys. In addition, management can identify not only a general result but and the reasons behind it by measuring individual processes, structures, specific qualities, seasonality, and even compare at retail level department stores and to cover certain groups of consumers.

Regarding the role of MC research as part of an integrated evaluation and rewarding system, the role of MC research should be communicated internally, what are the parameters under review, and the exact results of these. Accepting all of the above staff is essential in order for the evaluation system to work effectively, and to use the findings to improve processes, functions and, of course, staff. Therefore, the internal communication of the way and the objectives of the MC research should be completed before it starts. In the context of staff rewards, results can be used in both moral and material compensation schemes. It has been recognized that such rewards have at least direct and short-term benefits in terms of quality of service. In the case of saturation of staff and the acquisition of the required level of service, different scenarios, targets and remuneration systems should be applied to renew employees' interest and to find new ways of improving the service provided.

Incorporating MC Research Into Its Corporate Evaluation and Commercial Operation

The MC survey may be carried out and utilized or defined by different parts of a business on a case-by-case basis. The Marketing department can leverage the findings by comparing them and utilizing them with customer satisfaction surveys. In most cases, the process concerns the human resources department, while the sales department is the one that draws on the findings directly, while recipients may also be the quality department or the internal audit department.

The fact is that there is a lot of room for the use of MC research combined with findings of customer satisfaction surveys, sales customer complaints, or customer retention, leading to an incomplete exploitation of all findings. There are not many examples of creating a balanced scorecard system (Kaplan & Norton, 1992) which will answer the following questions:

- How well the store achieves the required service levels (MC Survey)

- How do they understand the service at a particular point of sale (satisfaction surveys, complaints)
- How effective is the store in creating relationships with its customers (cross selling, customer retention)
- How effective is a sales point? (Sales department)

MC Investigations and Human Resources Management

MC search programs are defined as a tool for evaluating and improving customer service. Their development and implementation are also directly linked to the management of human resources and the involvement of employees of the company.

This process starts with the determination of the program's objective objectives, and ends with the reward of the employees and the implementation of the necessary corrective changes. Employee involvement is seen as a necessary success factor for the program at all stages of its implementation. This process prevents employees from treating clients as opponents or spies, undermining the same process that MC research is about to improve. Employees should therefore know what their expected levels of performance are, as well as the points at which the service provided is assessed. Research results should be used as tools to improve the service, not as a basis for assessing employee performance.

The purpose of linking with human resources issues is to encourage employee participation, create a positive climate that motivates, create a climate of teamwork, and serve to clarify the needs of education and evaluation of training programs. In the field of provision of service, its main purposes are to provide feedback on the operations of the store and to link the performance of employees with reward systems.

The Moral Dimension of MC Research

Although the relationships between competitors have been a bit concerned with the literature (Paine 1991), (Trevino & Weaver, 1997; Van Lujk, 2000; Spence et al., 2001) in some of publications argue that a competitor of a business has to be treated as "hostile" (Butterfield et al., 2000, pp. 991), while competitors compete with the general business environment and therefore have fair treatment (Green, 1994, pp. 457).

Green more specifically argues that intense competition is usually beneficial to customers, although at times in the name of competition some of the worst environmental offenses have been committed.

Compilation of data from competing companies is increasingly common in the modern environment, and is the result of increasing pressure on profitability, capital efficiency, and technology development.

Competitive knowledge thus emerges as a major competitive advantage and should not be questioned legally or ethically, according to Boatright (2000, pp. 139). As much pressure for the collection of such data increases, both the resistance for non-violation of moral norms decreases (Kahaner, 1996).

The collection of information can be derived from many sources, such as newspapers, magazines, internet, and sometimes industrial espionage and illegal methods such as data theft or telecommunication interception. It is true that illegal methods such as data theft or telecommunication

crimes are condemned by the majority of companies. On the contrary, legitimate methods of collecting and evaluating information on competition are widely accepted. Despite developments in the field, however, there are no guidelines defining legitimate and ethically acceptable practices. Even if industrial spying is considered to be illegal and immoral, much of the rest of the actions is a "gray zone".

The MC survey is also part of the wider context of collecting competition data. These figures, of course, are available to the public, so in some respects (Furtado, 1998) there are no issues of legality and morality. Although MC research is not considered a malicious but mild competitive practice (Butterfield et.al 2000, pp. 94), according to a UK study, its extensions should be looked at better.

According to Lynne, Sharp & Paine (1991), four basic rules govern the collection of data, and what the MC investigation should not include.

- Theft of documents or other material property
- Deceit or other form of false entry. An attempt to influence the judgment of people holding key positions or confidential information
- Secret or unauthorized monitoring
- Search for information from people who are not responsible for spreading it.

Some of the above rules can be applied directly to the MC survey. The researcher is presented as a customer, and records the seller or other employee of the competitor without his consent. Consequently, research may be questioned as to the honesty of the person using it (Boatright, 2000, pp. 141). According to Butterfield et.al (2000, pp. 982), misleading the researcher's identity (which is self-evident in the MC survey) raises important ethical dilemmas. In the UK, the Society for Competitive Advantage argues that researchers can not lie about their identity, but they can be as vague as possible (Curtis, 2000, pp. 28).

One last point to consider is whether a retail store can be described as a public space where the notion of privacy does not apply. According to Green (1994, pp. 269), a last resort concerns the recruitment of researchers who are new to the company who do not have the knowledge or experience to take a critical attitude in the commands they will accept or even be students who practice and therefore will receive the directions given to them effortlessly. These groups are reasonable to adopt more easily some controversial practices.

The mobile phone industry in the UK has experienced, as in other developed countries, rapid growth in recent years, both in its infrastructure and in the services it offers. This growth has increased competition between sales networks to gain an increasing customer base and consequently increased market shares. As a result, attracting new customers combined with the retention of existing ones, has created a highly competitive environment dominated by consumers' tendency to find new opportunities and more cost-effective bills to meet their needs.

Canceling / interrupting a mobile connection remains one of the most important issues for companies, which should deal with it properly in order to be profitable and competitive. The forecasts give the market a great deal of room for growth, so the battle is set to increase. Competitiveness is therefore a major success factor in the industry. The success of mobile operators is to attract new customers by providing improved services and offers. not only are they good for the consumer but also better than the services offered by the competition. A valuable tool in this effort is the MC survey that allows us to monitor competition and benchmark our services with him.

In a survey conducted in the UK, the results showed that: The results of a MC survey outweigh any negative effects or moral dilemmas, and its effects on competition are of particular utility. More specifically, through research, prices, offers, products and services can be compared with our competitors.

Although it is argued that through MC research, one of the benefits of consumer MC research is to increase competition, resulting in lower prices and the promotion of improved services. Indeed, through timely information of companies and between price comparison can improve the response time of competitors and their reflexes. However, when in practice a provider increases its tariffs, this process can be reversed. There is thus little chance of lower tariffs and higher for their growth. Consequently, we conclude that the driving force is to gain market share and new customers, not so much the supply of new services on the basis of purely incentive for consumers.

With regard to first-line salespeople, it was argued that the main benefits of MC research for them are to practice their skills in selling and to control their level of knowledge with positive feedback on the impact of research on the quality of their work. At the opposite and at the sellers level, the difficulty of servicing the researcher in case of a particular workload is a negative phenomenon.

A key point when discussing the moral dimension of a MC's research is to conceal the identity of the researcher, the deception he performs by representing the client. The way the researcher will behave determines the degree of deception he is doing and who can overcome the commonly accepted limits. So, as the researcher limits himself to observing, recording, and asking the average client, as the author claims, there is no problem. However, if he makes false impressions of high consumption or is asking questions to cheat the seller about the products he is going to buy, there is a serious version of cheating.

Monitor and Conduct a Survey Without Consent

A MC investigation is conducted without the organization under review, as long as the vendor does not know that it is subject to monitoring and evaluation for purposes other than the alleged market. According to good research practice, the subject should be protected through practical informed consent (Silverman 2000, pp. 201). The evaluators should be informed about the purposes of the survey, about their role in it and finally have the opportunity to give their consent. Such consent is not sought in the case of a MC investigation because it is natural for the respondent to refuse such inquiry. Therefore, there is de facto questioning the moral aspect of the research.

Evaluating MC Research With Regard to the Quality of Its Results

The evaluation of business performance (especially retail chains) is increasingly receiving the attention of the academic community. Initially, a disproportionately large volume was devoted to assessing the quality of service provided by clients using the SERVQUAL theory (Parasuraman, Zeithaml, & Berry 1988) and the investigations that used it (Carman, 1990).

Later, academic research proceeded in a broader context with the application of G-Theory (Finn & Kayande, 1997) through which the efficiency of surveys was assessed and improved using business benchmarks. More and more companies in recent years have also been using MC surveys to assess the quality of service provided. This is due to different reasons: First of all, the nature of certain services (fast food restaurants, gas stations, etc.) does not favor the customer satisfaction survey (especially the interview process), or even their recognition and recording to conduct the research on second year.

Another reason is that the customer cannot easily recall all the necessary details of his purchasing experience with the company (Wilson, 1998). In contrast, the researchers record not only the result but also all details of the purchase experience and the transaction process in general. Finally, the MCs can also record other items beyond those of purchasing experience, such as store image, queues, customer complaints, and more.

G - theory is now also used for the evaluation of MC research, utilizing research data on the quality of customer service and the quality of the sales environment (Finn & Kayande, 1997). The results show that MC research data is of better quality than customer satisfaction surveys when both methods use common evaluation tools.

Despite the higher cost of a MC survey, the cost-to-performance ratio is far better with regard to the findings to be used for decision-making. The optimal number of visits by researchers depends on the type of research we will carry out. Thus, when evaluating a relatively simple and consistent factor such as the store environment, 10 visits per point are considered necessary when evaluating the service provided can reach 40. However, in reality decision making or administrative changes are commonplace yet and after 3 or 4 visits.

Assessment of Personal Sale

When it comes to the retail sector, the skills of the vendors of a company play an important role in its success (Naylor & Frank, 2000). Next, there are several academic and other theorists to publish work on personal sales (Futrell, 1997). At the same time, this issue is particularly important in the literature (Levy & Weitz, 1998). Extensive reference is made to quantitative and qualitative sales performance indicators. However, no similar effort has been made to evaluate retail sales from the perspective of the retail customer.

Noteworthy are the investigations by Chonko, Caballero and Lumpkin (1990) which have conducted a survey in which student researchers had the role of candidate customers to look at the level of vendors and identify points that need staff training. Also, marketers were rated on a 19-point scale.

Bust et al., (1990) developed a matrix of 5 categories and 22 points according to which the supervisor can evaluate vendors under his / her responsibility. The five categories were:

- Merchandising procedures
- Customer service capability
- Ability to sell
- Knowledge of corporate processes
- Knowledge of product

Of the above categories, only one is eligible to be applied to a MC survey.

Other research such as that of Saxe & Weitz (1990), applies the SOCO scale used by customers to evaluate sellers and products and which was also used by Michaels & Day (1985), to evaluate vendors-suppliers by purchasing department employees. Spiro & Weitz (1990), developed the ADAPTS scale while Ramsey and Sohi (1997) showed a measure of the vendor's ability to listen to the customer's needs according to the latter's opinion. Futrell (1997), developed six evaluation categories within which there was a further set of parameters. These six categories, which Finn (2001), further developed and on which the present work was largely based, are as follows:

1. *First impression, staff image, time passed to customer service* (3 parameters). This category aims to capture the first impression of the customer in the early stages of the sales process.
2. *Evaluation - needs analysis* (4 parameters). Respond to needs, e.g. if the salesperson hears the correct questions that will allow him / her to diagnose the customer's needs. This class evaluates the seller's needs by asking questions and actively listening to the customer's needs.
3. *Product knowledge communicated to the customer* (6 parameters). For example, if there is a stock in the store from a particular product. This category allows us to control the extent to which the seller supports the solution he proposes to meet the customer's needs.
4. *Interpersonal communication* (4 parameters). Such as voice clarity (orthophony), eye contact. Through this category we can check whether the seller's communication is accepted by the customer.
5. *Product support* (4 parameters). That is, the arguments used to support the proposed solution. This factor is considered in order to assess the seller's ability to identify and communicate to the client the appropriate proposal to meet his / her needs.
6. *Sales tactics* (5 parameters). For example, enthusiasm and promptness in the presentation, prompting the purchase of a product / service. This category is reviewed to evaluate whether the customer has been positive in the way the seller faced the objections and handled the client's pressure to close the sale.

MYSTERY CUSTOMER PROCESS

When a MC investigation takes place, the data is collected by internal or exterior contributors of organisations, called MCs, suitably trained to observe, estimate and measure the customer service process, acting as potential customers. In order to extract the desirable information about the service process, MCs follow a line of predetermined and well-structured actions, and after the completion of questionnaires they come up with results concerning the service process (MRS, 2009).

Moreover, Parker (1988), supports the view that insofar as the employees know which criteria are important in the model of evaluation of the investigator, their efficiency is increased. In addition, Cramp, (1994), confirms the importance of the mixture of employees in the process of measurement. In short, the more involved the participant's feel in the programme, the greater the acceptance of the results will be (Rubel, 1995).

The steps of MC investigation involve the selection and training of personnel; the actual investigation and evaluation procedure; the collection of market information and the development of a report (Chen, 2005; Stefanelli, 1994). The specially trained MCs visit sub-units of organizations as customers attempting to detect the real behaviour of employees in their professional environment. Thus the probabilities for reception of false results and impressions are minimised (Beiting, 2001; Buxton, 1999).

Moreover, as Bryson (1991), points out, this method has major advantages: it is an excellent and reliable measurement of operational procedures; and it checks whether organisational standards (e.g. dress code, etc) are applied. However, the same author supports that this particular technique meets some weaknesses as well.

Firstly, the program is relatively expensive if there are many locations to be examined. Secondly, it might happen that the MCs deviate from the real behaviour of the employees. Thirdly, a lot of facts that occur in the work environment are not possible to be transcribed with a simple, short observation. This means that it is not easy to transcribe all the aspects of the problem but only an aspect of the situation. Finally, according to Bryson, (1991), the most important disadvantage of the method is the weakness to transmit literally and precisely the words and the behaviour of employees. This can be based on a number of factors such as the educative level of MCs, the subjectivity in their judgement and different mentality of people. For example, the opinion of two different people, regarding the level of service of the same shop may clash. This happens because not all individuals share the same level of sternness on the level of service (Bryson, 1991).

The above disadvantages lead the conclusion that a MC program should be designed with detailed attention so that it satisfies specific standards. According to Erstad (1998), the basic stages for the design of MC process are: the objectives, the evaluation form, the MC, the evaluation, the analysis, and the action needed. The objectives should be related to having satisfied customers as well as satisfied employees.

MC is meant to reinforce positive behaviour and modify improper behaviour but not to punish. The main goals of MC are the measurement of service quality that is delivered to the consumers (Hesselink & Wiele, 2003), the measurement of efficiency of the training programs (Morrison et al, 1994) and the evaluation of whether customers are treated equally (e.g. testing against discrimination) (Morral, 1994; Tepper, 1994). In addition, Spooner (1985) argues that the objectives of MC programmes are to highlight front-line customer contacts, to enable marketers to scrutinise and fine tune the human element; to track the progress of training; to reveal how positive the employee contact with customers is; and to identify areas that need training or further training.

The evaluation form (questionnaire) should be designed in such a way so as to provide objective observations, which are easily measurable (Cramp, 1994). It is also important to be explicit and allow the direct application of results (Leeds, 1995).

Usually, the questionnaires include binary type questions (yes or no) while in specific issues use the Likert scale (Gofton, 1995). Moreover, Spooner (1985), emphasises the importance of open-ended questions that encourage the MCs to transcribe their observations that seem interesting and important. Erstad (1998), underlines that knowledge from employees' experience is useful on defining and setting measurable standards. Additionally, he suggests finding out what the customers' value is and incorporate these in the evaluation form.

The choice, the information and the training of the MCs with regard to the objectives of the organisation are other important factors for the success of the program and should therefore be done with the necessary attention (Erstad, 1998). Furthermore, the evaluation should take place with impartiality, precision and should be based mainly on objective and realistic criteria. However, it is suggested that a limited amount of subjective information should be included as well (Erstad, 1998).

In the analysis stage, it is useful for the results to be analysed in combination with previous secret visits. Analyzing the data in a more suitable and better way, facilitates the identification of differences between service deliveries, behaviour of the employees, customer satisfaction and when the source of their origin is determined, efforts of minimisation and resolution should be made (Erstad, 1998).

The final stage is related to the action that needs to occur afterwards. Parker (1988), claims that the positive HR approaches like the re-education of employees, are likely to achieve a quite good level of improvement in the quality of customer service. In other words, this stage is critical to develop reward and incentive schemes related to employee performance (Dorman, 1994). Moreover, it is substantial to develop further employees' technical and behavioural skills by providing coaching and mentoring.

USE CASE IN A LARGE RETAIL CHAIN

The target of this research is the utilisation of recent information concerning the performance during the process of a sale to educe specific suggestions to improve the quality of service. The present research implements this method in a retail network of computers sales. It aims to classify chain shops according to the performance level of customer service. Moreover, the additional objective is to determine the factors that need direct improvement and the ones that need improvement in the second and third phase according to their importance.

The determination and the hierarchical classification of these factors could support, that MC is a method to identify improvement areas in order to increase the level of service quality.

The application field of the present study constitutes a large Greek network organisation. Its retail shops expand to four cities. Specifically:

- Athens and the wider region of Attica (12 shops) [presented as A1, A2, …, A12 herein]
- Thessaloniki (2 shops) [presented as TH1, TH2 herein]
- Patra (1 shop) [presented as P1 herein]
- Heraklion (1 shop) [presented as H1 herein]

The survey conducted on all retail stores of the firm. Furthermore, all and each one of the personnel surveyed because the need of the firm was to evaluate all the retail chain. The survey was conducted in Athens and the rest region of Attica prefecture, Thessaloniki, Patras and Heraklion. In total, there were 5619 visits in a period of thirteen months, from January 2014 to February 2015. There was no fixed and steady number of secret visits per day, but instead the size of the visits varies (Finn, 2001). Lastly, all working shifts at all retail stores were visited for the purpose of the survey (Finn, 2001).

Furthermore, data was collected and analysed to compare (i.e. Benchmarking) the departments of the shops, but also the localities classified in the Region of Attica (A1, A2, A3, A5, A6, A9, A12), the Region A4-A7-A8, the Region of the centre of Athens (A10, A11) and the Region outside Attica (H1, P1, TH1, TH2).

We decided to consider this particular classification because the regions we chose may have differences between customers. For example, the Region of Attica consists of suburbs, whereas the region outside Attica includes shops in cities with a population of two hundred thousand and potentially with a different lifestyle and culture from the suburbs of the capital of Greece. We also extracted and reflected on “action diagrams” (Grigoroudis & Siskos, 2002) for each department so as to reach certain conclusions with regard to the level of service that we studied. To sum up, action diagrams combine weight and performance indices to depict the improvement areas. The frame of “action diagrams” is described at the Appendix.

The present research covers the total of products and services of the organisation. The departments of shops, which have been evaluated, are Laptop, Desktop, Software Upgrade, Regionally product department, Digital product department, Telephony, Consumables department (Cd-rom etc.), Stationery department, Technical service and Cashier’s desk. However, for the economy of space, in the present chapter we will refer only on the Laptop department.

Additionally, all the basic principles of the MC method were observed. The employees do not know that MC research is being carried out because the view is to use the results to improve the provided products and services. MCs are acting as potential customers and we assume that in this case, where the employees are not informed, the results will be more objective and realistic (Buxton, 1999, Beiting, 2001).

The implementation of the research took place in the cities and the regions that were mentioned above. Globally, the Laptop department was visited 526 times, over a period of thirteen months, from January 2014 to February 2015. The frequency of visits per day varies. Moreover we tried to implement the visits in all the time intervals for all the shops. The number of visits was decided according to specific articles that deal with the subject (Finn, 2001).

Given that that the processes of service quality measurement are supported by certain specifications that have been predetermined, we collected real world data aiming to achieve a realistic representation, the measurement included a number of subjective questions, which help us to conclude whether the customers felt excessive pressure to buy something or when the MCs consider that the personnel seems to be aware of the philosophy and policy of the organization.

Consequently, a realistic script was developed in order to measure the efficiency of services, the behaviour and customer service. The MC dedicates 1 – 2 minutes on the exploration of the exterior space before entering the shop. Then, he enters the interior and when an employee approaches him, the process of sale begins. At this stage, the MC is not expected to ask about specialised services, he is supposed to leave the employees relatively free in their choices in order to convince the MC to buy something.

In case the MC judges that he has not met the requirements of the evaluation form, he is then supposed to ask questions with a view to focus the salesman on further analysis.

Directly after the visit, and outside the shop's area, the MC completes the evaluation form and rates each sub criterion (Figure 2) from "very bad" to "very good" (1 to 5). We point out that the MC can proceed in the purchase, because of the evaluation of the Cashier's desk. It depends on the scenario as a visit in a shop is divided into three stages:

Stage 1: Control of exterior and interior area
Stage 2: Demonstration of interest for the product
Stage 3: Final stage of evaluation

As mentioned above, a typical retail MC questionnaire covers *greeting, customer service, facility cleanliness and orderliness, speed of service, product quality and employee product knowledge*. The questionnaire was designed according to the objectives of the proposed research. However, we included *facility cleanliness and orderliness* to another section of the questionnaire, which corresponds to the total view of the shop.

Concerning the service quality at the Laptop department, we included four primal criteria that are, “First Contact”, “Presentation” “and Closure of sale and Suggestion of Product X” (Figure 2).

The weights (Table 5) are defined by the project team (organisation experts, researchers) to correspond with the sub criteria, using the experience of the service process and several customer surveys. At this point it is important to mention that these weights are necessary for the determination of the weak and strong points of the network. However if these weights had not been included to the calculation of the performance level the results wouldn’t change greatly.

Figure 2. Criterion and sub criterion assessment space. The top level includes the main objective: Service quality at the Laptop Department. The second level captures the four main criteria, and the third level captures the sub criteria.

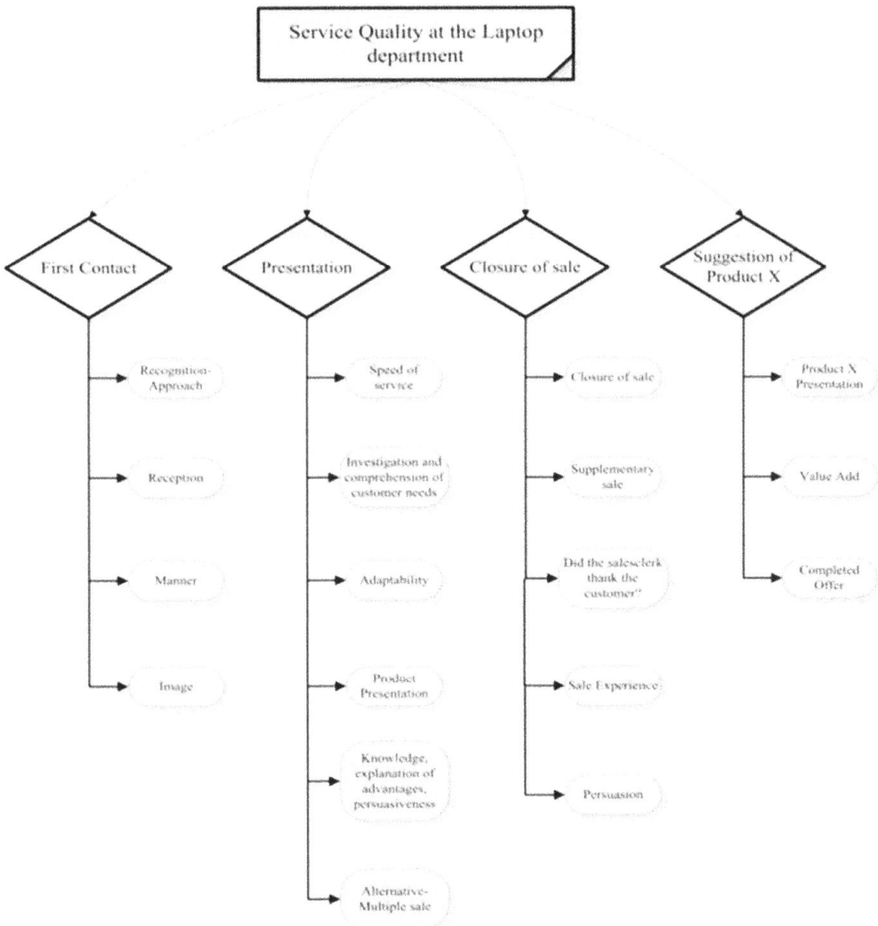

In each question a Likert scale, from 1 to 5 was followed.

1 = very bad performance.
2 = bad performance.
3 = fair performance
4 = good performance
5 = very good performance.

In addition a "very bad" rating signifies a very bad performance of the sales assistant and "very good" a delighted MC.

The criterion "First Contact" includes the "Recognition-Approach" of the customer. The "Reception" and "Manner", specifically at which level the sales assistant wants to service the customer and the "Image" of the sales assistant. The Presentation, that is to say the beginning of the sale process, we have "Speed of service", "Investigation and comprehension of customer needs", "Adaptability", which means how tangible the sales assistant's words were, "Product Presentation", "Knowledge, explanation of advantages, persuasiveness", which means how convincing the clerk was with his knowledge about the products, and "Alternative sale", which means multiple sale when he/she services other customers simultaneously.

The criterion "Closure of sale" concerns the ability of the sales assistant to close the purchase, namely to convince the customer to make a purchase and leave the shop feeling satisfied. The sub criteria that were included are "Closure of sale", "Supplementary sale", "Did the sales assistant thank the customer?", "Sale Experience", "Persuasion". The last criterion is the "Suggestion of Product X". We include it because it is a series of new products that the organization promotes and it is very important to be presented. The sub criteria that were included are "Product X Presentation", related with whether the products were reported, "Value Add", namely to promote the Product X series, and "Completed Offer".

RESULTS

Each shop has eleven departments where the survey took place, but for the economy of space we present only the results of the Laptop department. Concerning the descriptive elements of research, in figure 3 we observe the number of visits in the Laptop department for each shop separately. Most visits, therefore, were paid in the shops A7 and A11 where they reached 37. One less was at A1, A3, A4, A5, A6, A8, A9, A10 and A12. Outside Attica, most visited shops were TH1 and TH2 where they reached 24, while there were fewer visits in the shop H1. Shop P1 had the fewest visits.

Figure 3. Mystery customer visits in the laptop department per Shop

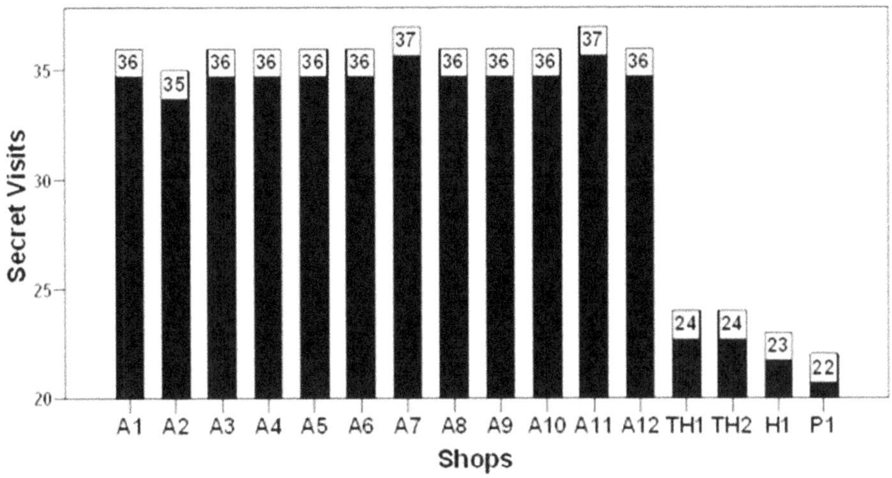

In figure 4 we observe the number of visits in the Laptop department for each region separately. Most visits, therefore, were in the shops of the region of Attica where they reached 251. Far fewer were realised in the rest of the regions. In the region A6-A10-A11, which is 3 big cities in Attica and we examine them separately, visits reached 109. In the shops of the regions outside Attica reached 93, while 20 fewer, that is to say 73, were shops in the of the region of the Centre of Athens.

EVALUATION RESULTS

Data was coded and statistically processed using SPSS version 16.0 (IBM SPSS, 2015). For decongestant of the action diagram, subcriteria are coded (Table 1).

Evaluation of internal consistency of user judgments we used Cronbach's alpha estimates (Frankfort-Nachmias and Nachmias 1992), which are summarised in Table 2 and suggest that experimentation achieved a high degree of response reliability and the order of sub criterion presentation did not bias user assessment.

In this section we will check statistically two fundamental hypotheses:

1. Is there a relation between familiar departments of the retail chain? In other words we are aiming to calculate the level of quality service in the process of sale within the Laptop departments and conclude if the performance levels are distributed uniformly at the shops.

Figure 4. Mystery customer visits in the laptop department per region

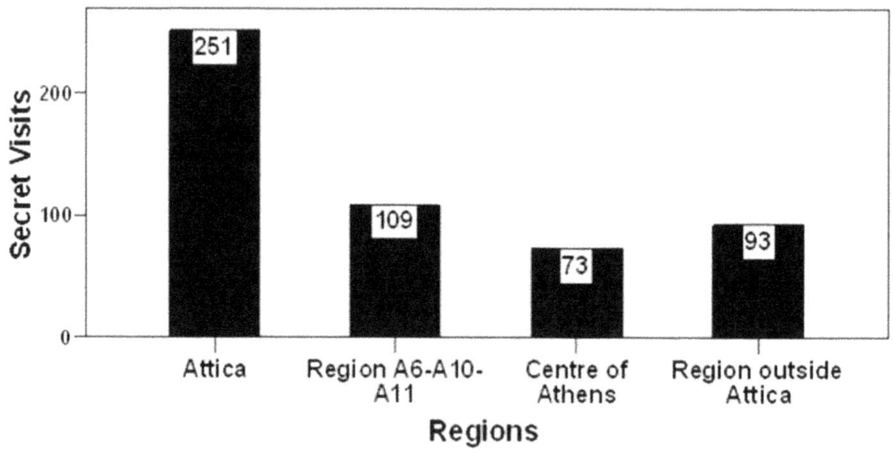

Table 1. Codification of the sub criteria

1: Recognition-Approach	**10: Alternative sale**
2: Reception	11: Closure of sale
3: Manner	12: Supplementary sale
4: Image	13: Did the sales assistant thank the customer?
5: Speed of service	14: Sale Experience
6: Investigation and comprehension of customer needs	15: Persuasion
7: Adaptability	16: Product X Presentation
8: Product Presentation	17: Value add
9: Knowledge, explanation of advantages, persuasiveness	18: Completed Offer

Table 2. Estimation of Cronbach's alpha values with respect to study datasets. A Cronbach's alpha value greater than 0.70 means that data are reliable and can be readily used to support analysis (Frankfort-Nachmias and Nachmias 1992)

Dataset	No. of Items	Responders	Alpha value
Subcriteria level	18	510	0,909

2. Are there significant differences between the service quality of the shops and of the regions?

The overall scores were classified in 5 levels for the facilitation of calculations.

1: ¨0 – 0,20¨ very bad
2: ¨0,20 – 0,40¨ bad
3: ¨0,40 – 0,60¨ fair
4: ¨0,60 – 0,80¨ good
5: ¨0,80 – 1,00¨ very good

Scientific Literature reports that Chi-Square test can be used as an interdependence test between variables (Dafermos, 2005). We used Chi-square test to trace the interdependencies between variables (Dafermos, 2005). SPSS's Chi-square is using Fischer's exact test, or Kruskal-Wallis test (Pallant, 2001). Obviously, high performance is defined when high mean score combining small standard deviation is observed.

Figure 5 depicts that generally the total level of the Laptop department varied round good levels. Additionally, MCs rated "good" 59,89% and "very good" 18,25% (Figure 6) of the visits.

Results in table 3 establish the independence between scores and variables. Namely Fisher's Exact Test, Sig. = 0, 083 > 0,05, means that the performance levels in the shops are distributed uniformly. In other words the sales assistants do not perform in various ways but in similar ways. Shops A6, A9 and TH2 (figure 5 & table 4) have the best scores. They were followed by H1, A10 and A3 (figure 5 & table 4). Inferior scores to the above were observed in A8, A2, A4 and A1. With more inferior scores than the previously mentioned and with an obviously mediocre performance are the A12, P1 and A5. Last in the rank are shops A11, A7 and TH1. The mean scores and the standard deviation in each shop are described on Table 4 and Table 5.

Table 4 presents the overall scores of the shops where A11, A1 and A5 are characterized by a high variability/fluctuation in their scores. Variability refers to how a group spreads out of scores. On the contrary, shops H1 and TH2 present the low variability among scores of the network shops within the Laptop department. That means that they are more unchanging in their performance than the remainder shops. Table 3 presents that there is an important difference between the mean scores of the shops (Table 3, Kruskall – Wallis test, Sig. = 0,001 < 0,05). Extreme values were observed and especially, scores with zero (0) value correspond to visits when nobody serviced the MC.

Figure 5. Overall scores of evaluations per Shop

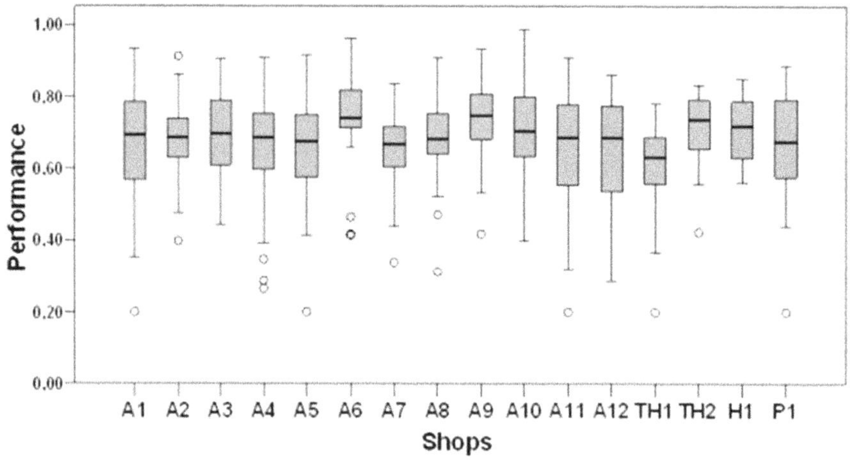

Figure 6. Percent frequencies of performance of the laptop department globally

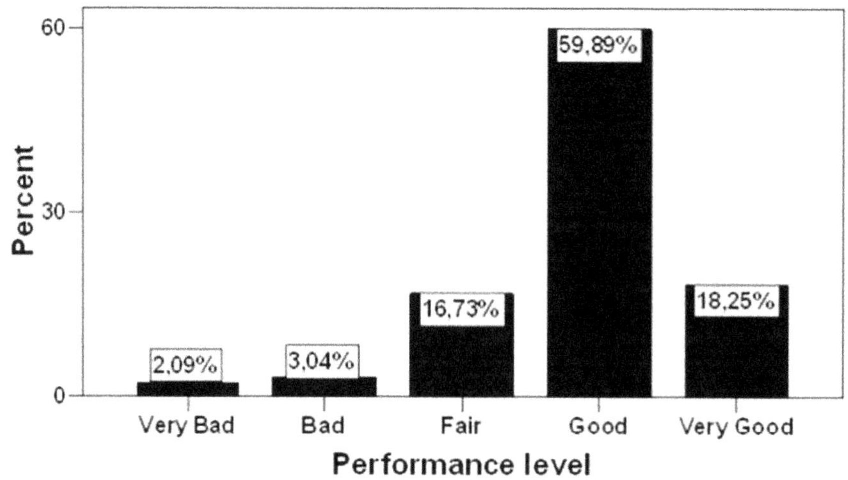

Table 3. Results of One-Sample Kolmogorov-Smirnov test for Normality, Homogeneity of Variances, Fisher's Exact test for the relation between the shops and the regions and Kruskal-Wallis test for the importance of mean scores differences

	One-Sample Kolmogorov-Smirnov Test (Asym. Sig.)	Homogeneity of Variances Test (Sig.)	Fisher's Exact Test (2-sided Sig.)	Kruskal-Wallis Test (Sig.)
Shops	0,0005	0,0005	0,083	0,001
Regions	0,0005	0,008	0,264	0,239

Table 4. Overall shop scores. Entries correspond to mean and standard deviation values. The last line reflects overall organisation score averaged (mean ± standard deviation).

SHOPS	Overall scores
A1	**0,65 ± 0,18**
A2	**0,68 ± 0,11**
A3	**0,69 ± 0,12**
A4	**0,65 ± 0,16**
A5	**0,65 ± 0,16**
A6	**0,75 ± 0,12**
A7	**0,65 ± 0,11**
A8	**0,69 ± 0,12**
A9	**0,74 ± 0,11**
A10	**0,70 ± 0,14**
A11	**0,62 ± 0,22**
A12	**0,65 ± 0,15**
TH1	**0,61 ± 0,13**
TH2	**0,71 ± 0,10**
H1	**0,72 ±0,09**
P1	**0,66 ± 0,15**
Total Laptop Department	**0,68 ± 0,15**

It was found that the overall scores and the regions are independent (Table 3, Fisher's Exact Test, Sig. = 0, 264> 0,05), meaning the performance levels are distributed uniformly in the regions. The region of Attica has the best scores, followed by the region outside Attica. Inferior to the above were observed in the region of the centre of Athens. Last in the ranking is the region A6-A10-A11. Table 6 presents the overall scores and the standard deviation per region.

Figure 9 presents the overall scores of the evaluation per region. The thick line is the median, the box is the inter-quartile range and the circles are the extreme values. The region of the centre of Athens is characterised by a big variability – fluctuation in its scores (Figure 7). On the contrary, the region outside Attica presents the smallest variability between the regions of the network in relation with the Laptop department.

This shows that it is more invariable in its performance than the remainder regions. It was found that there is no important difference between mean scores of the shops (Table 3. Kruskall – Wallis test, Sig. = 0,239 > 0,05).

Table 5. Organisation and shop ratings across evaluation sub criteria and sub criterion weight values. Sub criteria are described in figure 2. Entries correspond to mean and standard deviation values. The second column reflects the overall sub criterion weight value (0-1) normalized. Last line reflects the overall shop score averaged over the 526 MC visits—overall scores are also reported in Table 8. Sub criterion weight values and overall organization scores are normalized with respect to the overall objective.

Subcriteria	Weights	A1	A2	A3	A4	A5	A6	A7	A8	Total
1	0,06	0,77±0,32	0,76±0,29	0,82±0,27	0,70±0,33	0,74±0,34	0,82±0,28	0,77±0,26	0,82±0,31	0,78±0,29
2	0,03	0,77±0,22	0,77±0,15	0,79±0,16	0,77±0,16	0,76±0,22	0,85±0,15	0,80±0,13	0,82±0,14	0,78±0,19
3	0,06	0,79±0,21	0,77±0,14	0,80±0,14	0,72±0,16	0,78±0,20	0,84±0,15	0,77±0,15	0,85±0,14	0,79±0,18
4	0,04	0,80±0,19	0,83±0,16	0,82±0,12	0,812±0,13	0,82±0,19	0,86±0,13	0,83±0,13	0,82±0,13	0,82±0,16
5	0,06	0,69±0,22	0,75±0,17	0,77±0,12	0,70±0,18	0,69±0,22	0,80±0,17	0,70±0,13	0,76±0,16	0,73±0,18
6	0,06	0,69±0,21	0,75±0,14	0,73±0,16	0,74±0,20	0,71±0,21	0,77±0,16	0,74±0,16	0,72±0,16	0,72±0,19
7	0,04	0,68±0,21	0,71±0,14	0,71±0,18	0,68±0,19	0,63±0,21	0,74±0,16	0,67±0,14	0,72±0,17	0,70±0,19
8	0,04	0,65±0,25	0,71±0,20	0,72±0,20	0,63±0,20	0,64±0,22	0,81±0,19	0,68±0,15	0,66±0,17	0,70±0,21
9	0,06	0,64±0,23	0,75±0,19	0,73±0,23	0,66±0,21	0,66±0,24	0,81±0,17	0,69±0,16	0,69±0,18	0,72±0,21
10	0,06	0,61±0,23	0,62±0,23	0,60±0,27	0,54±0,23	0,58±0,25	0,64±0,23	0,59±0,25	0,64±0,19	0,60±0,24
11	0,06	0,51±0,25	0,47±0,23	0,50±0,25	0,49±0,24	0,47±0,23	0,60±0,25	0,47±0,26	0,50±0,19	0,51±0,24
12	0,04	0,27±0,14	0,27±0,12	0,37±0,23	0,36±0,20	0,27±0,17	0,37±0,25	0,34±0,19	0,29±0,19	0,32±0,20
13	0,06	0,90±0,23	0,89±0,18	0,93±0,11	0,88±0,20	0,89±0,21	0,93±0,17	0,84±0,23	0,93±0,15	0,90±0,20
14	0,06	0,68±0,23	0,69±0,16	0,66±0,12	0,66±0,20	0,63±0,21	0,82±0,19	0,69±0,19	0,71±0,18	0,70±0,19
15	0,08	0,59±0,23	0,61±0,19	0,64±0,21	0,59±0,21	0,57±0,21	0,77±0,24	0,59±0,19	0,64±0,21	0,64±0,22
16	0,08	0,64±0,30	0,68±0,28	0,74±0,25	0,66±0,29	0,69±0,30	0,75±0,29	0,65±0,32	0,66±0,28	0,68±0,29
17	0,08	0,514±0,30	0,60±0,29	0,59±0,29	0,58±0,28	0,58±0,30	0,62±0,30	0,49±0,26	0,58±0,26	0,56±0,29
18	0,04	0,58±0,35	0,62±0,33	0,57±0,31	0,65±0,31	0,63±0,35	0,67±0,30	0,48±0,32	0,62±0,31	0,59±0,33
Total	1	**0,65±0,18**	**0,68±0,11**	**0,69±0,12**	**0,65±0,16**	**0,65±0,16**	**0,75±0,12**	**0,65±0,11**	**0,69±0,12**	**0,68±0,15**

continues on following page

Table 5. Continued

Subcriteria	Weights	A9	A10	A11	A12	TH1	TH2	H1	P1	Total
1	0,06	0,87±0,25	0,85±0,23	0,65±0,34	0,78±0,31	0,86±0,30	0,83±0,33	0,80±0,18	0,75±0,24	0,78±0,29
2	0,03	0,78±0,18	0,80±0,19	0,73±0,27	0,74±0,18	0,71±0,28	0,75±0,24	0,79±0,07	0,77±0,17	0,78±0,19
3	0,06	0,80±0,17	0,79±0,16	0,74±0,29	0,78±0,22	0,70±0,20	0,88±0,12	0,81±0,11	0,75±0,16	0,79±0,18
4	0,04	0,84±0,11	0,86±0,10	0,75±0,25	0,84±0,11	0,81±0,20	0,81±0,16	0,79±0,07	0,73±0,17	0,82±0,16
5	0,06	0,79±0,16	0,78±0,16	0,66±0,25	0,69±0,21	0,70±0,17	0,74±0,18	0,70±0,19	0,67±0,15	0,73±0,18
6	0,06	0,78±0,17	0,79±0,15	0,68±0,26	0,66±0,22	0,57±0,21	0,74±0,19	0,72±0,13	0,65±0,24	0,72±0,19
7	0,04	0,78±0,13	0,74±0,16	0,61±0,25	0,65±0,23	0,75±0,25	0,78±0,18	0,73±0,11	0,69±0,19	0,70±0,19
8	0,04	0,76±0,18	0,74±0,19	0,64±0,26	0,69±0,23	0,59±0,23	0,77±0,20	0,78±0,10	0,68±0,16	0,70±0,21
9	0,06	0,79±0,16	0,77±0,17	0,66±0,27	0,67±0,23	0,69±0,23	0,83±0,18	0,73±0,14	0,68±0,16	0,72±0,21
10	0,06	0,69±0,22	0,66±0,26	0,51±0,26	0,63±0,20	0,54±0,22	0,61±0,25	0,60±0,22	0,51±0,24	0,60±0,24
11	0,06	0,58±0,21	0,49±0,21	0,49±0,25	0,46±0,25	0,41±0,23	0,53±0,31	0,57±0,21	0,57±0,14	0,51±0,24
12	0,04	0,33±0,20	0,41±0,28	0,32±0,20	0,24±0,12	0,27±0,14	0,28±0,17	0,53±0,23	0,34±0,19	0,32±0,20
13	0,06	0,96±0,11	0,88±0,21	0,78±0,30	0,89±0,23	0,89±0,26	1,00±0,00	0,90±0,12	0,85±0,18	0,90±0,20
14	0,06	0,78±0,14	0,74±0,19	0,66±0,25	0,68±0,22	0,68±0,15	0,76±0,12	0,72±0,10	0,69±0,16	0,70±0,19
15	0,08	0,76±0,17	0,67±0,25	0,61±0,26	0,58±0,23	0,58±0,20	0,73±0,21	0,75±0,20	0,63±0,17	0,64±0,22
16	0,08	0,74±0,28	0,68±0,28	0,61±0,34	0,67±0,29	0,62±0,33	0,70±0,27	0,70±0,23	0,67±0,26	0,68±0,29
17	0,08	0,61±0,29	0,54±0,31	0,53±0,30	0,51±0,31	0,39±0,29	0,60±0,34	0,67±0,22	0,64±0,27	0,56±0,29
18	0,04	0,74±0,27	0,54±0,33	0,57±0,34	0,55±0,36	0,39±0,31	0,52±0,38	0,61±0,27	0,58±0,28	0,59±0,33
Total	1	**0,74±0,11**	**0,70±0,14**	**0,62±0,22**	**0,65±0,15**	**0,61±0,13**	**0,71±0,10**	**0,72±0,09**	**0,66±0,15**	**0,68±0,15**

Figure 7. Overall scores of the evaluation per Region

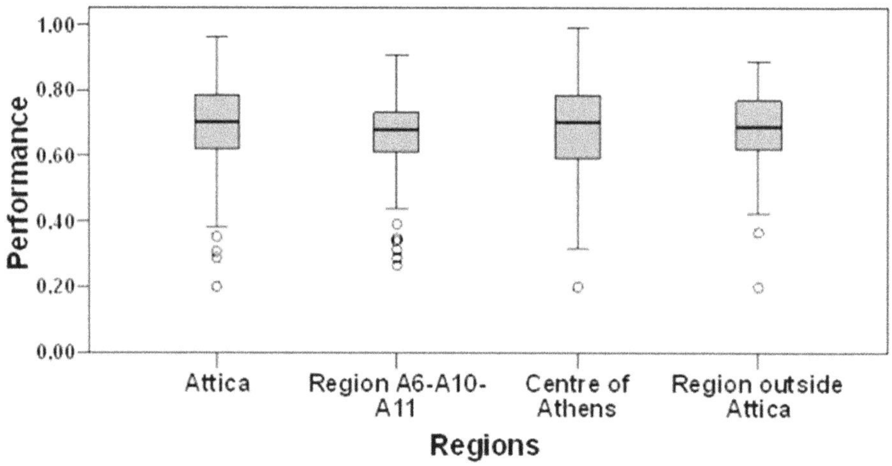

Table 6. Overall scores per region. Entries correspond to mean and standard deviation values. The last line reflects the overall organisation score averaged (mean ± standard deviation).

Regions	Overall scores
Attica	**0,69 ± 0,14**
Region A6-A10-A11	**0,66 ± 0,13**
Centre of Athens	**0,66 ± 0,19**
Region outside Attica	**0,68 ± 0,13**
Total Laptop Department	**0,68 ± 0,15**

At this point we would like to remind that an objective of the present chapter is to suggest specific actions to the organisation. Combining the previous results, where we ranked the shops and the regions, we present the following classification of the customer service elements. Eventually, the organization could concentrate on the lowest ranked shops and regions combining the actions we suggest, which are specified by the following Action Diagram, to improve the quality service.

Furthermore, we manage to identify the priorities of the organization by observing the Action Diagram (Grigoroudis and Siskos, 2002.). The relative Action Diagram (see Appendix at the end of the text) is applying for the comparison of the network shops. Action diagrams are also used for the determination of weak and strong points of the network.

Finally, action diagrams are able to identify improvement areas and extract the priorities of the organisation. Improvement areas can easily be found when the level of the performance is low. Additionally, the combination of the level of importance and the performance implies the classification of the priorities. Namely, high importance and low performance indicates first priority, high importance and high performance indicates the second, low performance and low importance indicates the third etc. Obviously, the sub criteria that belong to the right bottom quadrant represent the first priority because of the combination of high importance and low performance. Those that belong to the top right represent the second, those that belong to the bottom left represent the third and finally the last priority is represented by the sub criteria that cover the top left quadrant. For the economy of space sub criteria are coded (Table 10). From the Action Diagram (Figure 8) each sub criterion can be corresponded to the classification that is described in figure 9. Specifically,

Status quo (low performance/low importance)

- Supplementary sale
- Completed Offer

Figure 8. Action diagram of the laptop department

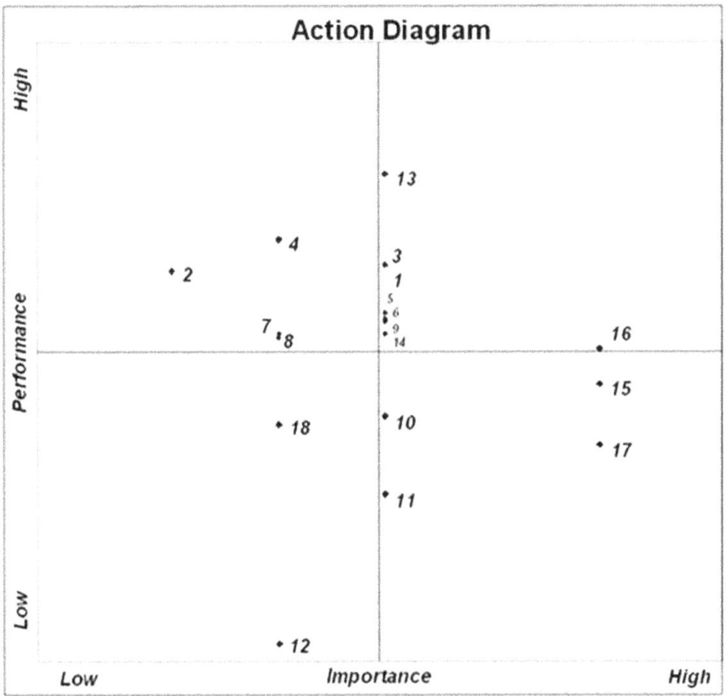

Leverage opportunity (high performance/high importance)

- Did the salesclerk thank the customer?
- Recognition-Approach
- Manner

Transfer resources (high performance/low importance)

- Image
- Reception
- Adaptability
- Product Presentation

Action opportunity (low performance/high importance)

- Value add
- Persuasion
- Alternative sale
- Closure of sale

Action diagrams are a similar technique to SWOT analysis. For instance, "Leverage opportunity" corresponds to Strengths, "Action opportunity" to Weaknesses, "Transfer resources" to Opportunities and "Status quo" to Threats. The Action diagrams indicate current and potentially critical satisfaction dimensions (Figure 8). MCs rate four important sub criteria with low scores. Specifically, "Persuasion", "Value Add", "Alternative sale" and "Closure of sale" are the sub criteria that are suggested to be improved by the organization. These sub criteria are defined as weaknesses of the organization and represent the first priority for improvement.

Additionally, the second priority may be given to the sub criteria in the top right quadrant, because there is room for improvement. Especially, "Product X Presentation", "Sale Experience", "Knowledge, explanation of advantages, persuasiveness", "and Investigation and comprehension of customer needs" and "Speed of service" are close to the X axis and obviously they are far from becoming clearly Leverage opportunities for the organisation. The strengths of the organisation became three sub criteria, "Did the sales clerk thank the customer?", "Recognition-Approach" and "Manner". In addition to this, there is room for improvement. However, they are close to the Y-axis, therefore we cannot say that they are of a high importance.

On the other hand, the third priority is the bottom left quadrant. The Status quo quadrant includes two sub criteria, which are "Supplementary Sale" and "Complete Offer". These became the threats to the organization because potentially their importance may be increased, and may become critical in the near future if customers give more importance to them. The last priority of the organization is the Transfer resources quadrant that includes four elements: the "Image" of the sales assistant, the "Reception", the "Adaptability" and the "Product Presentation". These are the opportunities of the organization and their resources can be used somewhere else more efficiently.

Finally, it was found that the sales assistants' performances are not distributed uniformly to the shops (Table 7, Fischer's Exact test Sig.<0,05) in relation to the majority of the sub criteria. The performances on "Product X Presentation", "Investigation and comprehension of customer needs", "Alternative sale" and "Speed of service" are distributed uniformly to the shops (Table 7, Fischer's Exact test Sig.>0,05), namely the sales assistants perform in various ways.

Table 7. Fischer's exact test and Kruskal-Wallis test per subcriterion

Subcriteria	Fisher's Exact Test (2-sided Sig.)	Kruskal-Wallis Test (Sig.)
1: Recognition-Approach	0,0005	0,006
2: Reception	0,015	0,598
3: Manner	0,003	0,010
4: Image	0,004	0,198
5: Speed of service	0,052	0,037
6: Investigation and comprehension of customer needs	0,163	0,006
7: Adaptability	0,002	0,018
8: Product presentation	0,0005	0,0005
9: Knowledge, explanation of advantages, persuasiveness	0,007	0,001
10: Alternative sale	0,579	0,074
11: Closure of sale	0,017	0,066
12: Supplementary sale	0,0005	0,0005
13: Did the sales assistant thank the customer?	0,0005	0,0005
14: Sale Experience	0,0005	0,001
15: Persuasion	0,001	0,0005
16: Product X presentation	0,304	0,869
17: Value add	0,023	0,113
18: Completed offer	0,0005	0,018

Additionally, it was found that the mean performances are significantly different (Table 7, Kruskal-Wallis test Sig.<0,05) in relation to the majority of the sub criteria. However, the mean performances on "Reception", "Image", "Alternative sale", "Closure of sale", "Product X Presentation" and "Value add" are not significantly different.

To sum up, primary attention must be paid to those sub criteria which belong to the priorities of the Action diagram (Figure 8) taking into account the uniformity of their distribution at performance levels and also the significance of their differences. If we observe significant difference between mean scores, then we have to pay attention firstly to the low performance shops.

The low performance shops, which distributed in Table 8, are the shops that have lower mean score than the global mean score of each sub criterion and additionally high level of variability. By the first priority of the action diagram we can conclude that, "Persuasion" is suggested to be improved firstly in the low performance shops (Table 8) and afterwards to the whole network, additionally "Alternative sale", "Closure of sale" and "Value Add" have to be improved directly in the whole network. Concerning the second priority, "Product X Presentation" it is recommended to be improved directly to the whole network. Regarding the other sub criteria of the second priority, it is recommended they be improved firstly in the low performance

Table 8. Low performance shops per subcriterion with significant difference between shops

Priorities	Sub criteria	Low performance Shops
First	15: Persuasion	A5, A12, TH1, A1, A4, A7, A11, P1
Second	14: Sale Experience	A5, A11, A4, A3, A1, A12, TH1, A7, A2, P1
	9: Knowledge, explanation of advantages, persuasiveness	A1, A11, A5, A4, A12, P1, TH1, A8, A7
	6: Investigation and comprehension of customer needs	TH1, P1, A12, A11, A1
	5: Speed of service	A11, P1, A1, A5, A12, H1, A4, TH1, A7
	1: Recognition	A11, A4, A5, P1, A2, A1, A7
	3: Manner	TH1, A4, A11, P1, A7, A2, A12, A5
	13: Did the sales assistant thank the customer?	A11, A7, P1, A10, A4, TH1, A12, A5, A2
Third	12: Supplementary sale	A12, A5, A1, TH1, A2, TH2, A8
	18: Completed Offer	TH1, A7, TH2, A10, A12, A11, A3, P1
Last	7: Adaptability	A11, A5, A12, A7, A1, A4, P1
	8: Product Presentation	TH1, A4, A11, A5, A1, A8, P1, A7, A12

shops (Table 8) and afterwards in the whole network. Concerning the third priority, "Supplementary sale" and "Completed Offer" it is recommended they be improved firstly in the low performance shops (Table 8) and afterwards in the whole network. Regarding the sub criteria of the last priority it is recommended to be improved in the same way as the third priority.

DISCUSSION

The original application presented in this chapter illustrates the implementation of a MC methodology in the retail sector, specifically in one of the biggest networks of computer stores. MC data shows where service performance is likely to cause dissatisfaction and enables proactive measures to be taken – during both the process design and maintenance stages (White, 2004). This chapter presents a completed MC survey that provides the level of quality service and classifies the improvement areas that have been determined by the analysis.

The results of the Finn & Kayande (1999) research conducted by A. Finn (2001) are that cost-effective customer-based research is cost-effective when assessing the retail environment in retail outlets. It is also a justifiable approach when individually assessing personal sales in these places, although the required visits are twice as many as those required in the first case. Consumer stores need more visits to make safe conclusions with regard to shops selling goods that meet long-term needs.

Also, MC research is more effective in evaluating individual retail points in relation to store chains when considering the ability to sell staff. The reason is the highest number of visits for safe deduction in the case of chains. The same applies to the assessment of the sales environment where the number of visits is double. This is due to the homogeneity of retail chains over individual stores, which implies more accurate measurements to draw meaningful conclusions. The above demonstrates that MCs can be a workable and cost-effective alternative to customer satisfaction surveys in retail point evaluation. It also concludes that 2 or 3 researcher visits - as usual - are not enough to evaluate retail outlets, especially those that sell fast-moving consumer goods.

MS data showed the points or issues that service activities performance is likely to cause dissatisfaction, enables proactive measures to be taken on employee performance and classifies the improvement areas that have been determined by the analysis and that can be used as a benchmarking tool (Zuhair et al, 2015). In spite of this, MS studies should not be identified as a mechanism that specifies customer satisfaction areas, but as an evaluation tool that indicates employees fine-tuning to firm's desired service process. Table 9 reveals five (5) points of differentiation between customer satisfaction and MC characteristics.

Table 9. Characteristics of CS and MS characteristics

	CS	MS
1	Can be used for understanding customer attitudes & perceptions	Can be used on measuring process performance and service standards
2	Emphasis on post purchased behavior and on measuring outcomes	Emphasis on service standards of the company in relative terms again competitors
3	Not identical to identify weaknesses in the service delivery approach. The management can't enforce easy the process to improve the service delivery process	Identical for understanding the service situation process before the outputs (the CS stage) of the service delivery process
4	Customers are often unable to remember the detail of a service experience. It gives an overall impression of the service experience	Useful for the description of service delivery inputs, processes and outputs
5	Respondents find it difficult to recall a recent service encounter	Can be a useful tool to describe the service encounters, the moments of truth, the contact points of the service delivery process(*)

Notes: Where (*), a. " Moments of truth" = are contact or interaction places between a customer and a firm (through a product, sales force, or visit) that gives the customer an opportunity to form (or change) an impression about the firm it is also called as "service encounter", b. contact points = are a persons or a department serving as the coordinator(s) or focal point of information concerning an activity or program.

The advantage is the extraction and the utilisation of the whole information that were gathered. The results solved practical problems of the network and specifically include the determination of the weak and the strong points of the organization and the performance evaluation of the network, as were defined by the objectives of the chapter. That is the reason why Action diagrams became the appropriate method to accomplish the objectives. Additionally, the most articles related to the MC data analysis focused only to the level of quality service. This chapter focused not only to this but moreover to add a tool at the interior benchmarking system. This tool is the relative action diagrams which determine the strong and the weak points of the organization, about the service quality. However, if the weights had not been defined this determination would not take place. At this way, from the data analysis only the level of quality service would be shown. At this point, it is necessary to mention that the results, without respect to these specific weights, would not have been changed notably, but without the weights we could not extract the Action diagrams.

The present study has expanded the understanding of MS application method by incorporating three more steps. Advanced statistical techniques incorporated for blinking up the current organizational status in terms of employee performance and the delivered service quality level. Action diagrams were used as a performance-importance map to indicate the strong and weak points in terms of criteria and to define the required improvement efforts. Results supported that MS is a useful tool to capture the actual service delivery process.

Furthermore, if there is a relation between the shops (and the regions) and their performance and if there is a significant difference between those performances, was examined. Therefore, the objectives of this chapter were achieved, according to the results, and finally supported the statement that MC is a very useful tool to improve service quality and determine specific elements. Finally, the results of the current chapter correspond to the objectives that Spooner (1985) has mentioned; because from the results there could be information extracted, concerning the performance in the sale process. Specifically, the results determined that 1) the status of customer contacts with the sales assistants (objectives 1, 3 & 4), 2) the strong and the weak points (objective 5) (and finally 3) the combination of them could be used to cover the second objective.

Another contribution is related to the improvement opportunities subject on predetermined criteria. The methodology classified these opportunities in order to be distinguished according to their level of significance and urgency. This resulted in pointing out the exact store or stores which need improvement in a specific area. The goal of similar studies, as pointed out from the literature review was just the comparison and/or the ranking of two or more competitive stores in a chain.

Another goal was to apply the method to specify the delivered service level in each store in terms of store image and to determine its weak points. This goal was accomplished by collecting and analyzing quantitative data. In order to avoid issues of bias, improvement opportunities were indicated (low score on a criterion). Improvement opportunities defined by using action diagrams on which every criterion placed according to the significance and the performance of each store per region. For example, when a store scores high significance on a criterion and low performance on the same criterion then this consist an improvement space. When a statistical difference occurs on the same subject criterion then a decision has to be made for improving the quality level of a store or the whole network. This is due to the fact that it is preferable and more effective to give improvement directions to similar performing network of stores than to a varying performance sample of stores.

The MC method was developed to measure the service quality, in addition to more often-used survey methods (Hesselink & Wiele, 2004) like customer satisfaction measurement. It should be noted that customer satisfaction is a dynamic parameter of the business organisation. Changes in the current market can affect customer's preferences and expectations. For example, some satisfaction dimensions may become critical in the near future if customers pay more importance to them, thus, the installation of a permanent customer service and customer satisfaction barometer is considered necessary. The main advantages are:

- The organization will have the ability to analyse customers' behaviour for different regions in the country, taking into account special regional characteristics.
- An interior benchmarking system can be established, based on customer satisfaction and MC evaluation in each branch. This way, the "weakest" stores of the network may be identified and improved.
- Competition analysis will be performed for different regions of the country.
- The effectiveness of marketing plans will be evaluated through customer satisfaction and service measurement.
- The establishment of a motivating system for employees may be directly related to customer service measurement. In this way, productivity may be improved and efficiently measured.

It is significant to notice that MS evaluates mainly the process rather than employees, although it also brings valuable information for front line employees as part of service process. However, more research is required about the insights of employees' low performance and in this perspective, it must be also examined the role and the importance of relationships between internal customers. Customer experience should not be only connected as the responsibility of the front office employees. As mentioned above, it could be an important mistake to consider customers only people and organizations external to a company. An integrative service- oriented way should support and provide the proper services for internal and external customers. Customer satisfaction- or else verbalized positive customer experience- is the result of interactions between external customers and employees who they are also internal customers to other departments and sometimes organizations. Consequently, internal customer relationships should not be ignored.

Since MS surveys function complementary to customer satisfaction surveys, the present study strengthens the effectiveness of MS in retail operations, since it has developed a sophisticated tool for accomplishing more strategic issues according to which specific improvement actions can be taken. In particular, although MCs' valuable offer can be used from high level managers in order to construct a more customer-oriented strategy. MS survey should be more considered as an evaluation vehicle of employee performance that could ameliorate service quality and enhance retailing company.

Seeking additional valuable information for customer satisfaction, it is essential to examine external "real" customers' perceptions and feelings about the delivered service quality and combine them with MS findings. Table 5 demonstrates a proposed procedure of six (6) steps to combine customer satisfaction and MC research tools.

Table 10. Proposal of a procedure to combine CS and MS research tools and studies

First step:	Identify CS requirements through a CS research study.
Second step:	Put the service standards based on a customer oriented study.
Third step:	Control the whole service delivery process with MS studies.
Fourth step:	Point out Strengths and Weaknesses in the service delivery process.
Fifth step:	Check the matching of CS requirements and the MS standards. Improve the contact points and the service encounters of the Service delivery process with the use of the 7 Ps (Product, Price, Place, Promotion, Physical Evidence and People).
Sixth step:	Make continuous CS research (longitudinal studies) to identify changes on customer requirements and match the MS standards to them. Create indices to improve long term the service quality process. Enforce it with CRM practices and technology, e.g. CRM software.

Therefore, in figure 9, for those combining both measurement tools, it is proposed a decision support matrix which, according the related CS and MS scores, indicates four different areas of actions described as below:

High MS Score Area

The organizational structure is evaluated with a high score based on firm's quality standard, but it doesn't consort with customer perceptions (Nikabadi et al, 2014). This structure is enough to satisfy only the supply side requirements and managers should give more emphasis to customers' desires and expectations. Investments on processes, qualitative tangible assets and perhaps, employees could make the organization more effective and customer oriented. Research is also suggested to identify more specific characteristics of the product or service offer.

Excellent MS and CS Score Area

Quality and current organizational status are in a higher delivered and customer satisfaction level. The organizational structure is sufficient, satisfying customers' desires and expectations. Investments on processes, qualitative tangible assets and employees are already effective and respond to customer needs. In this case, firms should enforce their position and keep investing.

Low MS and CS Score Area

The organization should make changes in order to build a better and more effective structure for qualitative services and to construct processes based on quality standards.

To occur satisfied customers, it is required the improvement of processes, assets and employees in long term basis. Besides, the organizational structure and the quality level can affect positively or negatively the level of customer satisfaction.

High CS Score Area

Customers are satisfied, although the quality standards may be too high and the MS controlling process scores in the lower level evaluation area of the matrix. In this case, organization should improve its quality level and if possible, to make changes to quality standards. It could be also proposed to check organizational cost-benefit analysis per action.

It has become clear that MC surveys are necessary at this time more than ever before, in order for a company not only to overcome its competitors but also to improve its position as well as to win and maintain its reputation, its credibility and above all its customers. In addition, businesses now have to use this technique for one more reason, the results of the survey are concrete, measurable and show the areas behind the business, i.e. the points that the business should focus on in order to achieve its goal. In addition, when you know exactly where it is, it's easier to find the solutions.

The MC method measures the process of delivering customer service (performance) while the customer satisfaction method measures the outcome of customer service performance (perceptions and expectations). Both methods are important to the

Figure 9. The proposed CS and MS decision support matrix

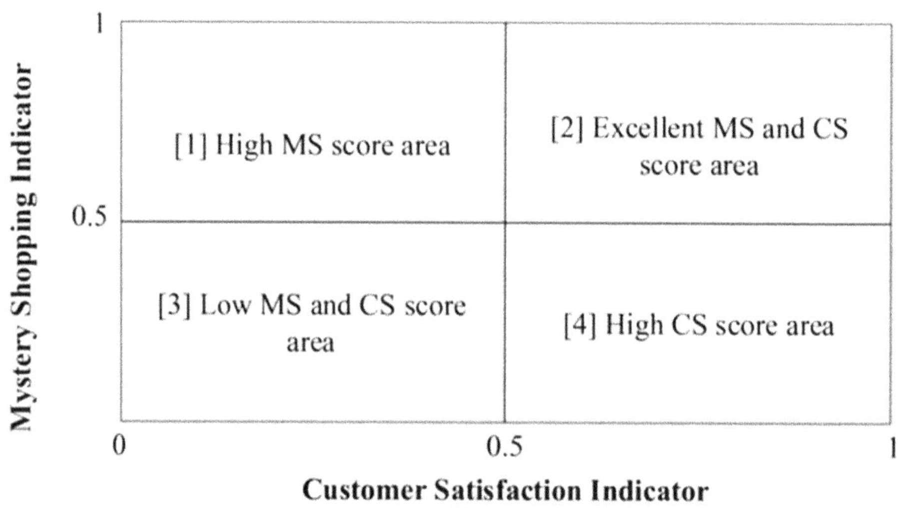

overall health of a customer-driven organisation. Both are part of an ever-repeating process of continuous improvement. In the end, there can be no customer satisfaction without customer service (Buxton, 2000).

The proposed methodology lied on Erstad's (1998) methodology. However, steps 4, 5 and 6 modified as described herein and constitute a proposal for enriching and improving the method. In addition, the proposed methodology was reinforced with 3 extra steps:

1. Monitoring and benchmarking the performance of chain stores by department and criterion. This step makes it possible to compare the departments as well as to identify the shop and the employee that needs to be improved.
2. Identify areas of improvement through action diagrams. Their use determines the organization's priorities by identifying the criterion that needs improvement.
3. Associate criteria per priority to the lowest performance yielding stores, in order to diagnose the weak points of the chain e.g. how important this weak point is and where its largest proportion located.

The proposed methodology does not have usage limitations. It can be applied to all sectors and types of businesses as all businesses are addressed to either external or internal customers.

Finally, the method applied above is useful to identify the strengths and the weaknesses of an organization and when you realize what your strong and weak points are it is much easier to deal with them. This way the customer will be satisfied and that is the very reason why an organization exists.

FURTHER RESEARCH

Concerning the further research, factor analysis and Multi-criteria Satisfaction Analysis (MUSA) (Grigoroudis & Siskos, 2002) is recommended. With the first, data volume would be decreased. Concerning the second, we could not apply it because of the deficiency of the global satisfaction question that is necessary for this particular analysis. However, a mixture of a customer satisfaction measurement method-tool and a customer service measurement method could be discussed. Apart from the statistical and multi-criteria analysis of the method which is very important, MUSA could be useful to identify the weights of the sub criteria which can be extracted and correspond to the MCs. Afterwards, these weights can be compared with the weights that were defined by the project team to gather important conclusions. Scientific literature identifies aspects in the comparison of derived and stated importance in customer satisfaction surveys (Grigoroudis & Spyridaki, 2003). Additionally, this way of comparison with regard to customer service surveys could be an interesting tool to extract further information.

Finally, regarding the service design and the recommended criteria for service excellence, first-line employees' perceptions should be acquired, using qualitative approaches, such as interviews with the personnel, attempting to investigate in-depth MS process and protocol. However, managers should necessarily discuss MS surveys results not only with their first- line personnel, but with internal customers too. Otherwise, MS surveys do not make sense both ways: as an organizational tool for a better employee performance and as a marketing technique for the evaluation of processes that target to the maximization of customer satisfaction. The use of MS and CS metrics can support also the Customer Service system of the Business. Weaknesses in services such orders delays and the corresponding low customer satisfaction are useful elements for the implementation of strategies to improve the Quality level of the service company processes. Monitoring of MS and CS indicators of measurement can be done through their integration into the company's CRM system (Kim et al., 2012; Sigala, 2008; Dimitriadis & Stevens, 2008). Therefore, a future work could be focused on an empirical research of first-line employees' perceptions regarding the service design and the recommended criteria used in the survey.

The Future of MC Investigations

In the future, MC research is expected to continue to grow as a tool to improve service and service delivery. The main reason is the growing sensitivity of businesses in these two areas and the recognition of their importance.

It may face competition from other applications such as customer focus groups and satisfaction surveys, although it may work with them. In order to achieve the best possible performance, it should be integrated into an integrated system and directly linked to customer satisfaction, staff performance, complaints, and, most importantly, to sales performance, a factor with which the present work.

REFERENCES

Aftermarket Business. (1994). *'Sell' ... you're on 'candid camera'.* Author.

Ahearne, M., Jelinek, R., & Jones, E. (2007). Examining the effect of salesperson service behavior in a competitive context. *Journal of the Academy of Marketing Science*, *35*(4), 603–616. doi:10.100711747-006-0013-1

Al-Hemyari & Alsarmi. (2015). Standards, benchmarks, and qualitative indicators to enhance the institutions' activities and performance: Surveys and data analysis. *International Journal of Knowledge-Based Organization, 5*(4), 37-61.

Anderson, D. N., Groves, D. L., Lengfelder, J., & Timothy, D. (2001). A research approach to training: A case study of mystery guest methodology. *International Journal of Contemporary Hospitality Management*, *13*(2), 93–102. doi:10.1108/09596110110381906

Armstrong, J., & Arvantis, C. (1996). *MC: Evolution of Large-Scale Programme within Financial Institutions*. ESOMAR/EFMA Conference on Banking and Insurance: Optimising Change, Barcelona, Spain.

Avlonitis, G., Hart, S., & Tzokas, N. (2000). An analysis of product deletion scenarios. *Journal of Product Innovation Management*, *17*(1), 41–56. doi:10.1111/1540-5885.1710041

Bateson, J. E. G. (1992). *Managing Services Marketing* (2nd ed.). Orlando, FL: Dryden Press.

Beiting J. (2001). Taking the mystery out of patient satisfaction. *Quirks Marketing Research Review*, 651.

Berry, L. L., & Gresham, L. G. (2001). Relationship retailing: Transforming customers into clients. *Business Horizons*, *296*, 43–47.

Berry, L. L., Parasuraman, A., & Zeithaml, V. (1988). The Service -Quality Puzzle. *Business Horizons*, *31*, 35–43. doi:10.1016/0007-6813(88)90053-5

Boatright, J. (2000). *Ethics and the Conduct of Business* (3rd ed.). Prentice Hall.

Boyd, M. (1995). A case for incentives. *Incentive*, *169*(2), 14.

Bruno, K. (1988). Today's mystery guest is … a food spy, hired by management, to check service. *Nation's Restaurant News*, *22*(11), F24.

Bryson, J. (1991). MC: Uses and abuses. *Quirks Marketing Research Review,* 456.

Burnside, A. (1994). In-store spies snuff out poor service. *Marketing, 32*(3).

Butterfield, K., Trevinäo, L. K., & Weaver, G. (2000). Moral Awareness in Business Organisations: Influences of Issue-Related and Social Context Factors. *Human Relations, 53*(7), 981-1018.

Buxton E. (1999). MC reports: Are quality and timeliness competing priorities? *Quirks Marketing Research Review,* 454.

Buxton, E. (2000). Similar but different. *Quirks Marketing Research Review*, 20000102.

Chanal, V., Martin, A., Breda, G., & Bergantz, C. (2017). From research to innovation, how to create value for an IT consulting and engineering service firm through its internal research activities. *International Journal of Knowledge-Based Organization, 7*(3).

Chen, H. Z. (2005). *The application of MC in measuring the service quality of hotel telephone reservation: A case study of a resort hotel. Master theses*. Taipei: National Dong Hwa University.

Cobb, R. (1995). Magical mystery lure. *Marketing,* 26.

Cramp, B. (1994). Industrious espionage and witnesses for the defence. *Marketing, 18*(17-18).

Customers Satisfaction Council. (1995). *Customer satisfaction assessment guide*. Schaumburg, IL: Motorola University Press.

Dafermos, V. (2005). *Social statistics using SPSS*. ZITI Publishing.

Dawes, J., Sharp, B. (2000). The reliability and validity of objective measures of customer Service: MC. *Australian Journal of Market Research, 8*(1), 29-46.

Dawson, J., & Hillier, J. (1995). Competitor MC: Methodological considerations and implications for MRS code of conduct. *Journal of the Market Research Society, 37*(4), 417–427. doi:10.1177/147078539503700406

Dimitriadis, S., & Stevens, E. (2008). Integrated customer relationship management for service activities: An internal/external gap model. *Managing Service Quality, 18*(5), 496–511. doi:10.1108/09604520810898857

Dorman, K. G. (1994). MC results can shape your future. *Bank Marketing*, *26*(8), 17–21.

Douglas, A., & Davies, J. (2007). *The Impact of Mystery Customers on Employees Disruptive behaviour*. Higher Education View project Lean Six Sigma in East Africa View Project.

Douglas, J. (2015). Mystery shoppers: An evaluation of their use in monitoring performance. *The TQM Journal*, *27*(6), 705–715. doi:10.1108/TQM-04-2015-0052

Dutka, A. (1995). *AMA handbook of customer satisfaction: A complete guide to research, planning and implementation*. Lincolnwood, IL: NTC Business Books.

Dwek, R. (1996). Magic of MC. *Marketing*, *17*, 41–44.

Eisman, R. (1993). Going undercover. *Incentive*, *167*(5), 70–71.

Erstad, M. (1998). MC programmes and human resource management. *International Journal of Contemporary Hospitality Management*, *10*(1), 34–38. doi:10.1108/09596119810199318

Evanschitzky, H., Iyer, G., & Caemmerer, B. (2008). Dimensions of satisfaction in retail settings: A research note. *Journal of Relationship Marketing*, *7*(3), 275–285. doi:10.1080/15332660802409621

Finn, A. (2001). MC benchmarking of durable-goods chains and stores. *Journal of Service Research*, *3*(4), 310–320. doi:10.1177/109467050134004

Finn, A., & Kayandé, U. (1997). Reliability Assessment and and Optimization of Marketing Measures. *JMR, Journal of Marketing Research*, *34*(2), 262–275. doi:10.1177/002224379703400206

Finn, A., & Kayande, U. (1999). Unmasking a phantom: A psychometric assessment of MC. *Journal of Retailing*, *75*(2), 195–217. doi:10.1016/S0022-4359(99)00004-4

Frankfort-Nachmias, C., & Nachmias, D. (1992). *Research methods in social sciences*. London: St. Martin's Press Inc.

Furtado, T. (1998). Ethical Dilemmas RegardingCompetitive Intelligence. In *Perspectives in Business Ethics*. London: Irwin McGraw Hill.

Gofton, K. (1995). *Service without a smile; customer service, marketing guides; customer loyalty. Marketing, 2(3).*

Gosselt, J., van Hoof, J., De Jong, M., & Prinsen, S. (2007). MC and alcohol sales: Do supermarkets and liquor stores sell alcohol to underage customers? *The Journal of Adolescent Health*, *41*(3), 302–308. doi:10.1016/j.jadohealth.2007.04.007 PMID:17707301

Green, R. (1994). *The Ethical Manager: A New Method for Business Ethics*. Englewood Cliffs, NJ: Macmillan.

Grigoroudis, E., & Siskos, Y. (2002). Preference disaggregation for measuring and analyzing customer satisfaction: The MUSA method. *European Journal of Operational Research*, *143*(1), 148–170. doi:10.1016/S0377-2217(01)00332-0

Grigoroudis, E., & Spyridaki, O. (2003). Derived vs. stated importance in customer satisfaction surveys. *Operational research. International Journal (Toronto, Ont.)*, *3*(3), 229–247.

Grove, S. J., & Fisk, R. P. (1992). Observational data collection methods for services marketing: An overview. *Journal of the Academy of Marketing Science*, *20*(3), 217–224. doi:10.1007/BF02723408

Hanke, E. (1993). Through the looking glass. *Credit Union Management*, *42*, 45-6.

Hesselink, M., Iwaarden, J.D. van & Wiele, A. van der (2004). *MC: A tool to develop insight into customer service provision.* ERIM report series research in management, ERS-2004-082-ORG.

Hesselink, M., & van der Wiele, A. (2003). *MC: In depth measurement of customer satisfaction*. Research Paper ERS-2003-020-ORG Revision, Erasmus Research Institute of Management (ERIM). Retrieved September, 2017, from http://repub.eur.nl/res/pub/281/

Hill, N. (1996). *Handbook of customer satisfaction measurement*. Aldershot, UK: Gower Press.

Hoffman, G. (1993). Customers can hone MC. *Bank Marketing*, *25*(8), 36–37.

Holliday, K. K. (1994). Mutual funds. *Bank Marketing*, *26*(7), 23–31.

Homburg, C., & Rudolph, B. (2001). Customer satisfaction in industrial markets: Dimensional and multiple role issues. *Journal of Business Research*, *52*(1), 15–33. doi:10.1016/S0148-2963(99)00101-0

Hotchkiss, D. A. (1995). What do your customers really think? *Bank Marketing*, *27*(3), 13–20.

IBM SPSS Statistics. (2015). Version 22. Retrieved from https://www.ibm.com/products/spss-statistics

Jankal, R., & Jankalová, M. (2011). Mystery shopping - the tool of employee communication skills evaluation. Business. *Theory into Practice*, *12*(1), 45–49. doi:10.3846/btp.2011.05

Kafchehi, P., Hasani, K., & Gholami, A. (2016). The Relationship between Innovation orientation and strategic typology in business firms. *International Journal of Knowledge-Based Organization*, *6*(2), 1-20.

Kehagias, J., Rigopoulou, I., & Vassilikopoulou, A. (2011). Linked mystery shopping inventory to customer-seller encounters. *Journal of Customer Behaviour*, *10*(1), 7–34. doi:10.1362/147539211X570492

Kim, M., Park, J. E., Dubinsky, A. J., & Chaiy, S. (2012). Frequency of CRM implementation activities: A customer-centric view. *Journal of Services Marketing*, *26*(2), 83–93. doi:10.1108/08876041211215248

Korhan, A. (2015). Effect of knowledge sharing and supply chain management on organizational performance. *International Journal of Knowledge-Based Organizations*, *5*(1), 19–32.

Leeds, B. (1992). MC offers clues to quality service. *Bank Marketing*, *24*(11), 24–26.

Leeds, B. (1995). MC: From novelty to necessity. *Bank Marketing*, *27*(6), 17–23.

McNerney, D. J. (1996). The link to customer satisfaction. *HRFocus*, *73*(9), 1–4.

Michaels, R. E., & Day, R. L. (1985). Measuring orientation of salespeople: A replication with industrial buyers. *JMR, Journal of Marketing Research*, *22*(4), 443–446. doi:10.1177/002224378502200409

Michelson, M. (2001). The difference between mystery shopping and marketing research. *Quirks Marketing Research Review*, 653. Retrieved from www.quirks.com

Miles, L. (1993). Rise of the MC. *Marketing*, *29*, 19–20.

Morral, K. (1994). MCs serve tests service and compliance. *Bank Marketing*, *26*(2), 13.

Morrison, E. (1994). Role definitions and organizational citizenship behaviour: The importance of the employee's perspective. *Academy of Management Journal*, *37*, 1543–1567.

Morrison, L. J., Colman, A. M., & Preston, C. C. (1997). Mystety Customer Research: Cognitive Processes Affecting Accuracy. *Journal of the Market Research Society*, *39*(2), 349–361. doi:10.1177/147078539703900203

MRS. (2009). *Market Research Society*. Retrieved October 2017, from www.mrs.org.uk

Naumann, E., & Giel, K. (1995). Customer satisfaction measurement and management: Using the voice of the customer. Thomson Executive Press.

Nikabadi, M.S., Hoseini, S.A.M., & Zangouri, S. (2014). The effect of total quality management on innovation performance with mediator role of organizational learning: (Case study: ABBAND mazandaran). *International Journal of Knowledge-Based Organization, 4*(4), 51-66.

Paine, L. S. (1991). Corporate Policy and the Ethics of Competitor Intelligence Gathering. *Journal of Business Ethics, 10*(6), 423-436.

Pallant, J. (2001). *SPSS Survival manual: A step by step guide to data analysis using SPSS for Windows Version 10*. Allen & Unwin.

Parker, L. (1988). MC for clues to improve service. *The Bottom Line (New York, N.Y.)*, *5*(1), 49–50.

Rubel, C. (1995). CSR works best when it also satisfies company needs. *Marketing News TM*, *23*, 38.

Schlossberg, H. (1991). There's no mystery in how to retain customers. *Marketing News*, *25*(3), 10.

Schlossherg, H. (1992, September). Latest image study finds mixed results for researchers. *Marketing News*, *8-9*. doi:10.1108/02634500010327917

Sigala, M. (2008). Service quality and customer relationship management: Interrelationships and implications. *Managing Service Quality*, *18*(5), msq.2008.10818eaa.001. doi:10.1108/msq.2008.10818eaa.001

Singh, R., Sandhu, H.S., Metri, B.A., & Kaur, R. (2014). Understanding organized retail supply chain performance indicators for business growth. *International Journal of Knowledge-Based Organization*, *4*(2), 64-76.

Spiro, R. L., & Weitz, B. A. (1990). Adaptive Selling: Conceptualization, Measurement, and Nomological Validity. *JMR, Journal of Marketing Research*, *27*, 61–69. doi:10.1177/002224379002700106

Spooner, L. (1985, July). MCs uncover keys to sales and service. *Savings Institutions*, 92.

Stefanelli, J. M. (1994). Using MCs to maintain hospitality company service standards. *Hospitality and Tourism Educator*, *6*(1), 17-18.

Stovall, S. A. (1993). Keeping tabs on customer service. *Bank Marketing*, *25*(6), 29–33.

Tepper, G.C. (1994). The merits of self-testing. *Mortgage Banking*, *54*(8), 76.

van der Wiele, A., Hesselink, M., & Van Iwaarden, J. (2005). Mystery shopping: A tool to develop insight into customer service provision. *Total Quality Management & Business Excellence*, *16*(4), 529–541. doi:10.1080/14783360500078433

Vavra, T. G. (1997). *Improving your measurement of customer satisfaction: A guide to creating, conducting, analyzing, and reporting customer satisfaction measurement programs*. Milwaukee, WI: ASQC Quality Press.

White, C. (2004). *Customer satisfaction research and MC: Why do both?* Retrieved September 2017, from http://www.marketresearchworld.net

Wilson, A. M. (1998). The use of MC in the measurement of service delivery. *Service Industries Journal, 18*(3), 148–163. doi:10.1080/02642069800000037

Wilson, A. M. (1998). The Role of MC in the Measurement of Service Performance. *Managing Service Quality, 8*(6), 414–420. doi:10.1108/09604529810235123

Wilson, A. M. (2000). The use of performance information in the management of service delivery. *Marketing Intelligence & Planning, 18*(3), 127–134.

Wilson, A. M., & Gutmann, J. (1998). Public transport: The role of MC in investment decisions. *Journal of the Market Research Society, 40*(4), 285–293. doi:10.1177/147078539804000401

Woodruff, R. B., & Gardial, S. F. (1996). *Know your customer: New approaches to understanding customer value and satisfaction*. Oxford, UK: Blackwell.

Yang, C. C. (2002). *Customer needs and satisfaction survey*. Taipei: SGS Taiwan Limited.

Yanli, Z., & Yao, Z. (2015). Challenges in organizational control: The economic and management perspectives. *International Journal of Knowledge-Based Organizations, 5*(1), 33–44.

Zeithaml, V. A., & Bitner, M. J. (2000). *Services Marketing*. New York: McGraw-Hill.

APPENDIX

Action Diagrams

Combining weights and average satisfaction indices, namely scores, a series of action diagrams can be developed. These diagrams indicate the strong and the weak points of customer satisfaction, particularly MC scores in this research, and define the efforts required for improvement (Grigoroudis & Siskos, 2002). These diagrams are also mentioned as decision, strategic, perceptual, and performance–importance maps (Dutka, 1995; Customers Satisfaction Council, 1995; Naumann and Giel, 1995), or gap analysis (Hill, 1996; Woodruff and Gardial, 1996; Vavra, 1997), and they are similar to SWOT analysis. Each of these maps is divided into quadrants, according to performance (high/low) and importance (high/low) that may be used to classify actions (Figure 9):

- **Status quo quadrant** (low performance and low importance): no additional action required, since the specific satisfaction criteria is considered of low importance by the customers.
- **Leverage opportunity** quadrant (high performance/high importance): the information revealed by this quadrant can be used as advantage against competition.
- **Action opportunity** quadrant (low performance/high importance): These are the criteria that that should be improved to increase the level of customer satisfaction.
- **Transfer resources** quadrant (high performance/low importance): resources and overall business endeavors relating to the specific characteristics of the product or service may be used elsewhere (e.g. improving satisfaction criteria of action opportunity quadrant).

The last step involves the improvement opportunities clarification. In this step, the first priority improvement opportunities of the action diagram were connected to the corresponding branch(s). Note that every improvement opportunity denoted by the subject sub-criterion (a).

For the construction of action diagrams, we used the relative variables $b_i^{'}$ and $S_i^{'}$ which can be calculated from the following equations:

$$b_i^{'} = \frac{b_i - \bar{b}}{\sqrt{\sum \left(b_i - \bar{b}\right)^2}}, \quad S_i^{'} = \frac{S - \bar{S}}{\sqrt{\sum \left(S_i - \bar{S}\right)^2}} \text{ and for i=1, 2, …, n}$$

where b is the criterion weight value and S is the satisfaction index value and $\overline{b}$ and $\overline{S}$ are the mean values of the criteria weights and the average satisfaction indices, respectively. Both axis importance which refers to the criteria weights and performance which refers to the average satisfaction indices take values in the range [0, 1].

This grid can be used in order to identify priorities for improvement. The bottom right quadrant is obviously the first priority, for the attributes are important to customers but organisation's performance is rated moderately low. The second priority may be given to the satisfaction criteria in the top right quadrant, especially if there is room for improvement. Issues of the third priority are indicated in the bottom left quadrant; although these issues are not terribly pertinent at the time of the analysis, they may be more important in the future if an organisation's performance is certainly not good. Finally, last priority for improvement should be given to the criteria in the top left quadrant because this category is the least important if an organisation's performance is relatively good. Apparently, priorities for improvement may vary among different companies, depending on the potential capabilities of improving the particular category. Relative action diagrams: these diagrams use the relative variables weights (b'i) and performance's scores (S'i) in order to overcome the assessment problem of the cut-off level for the importance and the performance axis. This way, the cut-off level for axes is recalculated as the centroid of all points in the diagram (Grigoroudis and Siskos, 2002). This type of diagram is very useful if points are concentrated in a small area because of the low variation appearing for the average satisfaction indices (e.g. case of a high competitive market).

Figure 10. Action diagram (Adapted E. Grigoroudis and Y. Siskos, 2002)

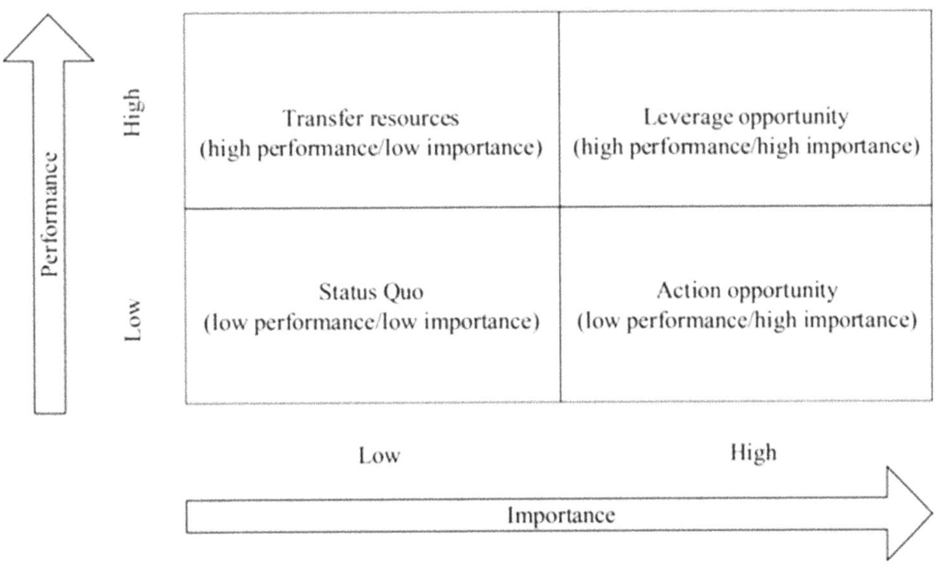

Chapter 11
A Case Study on Improving Learner Engagement by Incorporating ICT Tool Usage and Active Learning Strategies in Engineering Courses

S. Julius Fusic
https://orcid.org/0000-0001-9572-4025
Thiagarajar College of Engineering, India

Anandh N.
Manipal Institute of Technology, India

M. Thangavel
https://orcid.org/0000-0002-2510-8857
Thiagarajar College of Engineering, India

ABSTRACT

In the present scenario, the innovation in teaching is necessary to engage the students for the course of 36 to 40 hours. At the end of the course, the average student's strength will excel in performance and attains the course outcome with in-depth knowledge. To overcome the difficulties of slow and inactive learners for attaining the course outcome, it is necessary to renovate the teaching methodology. A few innovative learning methods like think pair share (TPS) activity, flipped classroom, online education, virtual classroom techniques, project-based learning, activity-based learning provoke the inactive or slow learners to be more active in learning the course. From studies, the current generation students are smart, and they feel conventional teaching methods like blackboard and PowerPoint presentations are monotonous. Here, the significance and impact of using ICT tools over conventional teaching methods is discussed on the fluid power automation course as a case study.

DOI: 10.4018/978-1-7998-2245-5.ch011

INTRODUCTION

In present days, the faculty members are practicing many active learning methods to make students to understand and remember fundamental concepts. The current teaching method needs innovation in delivering the lecture and to engage the students for entire course of 36 to 40 hours. At the end of the course, very few students attain the course outcome and excel in course depth knowledge. To overcome the hurdles of slow and inactive learners to meet the course outcome, it is necessary to renovate the teaching methodology. Initially, the course and its outcome are designed based on a consideration of ICT tools usage and skill (soft and creative) development approach engaging the students with innovative teaching practices. The addition of lab components in course content enriches the student practical knowledge. Few innovative learning methods like think pair share (TPS) activity, flipped classroom, online education, virtual classroom techniques, project based learning, activity-based learning stimulate the inactive or slow learners to be more active in learning the concepts. Learning through activities encourages the students concentration towards the course in South Indian education platform. Instead of conventional learning, the innovative ICT tools based learning dominates the current digitalized student's era. From studies, the present generation students are smart and they are uninterested in traditional teaching methods like black-board and power-point presentations. Traditional methods of learning reduced the interests among the students in learning the courses. As a consequence, presently there are many deliberations in upgrading the student education from Outcome-Based Education (OBE) to Conceive-Design-Implement and Operate (CDIO).

The CDIO syllabus focuses on developing personal, interpersonal and system building among students. This concept is new for all South Indian students and it emphasis on learner-centric education methodology. To create the revolution in an education system, the students should be well-equipped to face the challenges in learning methods. Nowadays, the students are experiencing the renovated teaching methods like simulation-based learning, project-based learning, demonstration-based learning, smart lab-based learning, flipping and inverse flipping classroom and gamifying based learning. By doing so, the students can learn more than the syllabus and think beyond the understanding category to apply category level as per the outcome-based education concept. Subsequently, students express more attention and self-motivation to apply the concepts from day to day life activities to enrich their skill from their knowledge. In this chapter, a set of engineering students from Thiagarajar College of Engineering (TCE) have experimented ICT tool-based learning approach for Fluid power automation course. A part of the students in TCE are from rural areas and they are lacking in English proficiency. The sudden implementation of a new education system for these students will not be much effective. It is a challenging task to successfully implement the ICT tools in courses for the rural based students. In this case study, grouping of students into two batches was done and in each batch there are slow and fast learning students, also rural and urban area based students equally in a proper ratio. Later on, by experiencing the learning approach individually a comparison was done for the enhancement of student creativity, depth knowledge and skills from course content. Simulation, learning through projects, flipped classroom, smart lab learning and gamifying techniques are few technical approaches. These approaches are not meant for all the chapters in that course. Certain topics like design and development of pneumatic circuits covered using simulation-based learning and smart lab learning techniques.

For example, in simulation-based learning, the theoretical explanation is done through power point presentation and practical demonstration is done using software tools like Automation studio and Fluid-sim. Initially, a demonstration was given through Fluid-sim software for the set of students experiencing ICT tool-based learning. Later, the students are allowed to the laboratory to practice the demo circuit and simulate the same. After completion of practice, an instructor will give certain industrial application and mentor the teams to use various methods like changing types of valves, change the sequence of operation and varying position of sensors. Finally, select a simulated circuit from a group randomly and interface it with hardware through LAN connection and run the operation for further understanding among students. Project-based learning and demo-based learning can be done through an application and industry based case study. These qualifying techniques made students to recollect the remember and understand concepts before continuous assessment test. This type of innovative approach made students to rely upon depth concept over the course. Many activities triggered the students to design and develop their own ideas into a product. The creativity will develop among students through system integration or sandbox concepts. These kinds of development create an opportunity to compete for the competition with a world-class education. The evaluation based on rubrics and data analytics of survey results and assessment results that shows the positive reflection of the activity-based learning than conventional learning method.

These ICT tool activities are not domain or discipline specific, so it will practice for all discipline students provided proper course design contents. Further, the chapter discusses the submission of assignment as the subtopic. Normally many students plagiarized the content and submit the assignment for grading purpose. The conventional assignments are for merely fetching marks but it does not provide depth knowledge in the course. Whereas the innovation in assignment like chart preparation, concept explanation and develop demonstration to rely practical knowledge over the topics, inverse flipped classroom made students to submit plagiarism free assignment and create opportunity to develop competitive skills like leadership qualities, team management, self-motivation, gestures development and also made students to think about knowledge-based approach. These assignments were assessed primarily as self-assessment and then indulged in peer assessment. At the end, the students realize the metrics in preparing assignments on their own way without plagiarizing other works that encourages the teamwork, self-motivation and leadership qualities. The impact of these innovations secures a better job placement in the future. Since the demo based assignment is requiring defending concept in which students develop gestures, free from stage fright, improve eye contact and communication. Also, these kinds of exposure prepare them to provide the solution for the needy peoples and become an entrepreneur. The significance and the impact of using ICT tools over conventional teaching method also discussed from the case study.

BACKGROUND

Christie Martin and Drew Polly (2017) explained the improvement in education among students by introducing co-teaching every subtopic. The co-teaching is defined as pre-service teacher which reforms the course as co-teaching by planning, teaching and assessing in a practical manner. Based on case study the pre-service teacher and co-teacher, co-planned to improve the student education through three pairs of comparative data. The author elaborates the collaborative learning based on group activity made students to practice and implement the theoretical knowledge effectively. The group activity improves the leadership quality, ethical approach towards problems, self-efficacy and so on. The student's perception of self and peer efficacy can be developed through the collaboration of universities with summer and winter academy self-teachers. The author explained the perception efficacy through survey assessment strategy. Oonk, Verloop, and Gravemeijer (2015) introduced the theory enriched with practical skills, as a role of a teacher who directs the students to design theory of each chapter along with practical experience or relate it with laboratory experiments. The major role in collaborative learning is group thinking. Group thinking among the peer was enriching the complicated task into more efficient and expandable one. The group learning was not only improving the student learning rather than peer and qualitative assessment, qualitative explanation among group students made the practical activity more efficient. Introduction activity and peer assessment develop collaborative learning strategy. The blended learning strategy was one of the efficient methods made faculty teaching method as simple as possible.

David Parsons (2017) approach provides an ample platform to engage students to learn course content with advanced technology and activities. The author introduced mobile learning approach to engage and interact student to student learning approach. The mobile learning approach in computer application explained the effectiveness in student affective traits among peers. The mobile assisted language learning practiced students to improve communication and student interactive skills. The student perspective over the language course might change and increase effective instructional strategies, training and professional practices in gestures and pronunciations. According to Owston (2013), the blended learning is not only technology implication but also learning strategy for students to students learning, peer learning and collaborative learning. Graham, Woodfield and Harrison (2013) stated blended learning with case studies in universities and three important methodologies like create awareness and exploration about pedagogical processes, adopt or implementation of course development processes and categorize data-driven instruction implementation among undergraduate students. David J. Nicol et.al (2006) demonstrated that the assessment of student activities using ICT tools was playing a vital role in education. To assess the higher education in students it was recommended to follow formative and summative assessment techniques. In relation with assessing the student capability based on knowledge and psychomotor skills, few researchers like Eleanor Hawe & Helen Dixon (2016) detailed about assessment for learning (AfL) technique for higher education students. In this assessment pattern the author explained that the students are not only assessed based on knowledge but also the thinking skills and feelings too. In another work, author Mike Mimirinis (2018) stated the assessment practiced by the course instructor would be electronic management of assessment (E-assessment). As the teaching style highly depend on digitalized component, teaching faculty can use E-assessment tools to evaluate the students skills and knowledge in online based learning methods like Learning Management System (LMS) and Massive Open Online Course (MOOCs). Guri Skedsmo et.al (2018) stated that assessment methods engrave the teacher's professionalism and promote the students learning process. Based on the assessed outcome course instructor would encourage the students by offering incentives, complementary gifts, which might trigger the student behavioral pattern and increase the effort of students in future.

CASE STUDY BACKGROUND

The department of Mechatronics Engineering offers B.E (Mechatronics Engineering) Programme since 2014 and M.E. (Mechatronics Engineering) since 2014. The department is supported by well experienced and qualified faculty members. The department focuses on imparting practical and project-based training to the students through competency-based curriculum. The department has different Special Interest Groups such as Automation and Control, Robotics and IOT. These groups work on promoting research, training to industrialist and development activities. The department has sponsored research projects supported by funding organizations like UGC, AICTE and DST. The Automation lab was well established in collaboration with Rexroth Bosch Ltd. The students can able to virtualize the smart industrial

Table 1. Course Outcomes

CO Nos.	Course Outcomes	Revised Bloom's taxonomy
CO1:	Classify the properties of pneumatic and hydraulic systems and their applications	Understand
CO2:	Classify and select the pumps and motors for the required applications	Understand
CO3:	Design the fluid systems with speed, pressure and direction control	Apply
CO4:	Design the hydraulic and pneumatic circuits for the given application	Apply

manufacturing systems like Pneumatics, Hydraulics, CNC technology, PLC based automation and so on.

The curriculum designed for IV semester fluid power automation students were based on the Choice Based Credit System (CBCS). The course outcome was listed as follows,

The course complexity is as follows: There are few complexities that fluid power automation syllabus deals with theory, analysis, design and application-oriented categories, where the slow learning students might feel difficult to understand. The syllabus framed based on prerequisite courses. But students from lateral entry category may directly join the course in the second year might not satisfy the prerequisite. So they don't understand the basics about the course. The syllabus might be hectic for students who don't qualify the prerequisite. Since it is an industrial application oriented syllabus, which makes students to relate theoretical knowledge earned in the classroom with an industrial application for better understand. The fluid power automation course plan is shown in below Table 2.

The assessment pattern for the course is shown in table below,

The technical education for the students at colleges will not provide deep learning skills over the course contents designed. In this work, the quality, reliability, standard of questions and assessment setting to evaluate student knowledge over the course can be analyzed using item analysis method. Also, the difficulty level and test effectiveness are measured to find the weakness of teaching and difficulty level of understanding in electrical machine course as a case study using Plicker software.

ICT TOOL USAGE AND PEDAGOGICAL PRACTICES

To overcome the complexity, the teachers who are handling the course must follow certain teaching methodology by introducing ICT tools as much as possible. In this chapter, the ICT tools like Think, Pair, Share (TPS) activity, Brainstorming, Peer

Table 2. Course plan

THIAGARAJAR COLLEGE OF ENGINEERING, MADURAI – 625 015.
(A Govt. Aided Autonomous Institution Affiliated to Anna University)
DEPARTMENT OF MECHATRONICS ENGINEERING
COURSE PLAN

Degree	: B.E (Mect)	Department	: Core Course
Branch	: Mechatronics	Semester	: IV semester
Course Name	: Fluid Power Automation	Course Code	: 14MT450
Staff in charge	: S.JuliusFusic		

S.No	Topic	No. of Lectures (Hours)	Methodology (BB,OHP, PP, Others)*	ICT activities
1	**Introduction to Fluid Power Automation**			
1.1	Need for Automation, Hydraulic & Pneumatic basics	1	BB	
1.2	Properties of hydraulic fluids – General types of fluids – Applications of Pascal's Law- Properties of air Kinetic theory of gases – Boyle's Law - Laminar and Turbulent flow – Reynold's number Selection criteria	2	BB& PPT	TPS activity
2.	**System Modelling by Bond Graphs**			
2.1	Hydraulic pumps and Motor	1	BB	Peer Instruction
2.2	Gear, Vane,and Piston motors	1	BB	
2.3	Gear, Vane,and Piston pumps	2	BB	
2.4	Selection and specification-Drive characteristics	2	BB/PPT	
2.5	pump performance – Variable displacement pumps	2	BB/PPT	
2.6	Compressors – Filter, Regulator, Lubricator Unit – Air control valves	1	PPT	Flipped Classroom
2.7	Linear actuator – Single acting, Double acting special cylinders	2	BB/PPT	
2.8	Cylinder mounting details	1	BB/PPT	
2.9	power packs – construction	1	PPT	
2.10	Reservoir , Accumulators	1	BB/PPT	
2.11	standard circuit symbols	1	BB/PPT	Flipped Classroom
Assignment –I submission /Activity-1	**CAT Test 1**			
3	**CONTROL AND REGULATION ELEMENTS**			
3.1	Direction control Valves 3/2 way valve – 4/2 way valve – 5/2 way valve. Shuttle valve – check valve	1	BB& PP	Flipped Classroom
3.2	Flow control valves - Fixed and adjustable	1	BB	
3.3	pressure control - Simple and compound relief valve ,pressure reducing valve, sequence valve, counter balance valve	1	BB	Flipped Classroom
3.4	Methods of actuation – types	1	Demo	
3.5	electro hydraulic servo valves- Different types- characteristics and performance	1	Demo/PPT	Peer Instruction
3.6	Electrical control of pneumatic and hydraulic circuits	2	Demo/PPT	Flipped Classroom
	Activity-2/ Assignment-2 **CAT Test 2**			
4	**Control Circuits**			
4.1	Speed control circuits	2	BB/PPT	
4.2	synchronizing circuit	1	BB	
4.3	Penumo hydraulic circuit, Fail-safe circuit	1	BB	
4.4	Sequential circuit design for simple applications using the cascade method, step counter method.	1	PPT/Demo	
4.5	PLC based Automation - use of relays, timers, counters	1	BB/PPT & Video	Discussion forum
Assignment – III submission/ DEMO based Defend Activity(Peer activity)			**CAT Test 3**	
	Total	**36**		

* - (BB- Blackboard; OHP – Over Head Projector; PPT-Power Point;)

Table 3. Assessment pattern

Revised Bloom's Category	Continuous Assessment Tests			Terminal Examination
	1	2	3	
Remember	20	20	20	20
Understand	20	20	20	20
Apply	60	60	60	60
Analyze	-	-	-	-
Evaluate	-	-	-	-
Create	-	-	-	-

Assessment, and Flipped Classroom were explained along with course. The group activities like Project-based learning, Peer Learning was also explained.

Problem Description

In Fluid Power Automation course students felt difficult to understand the technical terms and design the circuits for given applications. The terminal summative assessment results have many failures especially from lateral entry and slow learners category. As per Bloom's taxonomy, students may not understand how to apply the theoretical knowledge for a given scenario. Many students feel hard to relate the previous prerequisite course along with this course.

Optimal Solutions

Introducing ICT tools for each chapter may improve the student remembering capability and also encourage the lateral entry students and slow learners to understand the concept clearly. Introducing formative assessment in each sub-topic may improve the terminal results in a positive manner. The demo based or product based learning may improve the students to relate theoretical concept with industrial application. Bridge courses and capstone courses are introduced to make the students relate prerequisite course concepts with current course content.

Table 4. Think pair share activity

Domain: Fluid Power Automation Topic: Pneumatics & Hydraulics Introduction Target Students: Engineering IV year (Mechatronics) Think Pair Share - [5 minutes] Question: Before starting the Fluid power topic, teacher will ask students to write all real-time examples in relating with pneumatics and hydraulics application in day to day life and home appliances. Thinking Phase [3minutes]: It is like brainstorming session. The teacher will give some examples in the day to day life and ask students to think and respond. Pair Phase - [10 minutes] Question: Instead of pairing with neighboring students, faculty asks for a volunteer to come on stage. Later on the students will interact individually with the volunteer. The volunteer writes all the examples said by students on the black-board and ensures that the examples are not repeated and also the classroom is under control. Share Phase - [5 minutes] Now the knowledge is shared automatically with pairing phase itself. But in sharing phase students may justify their selected application based on certain criteria which will engage other students to share their knowledge as like peer learning. This will reduce the additional time for sharing this particular example alone. Finally the teacher will consider all examples and start the introduction about concepts of Pneumatics and Hydraulics.

THINK PAIR SHARE ACTIVITY

Initially, the student must refresh the prerequisite and other basic knowledge about the course in chapter 1. The ICT tool usage like Brainstorming, Think Pair Share (TPS) activity is to refresh the students mind and create a thirst to learn about fluid power automation. The TPS activity is explained as follows,

FLIPPED CLASSROOM ACTIVITY

Out of many ICT tool activities, the foremost activity will be assigning hands-on/ practicing work as an assignment for continuous assessment. In this the sub-topic is design a pneumatic circuit for given problems. The pneumatic circuit design will be developed, once the students understand about pneumatic system symbolic representations as shown in figure-1.

The out of class activity outcomes are as follows,

At the end of watching the videos, the students should be able to,

1. Explain the construction and working principle of valves, actuators and its symbolic representation. (Understand Level)
2. Design pneumatic circuit for sequential operation and cascading operation. (Apply Level)

Figure 1. Pneumatic system symbolic representations

Table 5. Out of class activity plan

CONCEPT	VIDEO SEGMENT	DURATION (in min)
Classification of Valves	V1 (0:00 to 3:40)	3 min 40 seconds
Valve construction and Working Principle	V2 (3:40 to 15:00)	11 min 20 seconds
Sequential operation	V3 (0:00 to 0:30)	30 seconds
Explanation using software	V4 (0:30 to 8:24)	7 min 54 seconds
Total Duration		23min 24seconds.

The flipped classroom video links are as follows,

- V1 & V2: https://www.youtube.com/watch?v=QPfeyugnN04
- V3 & V5: https://www.youtube.com/watch?v=UAjdFII37XU

The out of class activity plan are as follows,

Table 6. In-class activity plan

Learning Objective	Assessment Strategy	Expected Duration (in min)	Additional Instructions (if any)
Study about the construction and working of valves used in the pneumatic application (Understand Category)	Q1. Draw and explain the construction & symbolic representation of 4/2 DCV valve.	20 Min	Watch V1 & V2 and then provide a solution for Q1
Design various pneumatic circuits for given application.	Q2. Design pneumatic circuit for given electric furnace operation.	20 Min	Use software in laboratory or diagram using symbols shown in video V3 & V4.

Once the out of classroom activity is completed, the students will return to the classroom. Later, the teacher must initiate an In-class activity which relates the video such that it satisfies the bloom level assessment criteria.

The In-class activity outcomes are as follows,

At the end of the class, students will be able to,

1. Design any Pneumatic circuit for a given application. (Apply Level)
2. Solve real-world problems using sequential & cascading operations. (Apply Level)

The sample solution given by students is as follows,

- Design pneumatic circuit for given electric furnace operation
- Students will be taken to the laboratory and they are demonstrated with hardware and software. And then they deploy assignment as various applications to be simulated using software's like Fluid sim, Automation studio etc.

Thus they submit the simulated circuit in Automation studio as follows,

Step-1: Pneumatic Circuit for Electric Furnace

Step-2: Electrical circuit for operation of solenoid valves

Figure 2. Pneumatic circuit for electric furnace

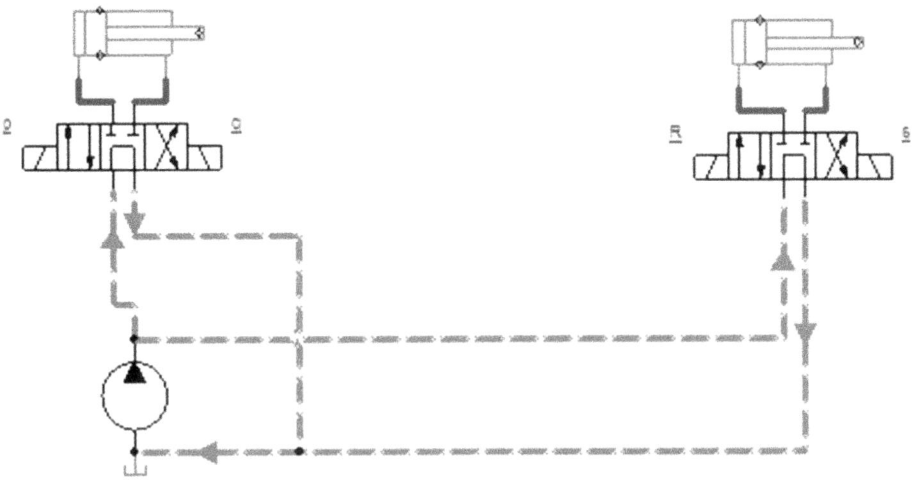

Figure 3. Electrical circuit for operation of solenoid valves

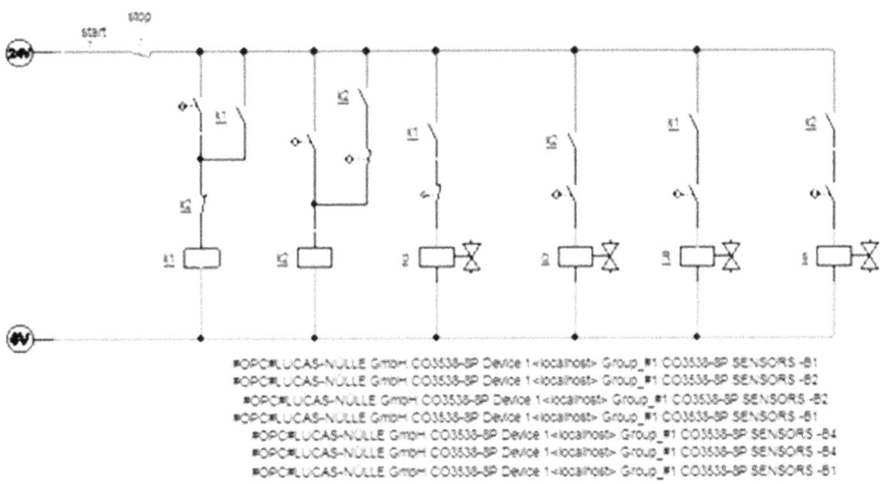

Table 7. Peer instruction activity - teacher

Give two objective type questions at the beginning of class and provide a summary of basic concepts: sample questions are shown below, In single acting hydraulic cylinders, the piston comes back to its original position due to__________ a)spring force b)self-weight c)the momentum of a flywheel d)all the above What is the difference between the pressure relief valve and pressure reducing valve? a) Pressure reducing valve is connected between pump and tank line while the pressure relief valve is connected between DCV and branch circuit. b) The pressure relief valve is always normally opened. c) Pressure reducing valve is connected between DCV and branch circuit while the pressure relief valve is connected between pump and tank. d) None of the above.

Procedure to do in Automation Studio

The procedure is as follows:

1. Automation studio E6.2 software to be opened and a new project were created.
2. The hydraulic components were dragged from the library and the connections are made as per the circuit diagram.
3. Here we use double acting cylinders and 4/3 DC valves with double sided solenoid actuation.
4. The variables assignment for the solenoids of both DC valves was done.
5. Electrical components such as contact switches, proximity switches, relays, solenoids were dragged from electrical component library.
6. Connections are made as per the circuit diagram.
7. The names and variables were assigned for the corresponding components.
8. To create OPC links, the following steps are used.
   Open Tools->Variable manager->OPC client->OPC Server-Add OPC server (Lucas nulle)
   Then open Groups->Add Group->Lucas Nulle.
   Open links->assign switches and valves.
9. Pneumatic kit connections are created by proximity sensors and solenoids.
10. The pneumatic kit was connected with pc by OPC cable.
11. The simulation was performed and the sequential operation was checked for Electric furnace.

Table 8. Peer instruction activity - student

For each question, the students will first answer individually. Then they will discuss with peers and come to a consensus and finally listen to instructor's explanation.

Figure 4. Plickers Sample Screenshots

Figure 5. Plickers class activities

Figure 6. Project-based learning sample

PEER INSTRUCTION STRATEGY (TEACHER PORTION)

The Peer instruction strategy explains the teacher to prepare sets of question from previous class lectures and they will be displayed in the classroom using projector. Once the students read the questions they will answer accordingly to their understanding. Later the teacher instructs the students to discuss with peers to find the appropriate answer. Finally students peer discuss about the correct answers obtained through peer instruction and conveys the conclusion to the teacher. In fluid power automation classes, these activities will be done using Plicker open source software.

PEER INSTRUCTION STRATEGY (STUDENT PORTION)

The sample screenshots of Plickers to support Peer Instruction activity are as follows,

Based on student, final display of Plickers card will be scanned by the teacher for final assessment and Plicker card samples are shown below.

Table 9. Project-based learning - Rubrics

Grading Rubrics for PBL based on blooms level.		
R's	**Description**	**% of Marks**
R1	Appropriate problem or application selection	5%
R2	Distribution of workload among teams	10%
R3	Literature review to identify problem gap.	10%
R4	Content delivery based on principle and working procedure as Chart presentation	20%
R5	A working prototype for the selected problem	30%
R6	Demo explanation and clarifications	20%
R7	Defend their work	5%

PROJECT BASED LEARNING

In this, the students will develop their own products and they will explain about the working model to peer group. The Project-based learning is another ICT tool which is implemented as assignment 3. Before initializing this assignment, the students will select their own team under their teacher guidance. After that, the teacher must float the rubrics for the PBL process based on Blooms criteria (as shown below). Once the students are aware of the rubrics they can select the topics based on criteria which satisfy all the taglines. The duration for the activity will be one month but every week there will be a formative assessment as reviews, where students prepare a chart presentation for ongoing work. Once the topic is freezed and the chart presentation satisfies the teacher, the student will proceed further to complete the work. At the end, the teacher who handles the course will act as internal examiner and invite other department teacher as external examiner for validating the peer work as like the main project. Thus the student can understand the concept clearly and relate the concepts with the industrial application and real-time application. Hence, the student satisfies the course outcome mapping at the end of the course. The student also understands the hands-on implementation of design and develops their own solution for the existing problem as a miniature.

The project has been evaluated based on the rubrics, as shown below,

At the end, the continuous assessment test 1 and 2 will be conducted and the results are assessed and compared with previous year results. The result focusses on the improvement in student learning and skills developed through the introduction of ICT tools along with courses. Finally, the syllabus for continuous assessment test 3 will be related to industrial application based chapters. So certain chapters are covered using a demo based peer teaching method and assign students to develop the group in a randommanner for final assignment as formative assessment.

ASSESSMENT

The education among Indian students is drastically improved from school level to college level during every academic year. Many students may fail to develop knowledge and skills in higher education in their domain. The change in academic pattern in higher studies especially engineering colleges are improving from common education to outcome-based education (OBE), choice-based credit system (CBCS) and conceiving-designing-implementing–operating (CDIO) and so on. Ali (2017) shown evaluation is an important tool to provide feedback to the teacher and it is an integral part of teaching-learning process. The development in the education system practically creates any impact over the student life is still an unsolved mystery. Item analysis makes use of evaluating and judging the test conducted to the students. Mostly, it identifies the performance of items to be considered over the test questions. Currently, many schools and colleges not delivering the lectures as learners centric rather they only exhibit the result oriented mode of teaching. The test conducted at the end of the course or like formative assessment test results will provide some important feedback to the teachers in many ways. The improvement in learning style among students needs some innovation and usage of ICT tools can be determined by the assessment methodology depends only on identifying mistakes in teaching method or distracters to the course. The post validation is better option to identify mistakes and paved a way to correct the mistakes or implement better teaching method, as shown by Popham WJ (2008). The assessment process is not a static method rather it has an impact on ongoing process and methods. Based on the assessment results, the activities of feedback and corrective measures are a dynamic process, as shown by Remesal (2011). This made the faculty realize that they are lack of innovative learning and lecturing skills. While comparing to European countries, many middle and high school assessment practices are regularly followed and in every term, they modified their methods based on measuring feedbacks. The test quality can be improved by assessing two methods like classical test theory by comparing facility value with discrimination index and item response theory can be applicable to MCQ's questions in the formative and summative assessment.

Academic performance of the students is more important in assessing their knowledge over the technical topics in the courses. The assessments are classified as summative assessment and formative assessment for measuring student's performance. In this work, the education measurement tools like item analysis measures the gap between educational objective and degree of learning. Item analysis is a process of

Table 10. Evaluation method index formulae

DI	FV		
	Less than 0.3	**0.3 to 0.6**	**More than 0.6**
Less than 0.2	Reject	Reject	Reject
0.2 to 0.3	Difficult	Improve	Easy
0.3 to 0.4	Improve	Accept	Improve
More than 0.4	Accept	Accept	Accept

collecting, summarizing responses received from the students and assess the quality of questionsbased on blooms taxonomy using Difficulty index (P), Discriminative index (DI), Facilitation value (FV), Effectiveness of Distracters (ED).

The study was conducted in the department of mechatronics engineering, second-semester electrical machines course as a part of the assessment. A total of 58 students in the class written the test for 50 marks (30 marks for remember, understand category and 20 marks for apply category questions). There is no negative marking for the test responses. Remember to understand category questions are multiple choice questions (MCQ's) remaining are essay type questions and time allotted was one and a half hour. Once the questions along with course outcome mapping submitted before the examination, the course designer along with the head of the department fix the pass mark as 25/50 mark (50%) through pre-validation of the questions. Then, after examination the response collected from the students are used for post validation using item analysis method. The post validation analyses required data or information based on total marks obtained by the individuals. The post validation has done in two phases, where MCQ's questions are analysed using item analysis and Apply category questions assessed by classical test theory. Then arrange the students based on merit or grade order. The upper 27% of students are grouped as upper graded students and lower 27% students are grouped as lower graded students. The remember category consist of four set of options where one is correct and the remaining three options are a distractor. The statistical data are converted into a mean, percentage, the standard deviation for n items. The relationship between P and DI, DI and FV for the items is determined by correlation analysis.

Block diagram for outcome-based education evaluation standard setting and Interpretation of DI versus FV for MCQ questions.

Figure 7. Block diagram for OBE evaluation

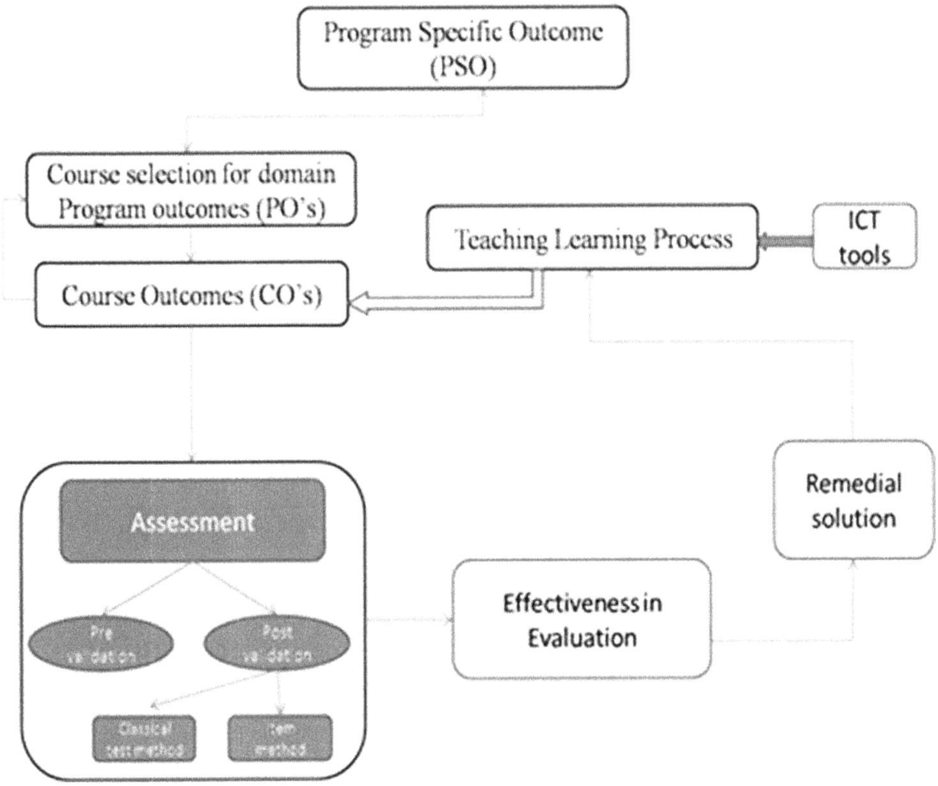

Table 11. Sample objective type remember CO's question analysis

Description	Formula	Explanation	Interpretation
Difficulty Index (P)	P=((H+L)/N)x100	H-No. of students correctly answer in upper-grade group L-No. of students correctly answer in lower grade group N- Total number of two groups including non-responder	P <30% -Difficult, P=30-70% -Accepted P > 70%= Not accepted (easy)
Discriminative index (DI)	DI=(Ri-Rj)/N)x2	Ri- ∑ (no. of upper grade students x total marks obtained), Rj- ∑ (no. of lower grade sttudents x total marks obtained)	DI is –ve then DI=0-0.19 poor DI=02-0.29- acceptable DI >0.3 excellent
Facilitation value (FV)	FV=(Ri + Rj)/N		
Distractor effectiveness (ED)	ED= (Nl-Nu/N)x2	Nl-no. of lower group students, Nu-no. of upper group students.	0-100% for MCQ's

Table 12. Evaluation method index formulae

<table>
<tr><th>ITEM-1</th><th>A</th><th>B</th><th>C*</th><th>D</th><th>ITEM-3</th><th>A*</th><th>B</th><th>C</th><th>D</th></tr>
<tr><td>UPPER 27%</td><td>3</td><td>1</td><td>11</td><td>0</td><td>UPPER 27%</td><td>9</td><td>4</td><td>0</td><td>2</td></tr>
<tr><td>LOWER 27%</td><td>6</td><td>3</td><td>5</td><td>1</td><td>LOWER 27%</td><td>6</td><td>5</td><td>3</td><td>1</td></tr>
<tr><td>ED</td><td>0.2</td><td>0.133</td><td>-</td><td>0.066</td><td>ED</td><td>-</td><td>0.066</td><td>0.2</td><td>-0.1</td></tr>
<tr><td>P</td><td>53%</td><td colspan="3" rowspan="3">Since the P,DI, functionality is within the limit, so this can be used without modification</td><td>P</td><td>50%</td><td colspan="3" rowspan="3">Since the P,DI, functionality is within the limit, so this can be used without modification</td></tr>
<tr><td>FV</td><td>0.53</td><td>FV</td><td>0.5</td></tr>
<tr><td>DI</td><td>0.4</td><td>DI</td><td>0.2</td></tr>
<tr><td>ITEM-5</td><td>A*</td><td>B</td><td>C</td><td>D</td><td>ITEM-8</td><td>A</td><td>B*</td><td>C</td><td>D</td></tr>
<tr><td>UPPER 27%</td><td>5</td><td>7</td><td>2</td><td>1</td><td>UPPER 27%</td><td>5</td><td>10</td><td>0</td><td>0</td></tr>
<tr><td>LOWER 27%</td><td>0</td><td>12</td><td>2</td><td>1</td><td>LOWER 27%</td><td>9</td><td>6</td><td>0</td><td>0</td></tr>
<tr><td>ED</td><td>-</td><td>0.33</td><td>0</td><td>0</td><td>ED</td><td>0.3</td><td>-</td><td>0</td><td>0</td></tr>
<tr><td>P</td><td>2%</td><td colspan="3" rowspan="3">The P value less than 30% so the question was mis-understood by the students and further modification to be done in the teaching method</td><td>P</td><td>53%</td><td colspan="3" rowspan="3">Since difficulty factor is inacceptable level the lower students mostly prefer A option. So that the discriminator can be modified</td></tr>
<tr><td>FV</td><td>0.022</td><td>FV</td><td>0.5</td></tr>
<tr><td>DI</td><td>0.133</td><td>DI</td><td>0.3</td></tr>
</table>

RESULTS AND DISCUSSION

The item analysis results for the course outcome, group of questions are analyzed and categorized under accepted level and modification need in discriminator or the teaching methodology are assessed and sample results are displayed in below table.

The essay category questions are also analyzed based on item assessment, complexity and remedial solutions are shown in below Table-12.

CONCLUSION

Based on the item analysis assessment results, 2018 academic council board of studies made few alterations in the 18MT230 - Electrical Machines course outcomes and also introduction of CDIO and innovative teaching tools were implemented. As per survey, only 10-15% of faculty was assessed the post-validation results and updated the remedial measures like introduction of ICT tools, MOODLE activities and product-based learning, flipped classroom and so on. Based on the classical analysis, the faculty designed system consists of high facility value and low discrimination index value among student's responses. This made an equal impact of faculty teaching methodology on fast and slow learners in the classroom. The reliability and standard setting over the course outcomes made improvisation in teaching plan and course objectives in a better way for ongoing and other batch members in a routine manner.

REFERENCES

Graham, C., Woodfield, W., & Harrison, B. (2013). A framework for institutional adoption and implementation of blended learning in higher education. *The Internet and Higher Education*, *18*, 4–14. doi:10.1016/j.iheduc.2012.09.003

Hawe, E., & Dixon, H. (2016). Assessment for learning: A catalyst for student self-regulation. *Assessment & Evaluation in Higher Education.*

Martin, C., & Polly, D. (2017). Handbook of Research on Teacher Education and Professional Development. IGI Global. doi:10.4018/978-1-5225-1067-3

Mimirinis. (2018). Qualitative differences in academics. *Springer Nature, 30*(3), 207-210.

Muhson, A., Lestari, B., Supriyanto, S., & Baroroh, K. (2017). The Development of Practical Item Analysis Program for Indonesian Teachers. *International Journal of Instruction*, *10*(2), 199–210. doi:10.12973/iji.2017.10213a

Nicol, D. J., & Dick, D. M. (2006). Formative assessment and self regulated learning: A model and seven principles of good feedback practice. *Studies in Higher Education*, *31*(2), 199–218. doi:10.1080/03075070600572090

Oonk, W., Verloop, N., & Gravemeijer, K. P. (2015). Enriching Practical Knowledge: Exploring Student Teachers' Competence in Integrating Theory and Practice of Mathematics Teaching. *Journal for Research in Mathematics Education*, *46*(5), 559–598. doi:10.5951/jresematheduc.46.5.0559

Owston, R. (2013). Blended learning policy and implementation. *The Internet and Higher Education*, *18*, 1–3. doi:10.1016/j.iheduc.2013.03.002

Popham, W. J. (2008). *Classroom Assessment: What Teachers Need to Know* (5th ed.). Boston: Allyn and Bacon.

Remesal, A. (2011). Primary and secondary teachers' conceptions of assessment: A qualitative study. *Teaching and Teacher Education*, *27*(2), 472–482. doi:10.1016/j.tate.2010.09.017

KEY TERMS AND DEFINITIONS

Bloom's Taxonomy: Bloom's taxonomy is defined as the classification of three different hierarchical domain of learning objective as cognitive, affective, and sensory domains. In education sectors, the outcome-based education (OBE) framework introduced Bloom's taxonomy in engineering education.

CDIO: The framework provides students by stressing engineering fundamentals in education in international standard through conceive, design, implement, and operate. Currently in all engineering education institutions, this framework is mandatory for obtaining ABET accreditation.

ICT Tools: The ICT tools are the latest technology or devices and concepts used in Information and Communication Technology among students to students, students to teacher interaction (e.g., flipped classroom, mobile apps, and clickers devices).

Chapter 12
Teaching Machine Design Curricula in Developing Countries:
A Case Study

Fredrick M. Mwema
Dedan Kimathi University of Technology, Kenya & University of Johannesburg, South Africa

Akinsanya Damilare Baruwa
https://orcid.org/0000-0002-6566-2705
University of Johannesburg, South Africa

Esther T. Akinlabi
University of Johannesburg, South Africa

ABSTRACT

In this chapter, a feasible approach to implement machine design curricula in developing countries is presented. The argument by the authors is that machine design should train engineering students in such countries to utilize local resources to solve practical societal problems. The approach illustrated here was used during 2015-2016 to teach machine design at Dedan Kimathi University of Technology, DeKUT, in Kenya. The approach involved grouping students of different interests and capabilities and tasking them to identify and study various problems in society. The groups were then required to propose machine design solutions to the identified problems. Finally, the groups were tasked to undertake the theoretical design and build CAD models for their projects. The students were monitored through individual weekly presentations to the instructor. The approach was seen to be successful to facilitate training in machine design.

DOI: 10.4018/978-1-7998-2245-5.ch012

INTRODUCTION AND BACKGROUND

The pedagogy in engineering courses has shifted from the deductive approaches to inquiry/investigative-based learning in which the students are encouraged to undertake their investigations to problems with the trainers (lecturers) acting as supportive figures to offer guidance and control (Wood et al., 2005). This way, the students are positively challenged to reflect critically on the problems and issues in their society and the impacts of their skills to the industry. Such approach cultivates perseverance, free thinking and adaptivity, which are necessary qualities for their future professions. These approaches are very useful when training engineering courses such as machine design (Royalty, 2018). For the approach to work effectively in teaching machine design, there should be enough and relevant facilities such as study books and equipment. Most engineering schools have extensively invested in books, journals and periodicals necessary to facilitate the teaching and training of machine design courses. However, while these materials are important to provide students with guidelines and challenging questions, the problems in those texts are mostly replicable, theoretical and do not sufficiently demonstrate the capability of the student to apply the concepts to real world cases. Hands-on experience is key to satisfactorily training of mechanical engineers and its importance has been underscored by Kolb's model of learning (Wood et al., 2005). There is therefore need for investment in industrial machines and facilities within the engineering schools to promote this mode of learning. This has been achieved, well, in developed countries and as such, the training of machine design in such schools is advanced, relevant and useful to the graduate students. The lack of training facilities in developing countries such as in Kenya leaves the trainers and trainees with the option of relying solely on the texts.

Most universities in developing world are underfunded and lack the basic training facilities in mechanical engineering. Another challenge with the training of machine design in developing countries is the 'curriculum'; the contents of the courses are formulated to match with those of the renowned engineering schools in the developed world. This makes it impossible for the instructors to deliver the core mandate of the course to the students and therefore compromising on the quality of the graduate engineers in those parts of the world. The syllabi of the machine design modules in the developing world should have realistic and achievable mandate. This can only be achieved through focusing on the issues of the society in which the universities are anchored (Seif, 1997). Therefore, the trainers of machine design in the developing countries must be creative to guide the students into utilizing their machine design concepts in proposing solutions to the local problems. This way, the students can gain deeper understanding of the underlying principles of the subject rather than just tackling theoretical problems from the developed world.

There are several publications focusing on innovative delivery of machine design course in engineering classrooms (Ranger and Mantzavinou, 2018; Andersson, 2012). Kanyarusoke (2019) demonstrated on the importance of integrative nature of training machine design to undergraduate students. The study demonstrated that rational approach to training machine design where the students are supposed to make products from the classroom designs enhances the creativity and understanding of concepts. Cohen and Katz (2015) noted that engineering undergraduate students should acquire practical skills during their training and demonstrated that one way to achieve so is through the use of 'design and manufacture laboratory' through which the students are supposed to produce components based on a taught design methodology. It was demonstrated that through such laboratory, students can gain professional skills on design procedure, teamwork and time management. Monterrubio and Sirinterlikci (2015) recommended for modifications of most machine design curricula to incorporate hands-on experiences to the students. Li et al. (2017) illustrated that using computer simulations for some of the machine elements during teaching enhances understanding and interest in machine design courses. Joyce et al. (2013) described a sustainable hands-on based training for machine design in which the students were supposed to implement their Capstone projects based on their two-year machine design curriculum. This way, it was concluded that students gain deeper understanding into concepts if they are engaged in projects right from the theoretical stage to the implementation. As such, machine design training should be based on sustainable projects through which they are actively engaged at all levels. Although there are several other publications discussing innovative machine design teaching methods to enhance hands-on training, there are limited researches on using societal problems (in developing world) to train machine design.

As such, the argument of the authors of this article is that, machine design, should not only focus on teaching students on just designing machines, but also on utilizing the local resources to solve the societal problems. As an illustration towards this proposition, the authors present an appropriate approach to train machine design (based on inquiry-based teaching methodology) in developing countries with a case study of Dedan Kimathi University of Technology, DeKUT (located in Nyeri County in Kenya) curriculum.

DESCRIPTION OF THE APPROPRIATE APPROACH

The machine design course at DeKUT is usually treated as a standalone module with the prerequisites of courses in mechanics, strength of materials, manufacturing, fluid dynamics, thermodynamics and engineering materials. The current curriculum of machine design is illustrated in Table 1. The overall purpose of the course is for

Table 1. Description of the machine design curriculum

Purpose of the course	To understand fundamentals of machines-machine component specifications and selection of standard parts. The joining of different machine components and their integration for machine design.
Learning outcomes	To design various machine components and standardization
Details of course outline	Shafts and axles
	Couplings
	Clutches
	Design selection and torque testing
	Bearings
	seals
	Cams and ratchets
	Joining, adhesion joining
	Weld design
	Standardization
	Use of design handbooks and catalogue
	Bill of materials
	Machine design project

the student to understand the fundamentals of designing machines and integration of engineering practice into machine design. The students are required to undertake laboratory work (machine design project), assignments and write examination towards the grading for the course. The course however does not describe the nature of the laboratory work although most of the relevant experiments related to the theory of machine design are usually covered in the prerequisite modules. The approach described in this study was employed to cover for the said experimental work (which most lecturers tend to ignore) for teaching of machine design in the department of mechanical engineering (DeKUT) during the 2015/2016 and 2016/2017 academic years. The uniqueness of the approach described in this paper is that the design projects were chosen based on the local problems rather than just the textbook projects.

The students were asked to identify problems within their communities and propose solutions to those problems related to machine design. The students were classified in groups of at least 8 individuals. The grouping was based on the various talents and interests of the students, that is, each group consisted of all individuals of different capabilities-those good in computations, computer aided design (CAD), design theory and software. This was necessary because the machine design module at DeKUT is an integration of nearly all the other modules in mechanical engineering curriculum and as such requires mixed understanding of different concepts. Additionally, the grouping was based on regional diversity of Kenya, i.e. communities in different parts of Kenya face different challenges in their daily lives. For example, communities living in highlands (central and rift valley) areas experience different problems from the pastoralists and those from arid and semi-arid regions. It was also considered that communities living in the cities have different problems from those in the rural areas of Kenya. That way the groups consisted of a mixture of individuals of different backgrounds, upbringings and skills. Such diversities are important in industrial design teams to harness full potential of teamwork (Heath, 2016).

The teams were required to identify solutions to the respective problems based on the facilities available in Advanced Manufacturing and training center (ADMATC) at DeKUT. The pictures of the typical manufacturing facilities in the ADMATC center are in figure 1. Prior to identifying their design projects and solutions, the student teams were requested to tour the center and remind themselves about the available facilities. Each group was tasked to identify common problem(s) in the society and formulate the problem statements and then propose mechanical engineering design solutions to the problems. Next, the students were required to design the projects based on machine principles and propose feasible manufacturing processes for possible implementation. To control the progress of the development and implementation of the design ideas, every individual of each team presented one-on-one (face-to-face) to the module instructor twice per week. In this way, the contribution(s) of each member was monitored, evaluated and enhanced. The teams were encouraged to share responsibilities based on the strengths of their members as it is the case of design teams in the industrial set up. In figure 2, a summary of the approach is illustrated.

The projects were assessed based on the template shown in Table 2. To further assess the success of the approach, the number performance of the students on the various projects was compared to the number of students who took up the same projects for their final (capstone) projects in both academic years.

SAMPLE PROJECTS

In this section, some of the projects developed by the design teams based on the instructional approach described in Figure 2 are illustrated. The problem statements, design solution and outputs have been captured from each project.

Figure 1. Some of the production facilities at ADMATC DeKUT. The design teams were tasked to tour the workshops to identify the equipment which can enhance implementation of their designs (Photos taken by Harrison Shagwira).

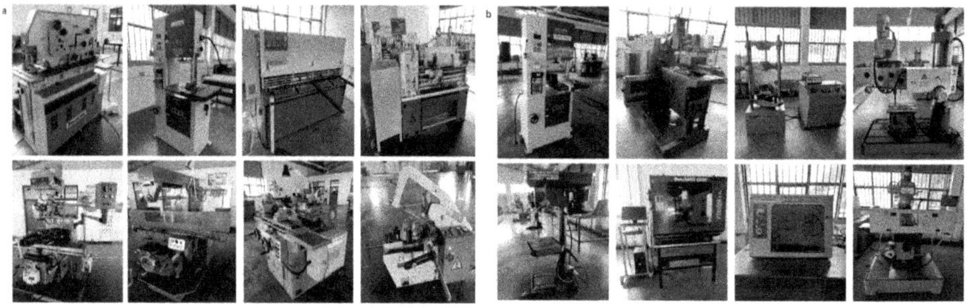

Figure 2. Summary of the approach

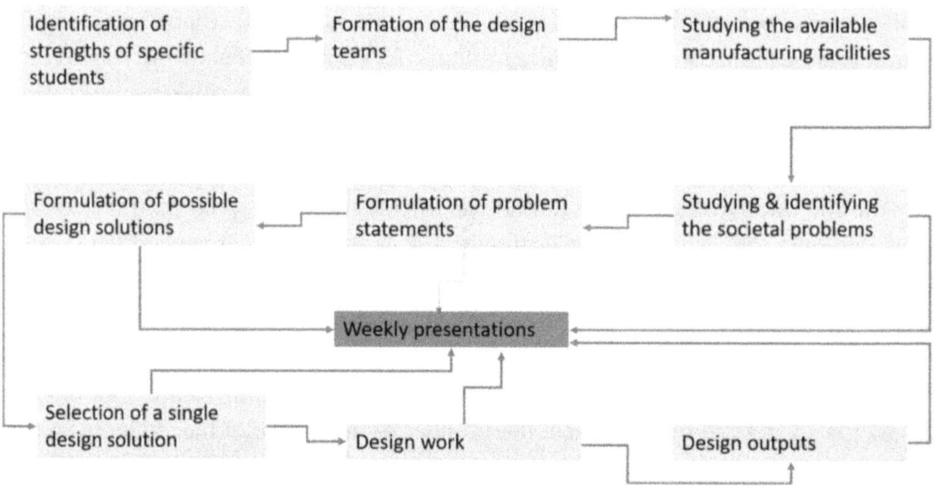

Project 1: Design of an Electrically Operated Sunflower/ Canola Oil Extractor for Household Application

Overview

Sunflower and canola seeds are food crops with both high nutritional and trade values and such they are consumed by humans and livestock. Their seeds are rich in edible/ healthy oil, which is extracted through various methods for human consumption. These crops are grown for both domestic and commercial purposes in rural Kenya.

Table 2. Assessment template for the projects

Assessment criteria	Requirement	Maximum marks (%)
Problem identification	The problem should be clear, local and related to the society. How has the student formulated the problem?	15
Problem solution	What project(s) can solve the problem? The student should illustrate the scientific guidelines of engineering design to choose the best solution. Is the solution viable?	5
Solution concept	Can the student visualize and sketch the project design, operation and how it will be manufactured?	15
Design theory	Has the student correctly utilized engineering concepts such as mechanics to design the project?	20
Computer aided designs	Has the student used CAD appropriately?	20
Project write up	Is the report technically written?	15
Weekly presentations	Is the student making progress or involved in the progress of the project?	10
Total		100

However, the value addition/processing is limited by the availability and cost of extraction facilities. Traditional, mechanical and chemical methods are used to extract the oil. Although the traditional methods are cheap, they are inefficient in extracting the oil (can extract only 20-30% oil). On the other hand, chemical methods are expensive and complex for adoption by local communities. Mechanical methods have been shown to extract a higher percentage of oils from canola and sunflower seed. In this work, design of mechanical oil extractor is described. The various components of the designed machine include the support table, hopper, auger screw, barrel, oil and cake drainage mechanism and metallic sieve. The machine runs on a power capacity of 14 kW with an average extraction capacity of 50kg/hr. The production cost of the machine is estimated at $461.00.

In this project, the machine is introduced, earlier works on similar machines reviewed, the various machine parts are designed and drawn using CAD software. The failure analysis of the machine, specifically the auger screw is done, and the allowable safety factor is determined. From the design analysis it is concluded that the failure of the auger screw component is mainly due to fatigue. The machine can be used for domestic sunflower/canola oil extraction among the communities in rural regions of Kenya.

The main objective of this project design is to improve the manually operated oil extractor machines currently in existence into electrically operated ones as well as perform the design aspect of the major parts of the machine so that it can be used for household applications. This was achieved based on the following specific objectives:

- To improve the operating mechanism of the machine to achieve a faster and efficient extraction process for small households.
- To design a low cost in material and operation oil extraction machine.
- To design a pressing screw that facilitates faster extraction of oil from the cake.
- To design a cake, choke mechanism that facilitates faster removal of pressed cake without clogging at the exit end of the machine.
- To design a more efficient machine that eliminates the use of solvent extraction by employing the use of a heating filament mechanism.
- To design a machine that is cheap to maintain and easy to clean when need be i.e. faster to dismantle and assemble.

Problem Statement

Traditional methods of oil extraction that mainly involves preliminary and hand pressing have proved to be less efficient requiring more manpower, achieve only 20-30% oil recovery, have longer processing times and produce low-quality oil. The development of mechanically, hydraulically and electrically operated oil extraction machines has seen most of these drawbacks surpassed and the realization of faster, more efficient and reliable oil extraction process.

Most sunflower and canola oil extraction methods that are in existence today are large scale-power or hydraulically operated machines. The machines make use of the screw and have the advantage of high capacity, small labor intensity and continuous production. With this development, the need for small/domestic scale machines has been ignored and the small-scale farmers in rural Kenya are left with option to sell their sunflower and canola produce without value addition. Most of the existing/available small-scale extractors are manual and hence inefficient to most rural farmers.

Outputs

The machine design considerations for this project were, high oil extraction, better efficiency, production of quality oil and low cost of machine. The main aspects of the machine structure were:

- Screw with Tapered Shaft and Variable Pitch: This type of screw was adopted because it gives higher rate of increase in pressure as compared to the other types of screws. Hence, it enhances efficiency of the extraction process.
- Conical drainage system: The system is cheap and practical.
- Drilled Holes as a drainage mechanism for oil since no extra parts for oil drainage system are required. It also lowers the overall cost of the machine.
- Heating filament and a temperature indicator on the inside of the barrel to lower the viscosity of the oil and to monitor the temperature of the machine while in operation respectively.
- A metallic sieve to filter even the smallest particles of the cake that might pass through the drilled holes.

The desired machine has the following specifications:

- Capacity of 45-50kg/hr. of extraction
- Speed range between 60 and 100 rpm
- Screw thread thickness of 3 mm, smallest pitch of the screw of 30mm, largest pitch of the screw = 70mm, length of the screw shaft is 500 mm with 300 mm threaded from one end, maximum and minimum diameter of 126mm and 26mm respectively, and a taper angle of 10°
- An extraction chamber with 130 mm internal diameter and 550 mm length
- Conical Oil drainage of 50mm to 150mm internal diameter.

The machine is to be operated by a motor of an approximate capacity of 14kW.

The 3D Computer Aided design (CAD) models were modeled and generated using Inventor software (Figures 3 and 4). Each component was modeled separately and then assembled. Components including, the bearing, bolts, nuts and washers were selected from the Autodesk Library.

Design Analysis

1. Hopper Design

The hopper is frustum-shaped and replaceable and to be fabricated from mild steel through sheet forming, welding and bolting. The design involves determining the volume of the frustum, V_f. The bottom part of the hopper takes the shape of a cylinder. We proceed as follows;

To determine the volume of the frustum, the concept of similar triangles (shown in figure 5) is used.

The height of the frustum, h is calculated as;

Figure 3. 3D CAD design of some of the major components of the oil extractor

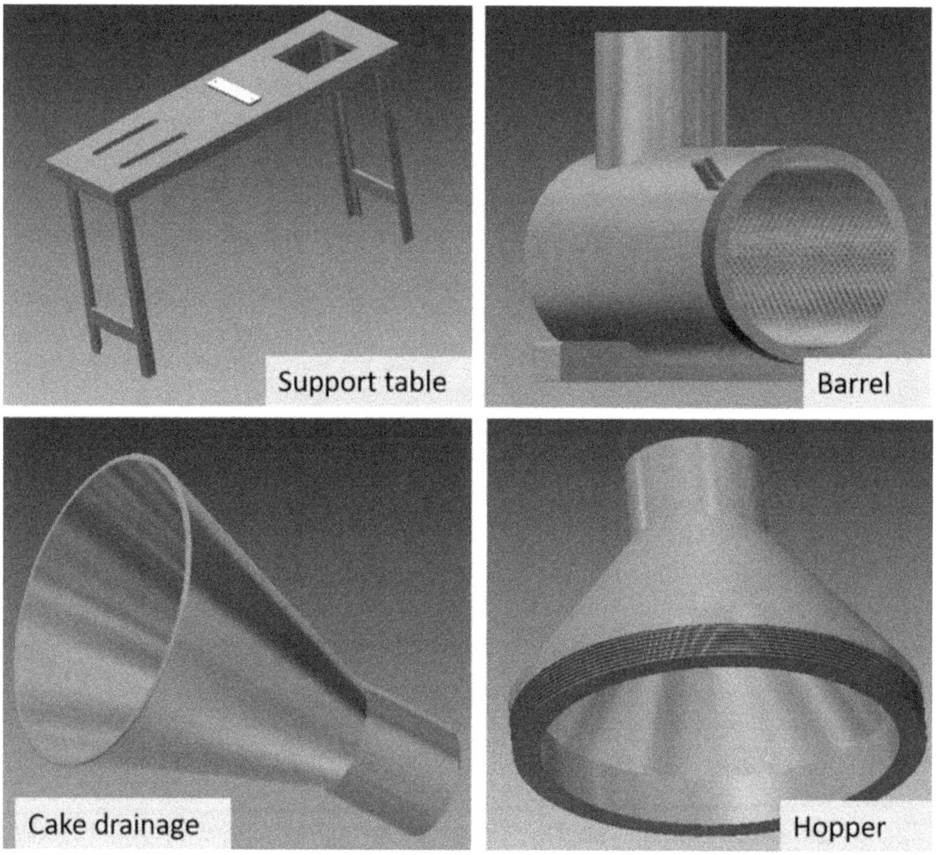

$$\frac{(250+h)}{h}=\frac{105}{35}$$

$$35(250 + h) = 105h$$

$$8750 + 35h = 105h$$

$$h = 125mm$$

Volume of the frustum of the cone is computed as;

$$V_f = \frac{1}{3}\pi(\mathrm{R}^2\mathrm{H}-\mathrm{r}^2\mathrm{h})$$

Figure 4. 3D CAD design of oil drainer, auger screw and the assembly of the machine

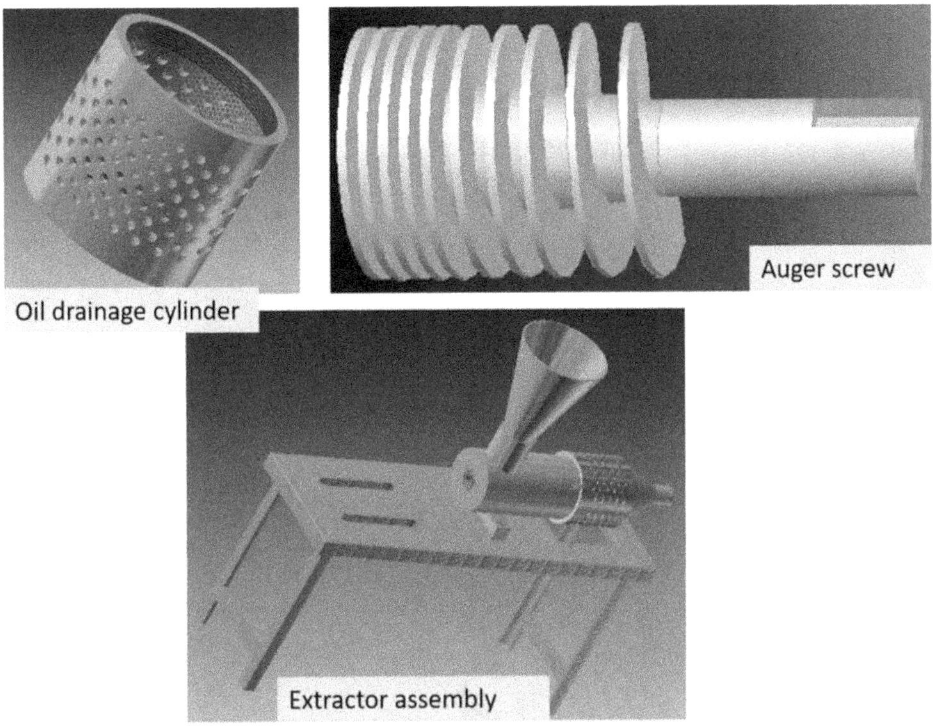

where;

R = Radius of the larger cone = 105 mm
r = Radius of the smaller cone = 35 mm
H = Height of the larger cone = 375 mm
h = height of the smaller cone = 125 mm

Therefore; $V_f = 0.004169\ m^3$
The volume of the cylindrical (V_c) part of the hopper is $3.838\times 10^{-3}\ m^3$
$\therefore$ Total volume of the hopper is $V_f + V_c = 0.008017\ m^3$

2. Screw Design

In designing screws for compression or extraction purposes such as the present case, compression ratio (R_c) is the most important parameter. The R_c is a function of cycle diameter of the screw (D_{cyl}), maximum (D_{max}) and minimum (D_{min}) diameters and it is computed as follows:

Figure 5. Illustrating some of the basic mathematical and engineering concepts applied in the analysis of the auger screw

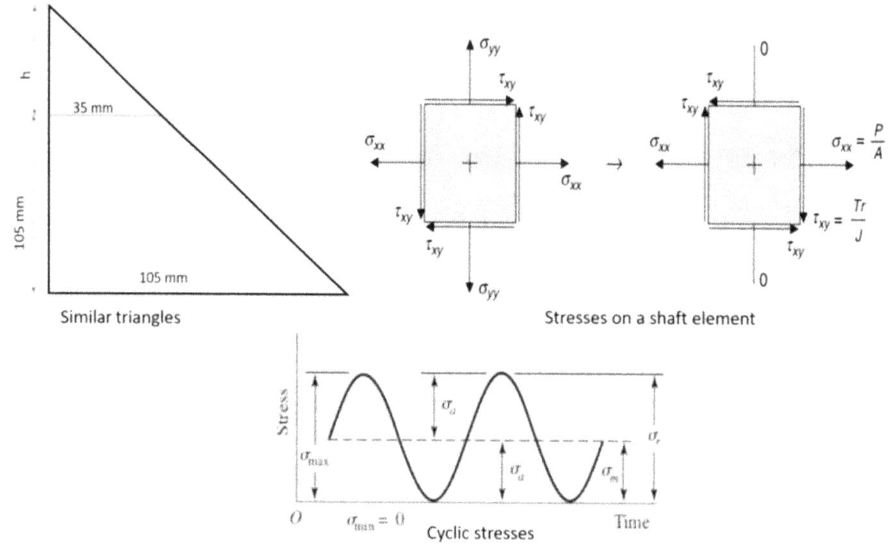

Compression ratio, $R_c = \dfrac{\left(D^2_{cycl} - D^2_{min}\right)}{\left(D^2_{cycl} - D^2_{max}\right)}$

Forces Acting on the Screw Shaft

There are two forces acting on the shaft: axial and tangential forces. The total axial force F_{axial} acting on the screw threads is given as;

$$F_{axial} = F\frac{(\cos\mu\lambda - \mu\sin\lambda)}{(\sin\mu\lambda + \mu\cos\lambda)}$$

where;

F is the force required to crush and push the seeds

λ, is the lead angle given by, $\dfrac{l}{\pi D_{\max}}$ with D_{max} as the maximum diameter and l is the length of screw.

μ is the friction coefficient (μ for steel is approximately =0.3 for SAE 1045 steel).

Given that the Maximum diameter of the shaft = 130 mm and the length = 300 mm

$$\text{Lead angle, } \lambda = \frac{l}{\pi D_{\max}} = \frac{300}{\pi \times 130} = 0.7346 rad = 42^{o}$$

Using the obtained value of lead angle and the coefficient of friction, axial force is:

$$F_{axial} = F \frac{\{\cos(0.3 \times 42^{0}) - (0.3 \sin 42^{0})\}}{\{\sin(0.3 \times 42^{0}) + (0.3 \cos 42^{0})\}} = 1.7574F \,.$$

In this case, it is assumed that F = 50kN. Hence,

$$F_{axial} = 1.7574F = 1.7544 \times 50000 = 87871.34N$$

Then to calculate the tangential force on the screw shaft, the following equation was used.

$$F_{tangential} = F_{axial} \frac{\{\mu \cos \lambda - \sin \lambda\}}{\{\cos \lambda + \mu \cos \lambda\}}$$

Thus, the tangential force acting on the screw shaft designed is given as:

$$F_{tangential} = F_{axial} \frac{\{\mu \cos \lambda - \sin \lambda\}}{\{\cos \lambda + \mu \cos \lambda\}} = 87871.34 \times \frac{\{(0.3 \cos 42) - \sin 42\}}{\{\cos 42 + (0.3 \cos 42)\}} = 40583.3N$$

The torque, T_s required to turn the screw shaft against the axial load is given by:

$$T_s = F_{axial} r_m \frac{(\tan \alpha + \mu / \cos \theta n)}{(1 - \frac{\mu \tan \alpha}{\cos \theta n})}$$

where;

F_{axial} = axial load resulting from crushing of seeds
r_m = mean thread radius

α = screw thread lead angle
μ = coefficient of friction; 0.3 (SAE G10200 steel)
θ_n = angle between tangent to tooth profile and a radial line measured in a plane parallel to thread helix (taper angle of the shaft 8^0)
θ =conical ejector lead angle (45°)

$$\tan\alpha = \frac{lead\ angle}{2\pi r_m}$$

where;

r_m is mean radius of conical screw shaft since it is very much in contact with the seed and does the actual pressing. Dimensions of the auger screw designed are as follows:
$D_{cyl'}$ = inner diameter of the circular barrel = 130mm.
D_{max} = maximum diameter of the shaft = 126mm.
D_{min} = minimum diameter of the shaft = 26mm.
Smallest pitch of the screw = 30mm
Largest pitch of the screw = 70mm
θ_n = taper angle of the shaft = 8°

From these dimensions, the mean radius and pitch of the screw were determined. The torque T_s, required to turn the power screw shaft is then calculated using equation 1.

$$T_s = F_{axial} \times r_m \left(\frac{(\tan\alpha + \mu / \cos\theta n)}{(1 - \frac{\mu \tan\alpha}{\cos\theta n})} \right)$$

$$T_S = 87871.34 \times 0.038 \left(\frac{(0.2094 + 0.3 / \cos 8^o)}{(1 - \frac{0.3 \times 0.2094}{\cos 8^o})} \right) = 1826.7 Nm$$

The power required to drive power screw shaft is determined by taking the average speed of the motor as N = 72 rpm. The angular velocity of the motor is given as;

$$\omega_n = \frac{2\pi N}{60} = 7.540 rad / \sec$$

The power required to run the motor is given by $P=T_s*\omega_n$. From which, the power, P, was determined to 14 kW.

Auger Screw Failure Analysis

The screw shaft is subjected to two types of forces:

i. Tangential force required to rotate the screw due to the motor, $F_T = 40583.3$ N
ii. Axial force, the force required to drive the cake out of the housing, $F_{axial}=$ 87871.34N

The tangential and axial forces acting on the auger cause torsional and axial stresses respectively. Therefore, the analysis will be based on the combined loads and considering the stresses on a shaft (figure 5). Normal stress, σ_{xx}, and the axial stress, τ_{xy} is the shear stress due to torsion.

The following **ASSUMPTIONS** are applied to the analysis.

1. 1. The shaft is of constant cross-sectional area whose radius is given by the average of maximum and minimum radii:

$$r_{mean} = \frac{r_{\max} + r_{\min}}{2} = \frac{126+26}{2} = 38mm$$

2. Only the maximum stresses acting on the shaft are considered. So, for this loading, elements on the surface of the shaft at radius (R) are of the greatest interest.

Determination of the maximum stresses due to a combination of axial and torsional loads:

Given that the axial force $F_{axial} = 87871.34$ N, Torque = 1826.7 N mean radius r_{mean} = 0.038 mm

The Cross-Sectional Area, A of the conveyor = $\pi r_{mean}^2 = 4.536 \text{ X}10^{-3}$ m^2

The axial Stress;

$$\sigma = \frac{F_{axial}}{A} = \frac{87871.34}{4.536\times10^{-3}} = 19.372MPa$$

The maximum shear stress is given by the torsion formula as;

$$\frac{\tau}{r} = \frac{T}{J} = \frac{G\theta}{L}$$

where; T= torque; J = polar moment of inertia; r = radius; τ= shear stress

For the solid screw auger, the polar moment of inertia J is given by;

$$J = \frac{1}{2}\pi r^4 = \frac{1}{2}\pi\times0.038^4 = 3.275 \text{ m}^4.$$

Substituting for J, r and torque, T in the equation,

$$\tau = \frac{TR}{J} = \frac{1826.7\times0.038}{3.275\times10^{-6}} = 21.2MPa$$

Calculation of the principal stress:

The maximum normal stress, i.e. principal stress (σ_1) is given by:

$$\sigma_1 = \frac{\sigma_{xx}+\sigma_{yy}}{2} + \sqrt{\frac{(\sigma_{xx}-\sigma_{yy})^2}{2} + \tau_{xy}^{\ 2}} \ ;$$

But $\sigma_{yy} = 0$; Thus;

$$\sigma_1 = \frac{\sigma_{xx}}{2} + \sqrt{\left(\frac{\sigma_{xx}}{2}\right)^2 + \tau_{xy}^{\ 2}} = \frac{19.37\times10^6}{2} + \sqrt{\left(\frac{19.37\times10^6}{2}\right)^2 + \left(21.20\times10^6\right)^2}$$

= 33.00 MPa

The principal stress σ_1 acts at an angle ϕ, given by the equation:

$$\tan 2\phi_p = \frac{2\tau_{xy}}{\sigma_{xx}-\sigma_{yy}} = \frac{2\tau_{xy}}{\sigma_{xx}} = \frac{2\times19.37\times10^6}{21.20\times10^6}$$

$\phi_p = 30.66°$

The failure of the conveyor is **mainly due to fatigue**. The shaft is subjected to **repeated stresses** since the stresses vary from zero to maximum value as shown for cyclic stresses in figure 5.

From figure 5, σ_m= mean stress and σ_a= alternating stress.

Fatigue Failure Analysis

From the standard tables, for the G10200 cold drawn steel, the deterministic ASTM minimum tensile and yield strength are:

Tensile strength = 470MPa
Yield strength = 390 MPa

Since it is a rotating element endurance limit, is denoted by $S_e^{'}$ and can be determined by:

For steel; $S_e^{'} = 0.5 S_{ut}$

$$S_e^{'} = 0.5 \times 470 = 235 MPa$$

Considering the endurance limit modifying factors:

The safety factor is given by;

$$\eta_f = \frac{S_a}{\sigma'_a} = \frac{158.696 \times 10^6}{41.5 \times 10^6} = 3.81$$

Using Langer criterion;

$$\eta_f = \frac{S_{yield}}{\sigma'_a} = \frac{370 \times 10^6}{41.5 \times 10^6} = 8.92$$

Thus, failure occurs due to fatigue according to Langer criterion.

Project 2: Design of A 200 Kg Animal Feed Mixer

Overview

The goal of this project is to design an animal feed mixer, capable of handling 200 Kg of feed constituents. Most of the local farmers, especially in rural and Rift Valley regions of Kenya where livestock farming is extensively practiced, utilize manual methods to mix the animal feed constituents. This machine will be designed with the objective of eliminating the reliance of manual operation. These manual methods are not effective and cumbersome especially where the farmer has more than ten animals or heavy feeding animals. The existing machines are also complex and expensive for small scale and domestic applications.

Problem Statement

Majority of the Kenyan population live in the rural areas and practice agriculture as the main source of livelihood. In rural highlands of Kenya such as Rift valley and central regions, controlled livestock agriculture is common due to lack of enough land spaces. Most of these farmers, small-scale, do not have access to affordable feed preparation machines; most of the machines are imported and expensive. Animal feeds are prepared by mixing different constituents to achieve a specific nutrient combination such as dairy meal, etc. The mixture should be uniform to ensure proper distribution of various nutrients in the feed. As such, a need for feed mixers to these farmers. Mostly, the mixing is undertaken through crude means, which are not effective in achieving a homogeneous mix of the feed. Additionally, such methods are cumbersome and time-consuming to the farmers.

The goal of this project is to design an animal feed mixer, capable of mixing animal feed constituents of up to 200 Kg. Various components such as the power/ drive mechanism, constituent delivery system (hopper), mixing shaft and housing were designed and specified.

Outputs

The working drawings and CAD assembly presented for final project of this project are shown below. The design involved determining the volumes of the hopper and mixing chamber, motor specifications, mixing shaft and supporting framework of the machine. Basic principles of mathematics, strength of materials and mechanics of machines were used just as illustrated in the previous project. As shown in the 3D CAD model, the cylindrical mixing chamber of 680 mm is directly anchored onto the frame through the centrally based horizontal mixing shaft on bearings.

Figure 6. Working drawings for the 200 Kg feed mixer as presented by the design team

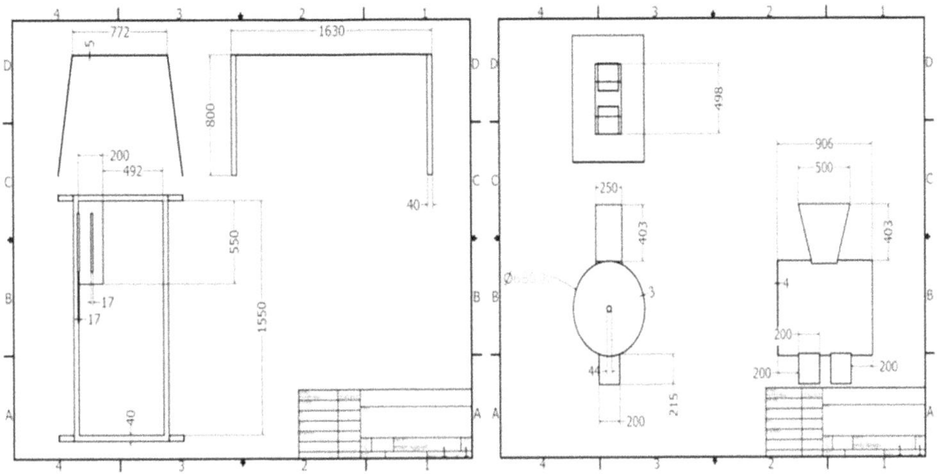

The mixing shaft has an auger screw configuration to facilitate mixing and flow of the constituents. The hopper has a volume of 312,500 cm^3 with major and minor widths of 500 mm and 250 mm respectively and a height of 400 mm. The machine consists of a wedge-shaped frame with general dimensions as 800 mm height, 1500 mm length and 680 mm width. The machine operates on a 3 kW and 700 rpm motor. The mixing is achieved through rotation of the mixing shaft.

Project 3: Design and Fabrication of a Pedal-Powered Washing Machine

Overview

Kenya does not manufacture washing machines and therefore the population depends on importation. This has rendered the machines expensive to acquire for low- and middle-class urban population with the cost of a 6 Kg washing machine going up to $1,000. As such, handwashing is the most common practice. The practice is tiring and inconveniencing especially for the working class in the urban and rural areas. The process is slow and some of the detergents used may affect the health of the person undertaking the handwashing. Additionally, the modern machines are electrically operated and cannot be used by those not connected to the national grid. In this project, an alternative manually operated machine is presented to serve those who cannot afford the modern machines and those living in rural areas without electricity.

Figure 7. 3D CAD assembly for the 200 Kg animal feed mixer

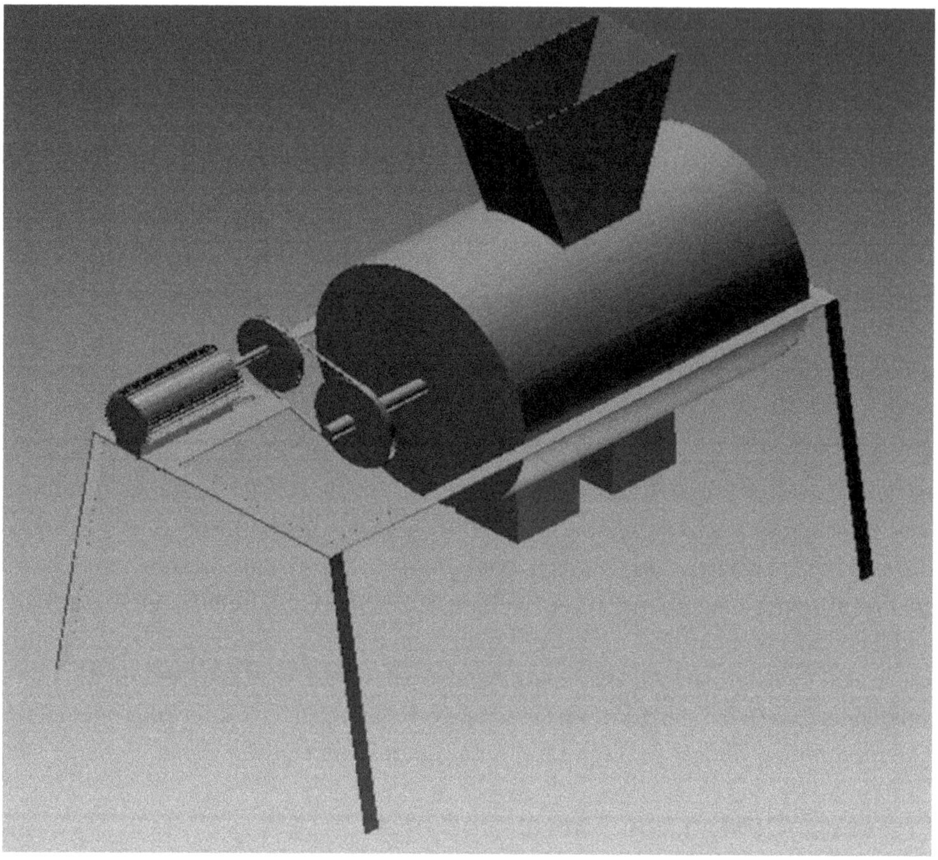

Problem Statement

In Kenya, most people wash their clothes manually. Human beings spent about an hour to hand-wash 20 Kg of laundry and detergents are mostly used to remove the dirt. This implies longer exposure to these detergents (especially for larger households); most of the local and market detergents have many chemicals such as enzymatic substances, which may be affect the human hands. As such, a growing need for washing machines in Kenya. Additionally, with the rising cost of electricity, electrically powered machines may be extra financial burden to most low- and middle-income earners. As such, there is a need to develop manually operated washing machines, which do not require electricity. Moreover, the existing electrically powered washing machines are considerably expensive to most, most owning a washing machine is seen as a luxury among the middle-income earners. The proposed one intends to cut down on this cost by approximately 35 percent since it does not have expensive

Figure 8. A domestic worker undertaking handwashing of clothes

parts such as electric motors, temperature sensors and the suspension systems and it is manufactured from locally available resources.

Therefore, the aim of this project is to design an affordable pedal-powered washing machine that can be used in both urban and remote areas in Kenya. The machine proposed here is easy to construct, maintain using locally available materials, it is also easy and comfortable to operate by both women and men compared.

This project work seeks to accomplish the following specific objectives:

- To design, fabricate and test a pedal-operated washing machine
- To increase efficiency in laundry by eliminating the physical stress of hand washing clothes and saving the amount of water and time wasted.
- To develop the most comfortable and effective pedaling mechanisms to operate the machine.
- To develop a low weight washing machine, that is easily movable within the washing chambers.

Outputs

The design of the machine was based on the following:

- A horizontal axis type of washing machine was preferred due to its compactness and flexibility.
- The intended capacity of the machine was 10 Kg, typically feasible for a medium-size family in Kenyan urban-rural setting.
- The framework was to be fabricated from angle lines (mild steel) of 3 mm thickness and joined through arc welding.
- The most important aspects of design were; the shaft holding the washing tub, the washing tub and ball bearing system. The design of the shaft and washing tub was based on the Guest's theory (due torsion loading during operation).
- The drive chain was designed using the theory of pulleys.

Based on the available resources at ADMATC, the manufacturing process was undertaken as follows;

- Sheet metal working: The washing and inner tubs were rolled and then welded on MIG.
- Cold forming: The framework was shaped in metal worker machine to obtain the various configurations and joined through welding.
- Machining: The shafts were machined on lathe, shapers and milling machines. Various holes were drilled using vertical drilling machines.
- Welding and joining: The welding of the various sheets and angle lines was undertaken using MIG welding operation. Spot welding machine was also used on the sheet joining. Bolts and nuts were used to join various parts together as well.

ASSESSMENT

The illustrated projects 1, 2 and 3 were among other many projects undertaken by the students during the module within the two academic years. In all the cases, the students defined the problem statement, objectives and justification of the project. The students demonstrated a significant understanding of the various problems affecting the society. Additionally, they demonstrated the application of mechanical engineering skills in solving some of these problems. All the projects presented (including those not mentioned here) were meant to address various issues in the

Figure 9. 3D CAD model and picture of the fabricated pedal-driven washing machine

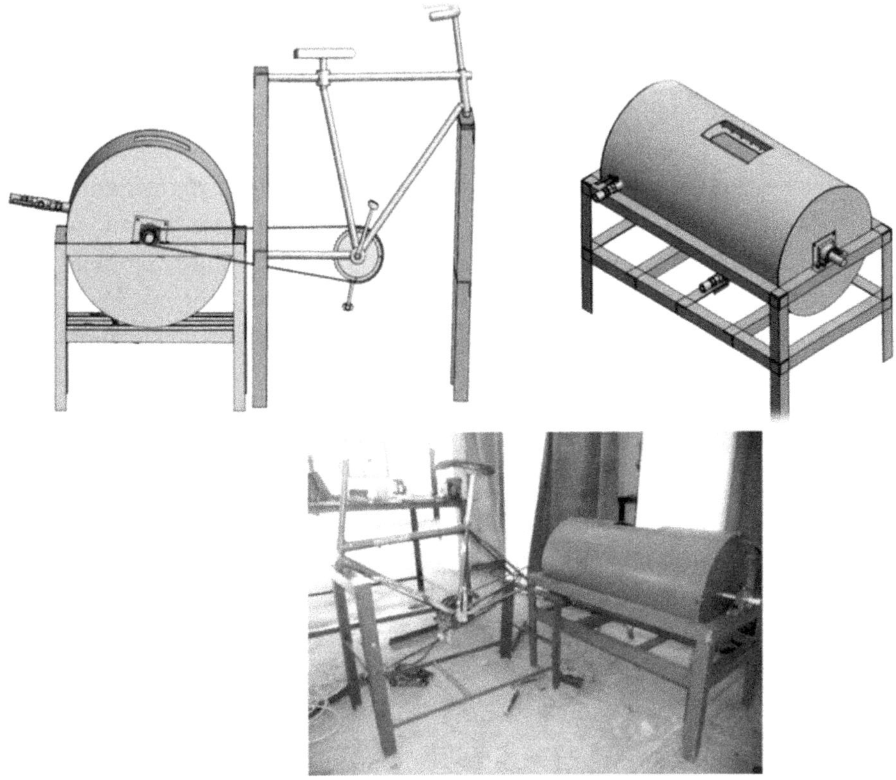

villages and cities. Projects 1 and 3 were the most successful since some of the students implemented those projects in their final (fifth) year capstone dissertation. The students in project 1 and 3 performed considerably better as compared to the rest of the class whereas those in project 2 were the poorest. The high performance was based on weekly presentations and presentation of the final report. The students who demonstrated effective use of engineering principles and CAD software got higher marks in the report. Additionally, a good link between the existing manufacturing facilities (ADMATC) and the project demonstrated a comprehensive understanding of the implementation of the projects. Although the students in project 2 demonstrated understanding of their problem, their theoretical and CAD designs were weak. The students who continued with these projects to their final year study performed better than those who worked on other projects for the two academic years.

CONCLUSION

The article has presented an approach to teach machine design and related modules in developing and poor countries to mechanical engineering students. The challenge of teaching machine design in developing countries is the lack of necessary facilities as it is the case in the modern world. As such, lecturers and instructors should advance feasible techniques, which challenge the students to utilize their engineering skills to solve practical societal problems such as poverty, etc. In this case, the students were grouped according to their various capabilities and tasked to identify problems in the society and propose a machine design project to address the problems. The students were required to study and state the problem (clearly), conceptualize the machine design project, design and build CAD models based on the available manufacturing facilities in our workshops. Each individual student made weekly presentations to the lecturer for weekly assessment and evaluation of their involvements in the overall progress of the projects. The approach was successful since all the groups identified a society problem and designed a machine project towards the problem. The projects compelled the students to revisit/learn CAD software and basic principles of engineering to design the various machines. This approach can be adopted in other schools in Kenya (and elsewhere) since most of the engineering schools grade machine design based on theory only (without experimental work as it in the case of other modules).

ACKNOWLEDGMENT

The authors acknowledge the DeKUT mechanical engineering students (2015-2016) for availing these project documents. Mr. Harrison Shagwira (a master's student in the department of mechanical engineering) is highly acknowledged for taking photos of the engineering facilities at the ADMATC, DeKUT.

REFERENCES

Andersson, K. (2012). Using Model-Based Design in Engineering Design Education. *ASME 2012 International Design Engineering Technical Conference & Computers and Information in Engineering Conference*. 10.1115/DETC2012-71081

Cohen, K., & Katz, R. (2015). Teaching Mechanical Design Practice in Academia. *Procedia CIRP*, *36*, 177–181. doi:10.1016/j.procir.2015.01.043

Heath, C. (2016). Why diversity matters: designing the right design team. *New Design Magazine, 6*(132). Retrieved from http://newdesignmagazine.co.uk/

Joyce, T., Evans, I., Pallan, W., & Hopkins, C. (2013). A Hands-on Project-based Mechanical Engineering Design Module Focusing on Sustainability. *Engineering Education*, *8*(1), 65–80. doi:10.11120/ened.2013.00008

Kanyarusoke, K. (2019). Mechanical Engineering Design: Going Over the Analysis-Synthesis Mountain to Seed Creativity. Engineering Design and Innovation Methods. doi:10.5772/intechopen.85174

Li, X., Wei, J., Ding, S., & Wang, Q. (2017). Teaching Mode of Mechanical Design Course based on Simulation Analysis Technology. *International Journal of Emerging Technologies in Learning*, *12*(7), 112–123. doi:10.3991/ijet.v12i07.7221

Monterrubio, L.E., & Sirinterlikci, A. (2015). A Hands-on Approach in Teaching Machine Design. *122nd American Society for Engineering Education.*

Ranger, B. J., & Mantzavinou, A. (2018). Design Thinking in Development Engineering Education: A Case Study on Creating Prosthetic and Assistive Technologies for the Developing World. *Development Engineering*, *3*, 166–174. doi:10.1016/j.deveng.2018.06.001

Royalty, A. (2018). Design-based Pedagogy: Investigating an emerging approach to teaching design to non-designers. *Mechanism and Machine Theory*, *125*, 137–145. doi:10.1016/j.mechmachtheory.2017.12.014

Seif, M. A. (1997). Design Case Studies: a practical approach for teaching machine design. In *Proceedings Frontiers in Education 1997 27th Annual Conference. Teaching and Learning in an Era of Change*. IEEE Explore 10.1109/FIE.1997.632754

Wood, J., Campbell, M., Wood, K., & Jensena, D. (2005). Enhancing the Teaching of Machine Design by Creating a Basic Hands-On Environment with Mechanical 'breadboards'. *International Journal of Mechanical Engineering Education, 33*(1), 1–25. doi:10.7227/IJMEE.33.1.1

Chapter 13
Implementation of Outcome-Based Education for a Course:
A Case Study

Tanya Buddi
https://orcid.org/0000-0002-9108-2704
Gokaraju Rangaraju Institute of Engineering and Technology, India

Anitha Lakshmi Akkireddy
Gokaraju Rangaraju Institute of Engineering and Technology, India

U. S. Jyothi
https://orcid.org/0000-0002-9302-855X
Gokaraju Rangaraju Institute of Engineering and Technology, India

ABSTRACT

Outcome-based education (OBE) is a learning theory based on objectives that derive outcomes for each portion of an instructional scheme. Every student has to accomplish the objectives at the end of the instructional experience. OBE is not limited to well-defined teaching or direct assessment strategies but involves indirect assessments to assist the learners in the attainment of defined outcomes. In this chapter, a case study on a course is described in all aspects of direct and indirect assessments. Initially, a correlation between programme outcomes (POs) and course outcomes (COs) is established duly analyzing the impact of CO on PO. The evaluation of COs using assessment tools are well-defined. The CO attainment percentage is evaluated using statistical methodologies, and the same is categorized to high, medium, and low-level attainments. The achieved level of attainments is correlated to Pos, and the same is adapted for all the courses in order to initiate the corrective action for further improvements in successive years.

DOI: 10.4018/978-1-7998-2245-5.ch013

Figure 1. A model for the curriculum emphasizing the importance of educational outcomes in curriculum planning

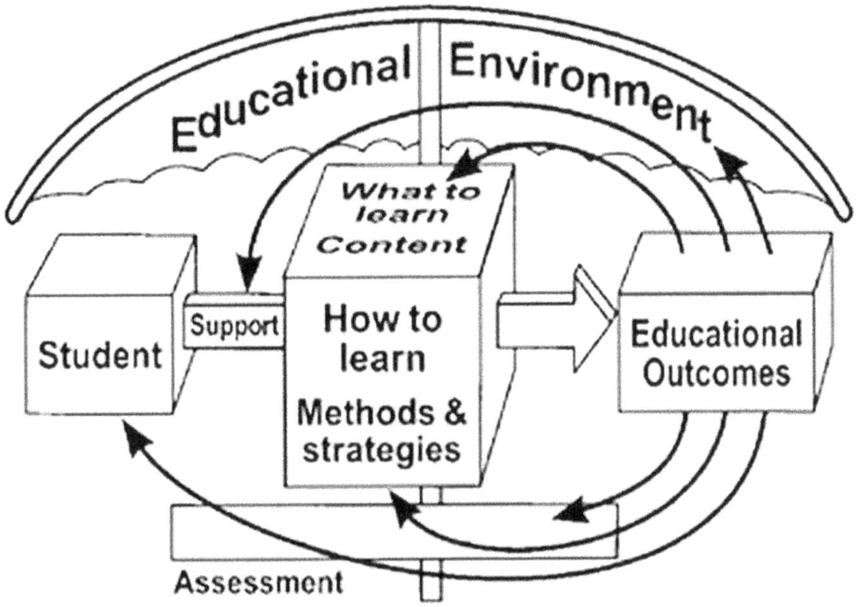

INTRODUCTION

OBE based on results has many inherent benefits that must make it an attractive model for curriculum planning with the assistance of curriculum developers, teachers, employers, students and the public. Despite the obvious appeal of OBE, research documentation and its effects are relatively low. Nevertheless, there are strong arguments for introduction of outcome-based education and assessment of its role in education. However, it represents that it is almost certainly a valuable education tool. Hopefully its adoption will foster a legitimate discussion about various kinds of educational outcomes are expected in and quantify the same (Harden, 1999). It is now accepted that learning outcomes should occupy a key position in curriculum plan and a model for the curriculum which recognizes as given in Figure 1

OBE meant to focus and organize everything in an educational system that is essential for all students to be successful at the end of learning experiences. It focuses on improvement in student learning and ultimate performance abilities to the highest possible levels at the end of graduation (Spady, 1994). In UK, the Quality

Assurance Agency for Higher Education includes Outcome-Based learning, as a component of the programme specifications to be reviewed. Institutions expect, to have a set of intended learning outcomes of the programme, teaching & learning methods that enable learners to achieve the outcomes. Further, the assessment methods are used to demonstrate the achievement (Harden, 2002). Malaysia has an experience of OBE paradigm shift, although it is not mandatory to follow the recommendations, the Malaysian Engineering Education Model (MEEM) paved the way for engineering schools to adopt OBE system. University Putra Malaysia (UPM's) experiences to develop the curriculum from first principles can be seen as an initiative and example of an institution on its way to OBE, but not without its own problems. In keeping with the ISO 9001 quality management system, the spirit of continuous improvement should be the driving force to keep institutions on the path to curriculum excellence (Aziz, 2005).

A lot of collaboration in organising and execution to implement the OBE system and requires through analysis. With this realization, UPM created an office automation system, which aids the staff of the institute to monitor the application of OBE (Jaafar, 2008). Appropriate application of OBE provides opportunities for fresh concepts and difficulties to create an approach to education that has resulted in enhanced learning outcomes. However, to effectively adopt OBE through tertiary schooling, both the educational staff and students need to know the roles goals. Education in tertiary institutions should not be a linear one-sided model, but rather an active and engaging process that is a transition for learners to prepare for the workplace (Rajaee, 2013). Institutions should focus on activity based education to encourage the learners to develop good attendance habits at the start of their studies (Taylor et al., 2017).

The essential elements for OBE to be successful is identification of learning material, student's achievement demonstrable measurables, multiple instructional and assessment strategies utilized to meet the needs of each and every student and adequate time and needed assistance to be provided so that each student can reach the maximum potential (Chandra, 2008). Tina has presented the consequences of academic performance of students related to their attendance, level of study, work shift, ethnicity, whether participants had any dependents, and how studies are funded (Tina et al., 2019).

OBE's application on self-examinable courses has strongly guided the teacher to enhance education and learning situations to generate more components, creative and qualified students with favourable values and attitudes as needed by the professors (Mansor, 2008). Sample information obtained from the Fluid Mechanics subject in the Fall of 2001 stated that the student's teaching efficiency is enhanced. The information gives us trust that the design of the evaluation instrument for the results-based engineering classes works in the beneficial direction. In order to improve the evaluation instruments, more information must be gathered. Collecting information is a long-term method. In order to improve student-learning efficiency and improve instructor-teaching performance, more information is required to statistically evaluate the instruments (Deng, 2002). It is proposed that the direction for future studies include a broader perspective on other possibilities to include the academic and the opinions and progression of the OBE system of the students. It is important to keep in mind the perspectives of system end-users against and orientation of regulators reform and strategy as witnessed in other areas of the globe (e.g. Western Australia and South Africa) that opposed OBE's concept (Akir, Oriah, Eng, & Malie, 2012).

The methodology of quantification includes a thorough articulation of the question document in which each issue is linked to a CO and its respective PO. Thus, one can assess the achievement of CO and the PO for a batch of learners by considering ongoing continuous assessment information. The amount of achievement can often be quickly varied depending on the student's type of education and performance (Ramchandra, Shivakumar, Samita Maitra, and K. MallikarjunaBabu 2014). In future research, the authors will study the causal relationship between the learning process and the learning outcome. They will also examine the effects of other features, e.g., collaboration and live broadcasting, and compare the remaining categories of the proposed Web learning taxonomy (Khalifa, Mohamed, & Lam, 2002).

PROGRAMME

Programme can be a multidisciplinary under/post graduate course which may be confined to be a specific area or integrated. It is designed to satisfy the demand for qualified technicians in different fields, combining elements of traditional engineering, medical sciences, liberal arts studies etc., Any programme is shaped in order to meet the demands of skilled works in various disciplines for solving wicked problems with design and development.

DEFINING PROGRAMME OUTCOMES

The outcomes of the programme are the statements that describes the expectations of the learners to understand and apply after graduation, which relate to the competencies, understanding and behavior acquired by learners through the programme. The POs should be framed such that they must meet the requirements of the stake holders namely Management, Industries, Parents, Students, Alumni, Employer, Professional bodies and Faculty. In the context of designing the POs for a Graduate, a set of generic Knowledge, Skills and Attitudes considered to be essential in the present

Figure 2. Graduate attributes for phrasing programme outcomes

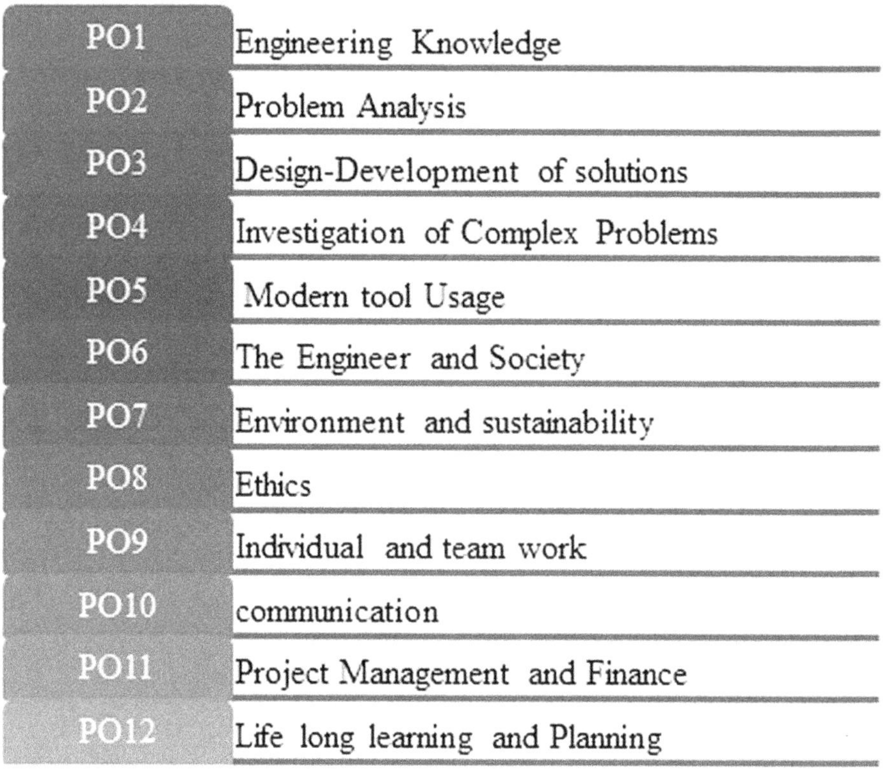

scenario. A programme is a combination of different courses relevant to specific discipline and the outcomes of programmes are quantified using COs which are evaluated based on the assessments of the course curriculum. The attributes for phrasing outcomes for a programme are shown in template Figure 2, which may be extended depending upon the specific programme requirements.

COURSE

A course is a part of programme which is defined based on the expected outcome of the student in a particular discipline to have adequate exposure for the concepts. The course curriculum, teaching and learning practice are the basis on which the programme is built. Curriculum typically relates to information and abilities that students are supposed to know, including teaching norms or teaching goals that are expected to fulfil.

Figure 3. Characteristics for framing course outcomes

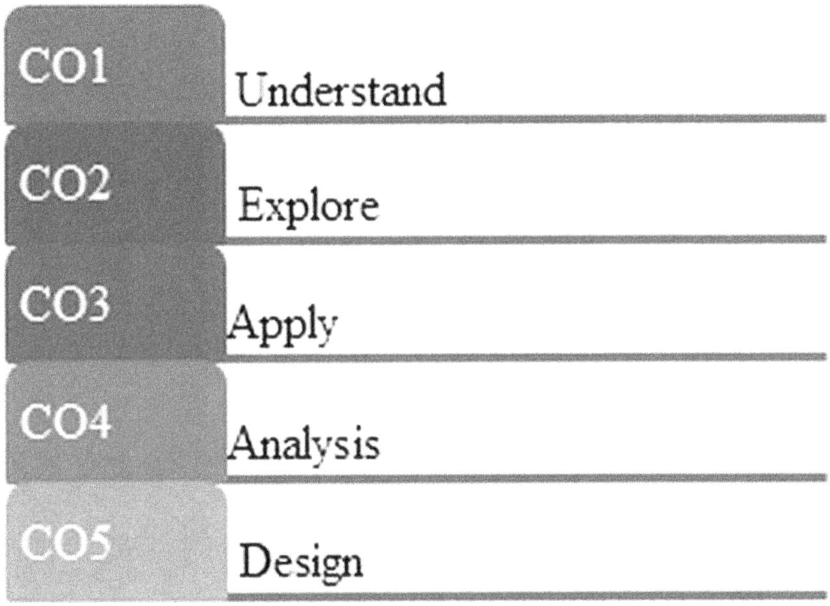

DEFINING COURSE OUTCOMES

Course Outcomes are narrow statements which describe the requirements of the learners, are able to apply the same at the end of each course. COs relates the student abilities, understanding, and behaviour while undergoing a particular course within a program. COs contribute to the results of the programme collectively and are mapped to the POs, not necessarily to a single one. The characteristics are shown in Figure. 3 which may be considered for framing the course outcomes.

MAPPING OF PO'S AND CO'S

As POs are attained through COs, from which the necessary information and evidences can be quantified to assess the level of achievement. In this regard, there is a necessity of preparing a matrix relating POs and COs as a plan for all associated courses in a programme. Every CO of a course need not meet all POs, but the entire COs must meet all the POS and extended POs. Usually a part of PO is embedded in CO, in order to assess the degree of matching. A predefined scaling is adapted by a

Figure 4. Planned mappings of COs vs POs

Planned mappings of CO vs PO													
Course Title	Course Outcomes	Program Outcomes											
		PO1	PO2	PO3	PO4	PO5	PO6	PO7	PO8	PO9	PO 10	PO 11	PO 12
Course 1	CO1	3	3	3	1	3	2	3	2	1	-	3	2
	CO2	3	1	3	3	2	-	2	1	2	3	3	3
	CO3	2	-	2	3	3	3	3	3	3	1	3	3
	CO4	2	3	2	1	3	3	1	3	2	1	-	3
	CO5	1	2	1	2	3	2	3	3	3	1	3	2

significant numbers like 3, 2, 1 and blank which means 3=Fully matched; 2=Partially Matched; 1= Poorly Matched; -- or Blank Not Matched as shown in Figure 4.

EVALUATION STRATEGIES

A detailed evaluation guideline allow the institution to assess itself and change/ modify the existing approaches for further development. The mapping of CO-PO enables to design the distribution of the course outcome ascertained by each component of the assessment and to view the relation between the results of the course and the program. The most efficient assessment in a teaching and learning society will be based on student learning outcomes, which promotes and favour the efficient teaching methods. Student learning assessments at its finest allow the learners to identify their own strengths and weaknesses indeed to correct their learning deficiencies and misunderstandings. When such assessment is employed correctly, learners can participate in self-assessment and ongoing performance improvement throughout their life.

Evaluation data is retrieved from different assessment tools from where identification of data, collection and consolidation towards the student's level of reaching course outcome as shown in Figure 5.

ASSESSMENTS

Educational assessment is the conventional process of documenting and using available data on the awareness, understanding, applicability towards the programme and to improve student learning aspect. Assessment information is acquired from the student's progress directly to evaluate the success of learning results or can be based on information from which learning can be inferred. The assessments may

Figure 5. Process involved CO assessment

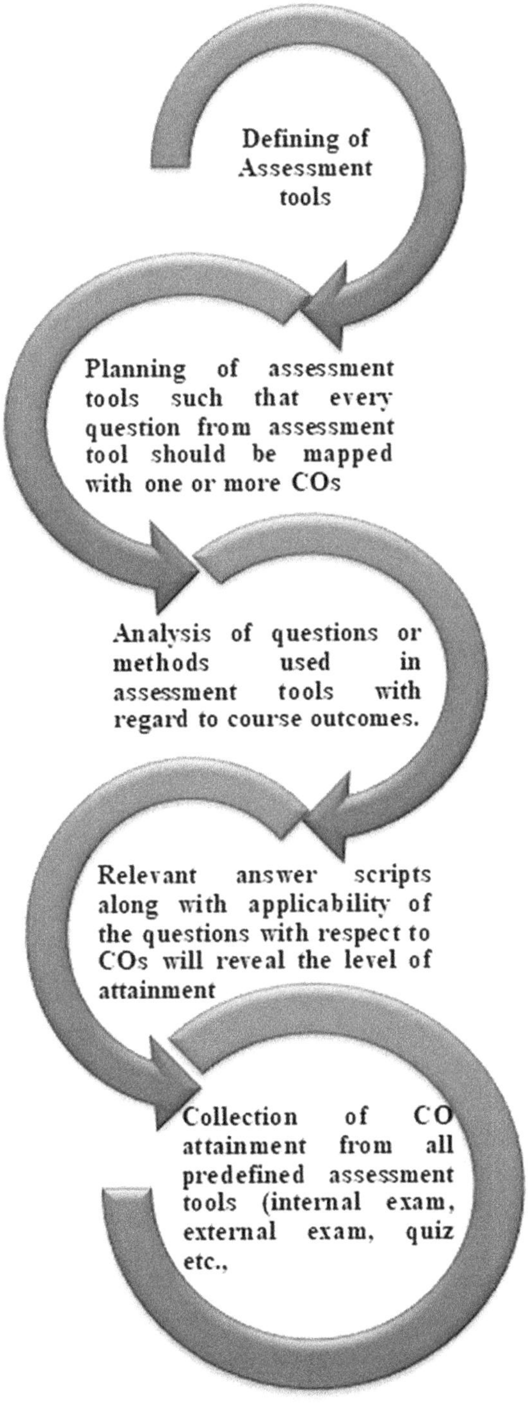

be categorized based on the nature of assessment such as Continuous, Formative and Summative. Continuous assessment is a form of educational examination that evaluates the student's progress throughout a prescribed course by means of activities. Formative assessments are in-process evaluations typically administered repeated times during a unit, course or academic programme. Summative assessments, evaluate student learning at the conclusion of a specific instructional period typically at the end of a course, semester or a programme.

As a continuous process, assessment establishes measurable and clear student learning outcomes for learning, provisioning a sufficient amount of learning opportunities to achieve these outcomes, implementing a systematic way of gathering, analysing and interpreting evidence to determine how well student learning matches expectations, and using the collected information to inform improvement in student learning.

In the present study, under Continuous assessment Flipped Classroom (FC), Think-Pair-Share (TPS) and Open Book (OB) activities, whereas in Formative assessments Internal- 1, Internal-2, Quiz-1 and Quiz-2 are adapted. External exam at the end of the course is considered as Summative assessment.

CONTINUOUS ASSESSMENTS

Continuous assessment is a constant process which shifts learners from passive learning to active learning, includes participation in team assignments, peer learning and problem-based learning.

FLIPPED CLASSROOM

Flipped learning is a pedagogical method in which the standard concept of graduate learning is inverted to introduce learners to pre-class teaching content, with classroom time then used to deepen knowledge through peer debate and teacher-friendly problem-solving activities. Students obtain understanding prior to class in the flipped classroom and use classroom time to exercise and apply concepts and thoughts through peer and teacher communication. Students reflect on the feedback they got after the class and use this to continue their teaching. Classroom time can be used to deepen learning and acquire higher-level cognitive abilities by offering learners with the material to obtain a fundamental level of knowledge and comprehension prior to school.

Some of the topics are given to the learners to attain the COs as a part of FC activity. The evaluation for all COs is done as per the template, shown if Figure 6.

Figure 6. Percentage of CO attainment with flipped classroom activity

S.No	Roll.No	CO1	CO2	CO3	CO4	CO5
		10M	10 M	10 M	10 M	10 M
1						
2						
3						
4						
5						
Total						
No of students attempted (NSA)						
Attempt % = (NSA/Total no of students)*100						
Average (attainment) = Total marks/NSA						
Attainment % = (average attainment/total marks for that question)*100						

THINK PAIR SHARE

Think-Pair-Share (TPS) is a collaborative learning approach in which learners work together in a pair or a group to fix an issue or problem to an allocated assignment. This method needs learners to think/understand the subject or solve a problem separately; and share thoughts with peers. Discussing a response with a partner maximizes involvement, focuses attention and involves learners. The TPS approach is a versatile and easy method to enhance the understanding of student's reading. It provides students time to reflect on a response and activates previous understanding. TPS enhances the oral communication skills of learners as they talk to each other about their thoughts. This approach enables learners become active learners and can also include writing as a manner to organize the activity.

The tasks that are assigned to the group attain the COs as a part of TPS activity. The evaluation for all COs is done as per the template, shown if Figure 7.

OPEN BOOK EXAM

The teacher chooses to read the text and develops the collection of issues or prompts that are aimed at important ideas of content. The teacher then defines the strategy's objective and gives discussion rules. As with all guidance on strategy, educators should model the procedure to make sure learners know how to use the approach.

Figure 7. Percentage of CO attainment with Think-Pair-Share activity

S.No	Roll.No	CO1	CO2	CO3	CO4	CO5
		10M	10 M	10 M	10 M	10 M
1						
2						
3						
4						
5						
Total						
No of students attempted (NSA)						
Attempt % = (NSA/Total no of students)*100						
Average (attainment) = Total marks/NSA						
Attainment % = (average attainment/total marks for that question)*100						

The problems that are given to the learner attain the COs as a part of OB activity. The evaluation for all COs is done as per the template, shown if Figure 8.

CONSOLIDATION OF COURSE ATTAINMENTS FROM CONTINUOUS ASSESSMENT TOOLS

All the activities under Continuous assessments are consolidated by considering the average for 10% and the entire COs are attained as per the template shown in Figure 9.

FORMATIVE ASSESSMENT

By means of two Internal Examinations the Formative assessment is evaluated for the present course study. The setting of the questions for internal exams will be such that each question should be mapped to at least one course outcome and all course outcomes should be covered in the pattern, which is shown in the template Figure 5 and Figure 7. For illustration purpose, it is presumed to allot marks of 10 for continuous assessment, 20 for formative assessment, 70 for summative assessment. In Formative Assessment 20 Marks is further distributed as 15 marks for descriptive and 5 marks for Objective. The weightage of marks is different for descriptive and objective questions as shown in template (Figure 10).

Figure 8. Percentage of CO attainment with open book exam activity

S.No	Roll.No	CO1	CO2	CO3	CO4	CO5
		10M	10 M	10 M	10 M	10 M
1						
2						
3						
4						
5						
Total						
No of students attempted (NSA)						
Attempt % = (NSA/Total no of students)*100						
Average (attainment) = Total marks/NSA						
Attainment % = (average attainment/total marks for that question)*100						

Figure 9. Consolidated percentage of CO attainment with continuous assessment tools

Continous Attainment					
Type of Assessment Tool	CO1	CO2	CO3	CO4	CO5
Flipped Classroom					
Think Pair Share					
Open Book Exam					
Assessments (if any)					
Average Attainment					

Figure 10. Template for internal-1 question paper (20 marks)

Descriptive **15 Marks**

Question No	Marks	COs
Q1. (a)	3M	(CO1)
Q1. (b)	2M	(CO2)
Q2.	5M	(CO3)
Q3.	5M	(CO4)
Q4.	5M	(CO5)

Objective **5 Marks**

Question No	Marks	COs
Q1.	1M	(CO1)
Q2.	1M	(CO2)
Q3.	1M	(CO4)
Q4.	1M	(CO3)
Q5.	1M	(CO1)
Q6.	1M	(CO3)
Q7.	1M	(CO5)
Q8.	1M	(CO4)
Q9.	1M	(CO2)
Q10.	1M	(CO5)

PARAMETERS USED FOR THE ASSESSMENT

1. Summing up all the Marks of students Obtained for each question which is related at least one CO which is treated as Total marks (Total).
2. The Question Paper is prepared either with choice or without choice for answering the questions. In case of choice opted papers (Descriptive), the

number of students attempted (NSA) for a particular question may not be equal to the number of students attended (Total attended), whereas for no choice question Paper (Objective) by default the NSA is equal to number of students attended.

3. The percentage of Attempt is based on Total and NSA which can be evaluated as

$$\%\text{Attempt} = \frac{\text{NSA}}{\text{Total attended}} \times 100 \quad (1)$$

4. Average attainment can be calculated by

$$\text{Average attained} = \frac{\text{Total}}{\text{NSA}} \times 100 \quad (2)$$

5. Attainment Percentage is calculated by

$$\text{Attainment}\% = \frac{\text{average attainment}}{\text{Maximum marks of a question}} \times 100 \quad (3)$$

Each CO attainment is consolidation of Descriptive and Objective attainment and the same may be evaluated using weighted average method

$$\%\ \text{Consolidated attainment of CO1} = \frac{(Q_1 \times M_1) + (Q_2 \times M_2) + \ldots\ldots (O_1 \times OM_1) + \ldots\ldots}{M_1 + M_2 + \ldots. + OM_1 + \ldots\ldots} \times 100 \quad (4)$$

where

Q_1= % attainment pertaining to CO1 in Q.No 1 of Descriptive.
Q_2= % attainment pertaining to CO1 in Q.No 2 of Descriptive.
M_1= Maximum Marks for Q.No 1 of Descriptive.
M_2= Maximum Marks for Q.No 2 of Descriptive.
O_1=% attainment pertaining to CO1 in Q.No 1 of Objective.
OM_1= Maximum Marks for Q.No 1 of Objective.

The same procedure is adopted for evaluating the % consolidation of remaining COs and the template in shown in Figure 11.

Figure 11. Attainment template for internal exam-1

		Questions are from first half of the syllabus														
		Descriptive Questions (Answer any Three)					Objective Questions									
S.No	Roll.No	Q1 (a)	Q1 (b)	Q2	Q3	Q4	Q1	Q2	Q3	Q4	Q5	Q6	Q7	Q8	Q9	Q10
		CO1	CO2	CO3	CO4	CO5	CO1	CO2	CO4	CO3	CO1	CO3	CO5	CO4	CO2	CO5
		3M	2 M	5M	5M	5M	0.5	0.5	0.5	0.5	0.5	0.5	0.5	0.5	0.5	0.5
1																
2																
3																
4																
5																
	Total															
	No of students attempted (NSA)															
	Attempt %=(NSA/Total no of students)*100															
	Average (attainment)= Total marks/NSA															
	Attainment % = (average attainment/total marks for that question)*100															

Similarly Formative assessment is done for Internal Exam-2 which consists of a pattern of both descriptive and objective as shown in the Figure 12. The attainment of the Internal Exam-2 is also calculated (Figure 13) as discussed above and each CO attainment is consolidated.

SUMMATIVE ASSESSMENT

External examination is adapted for Summative Assessment at the end of the course. As the entire curriculum of the course is covered the percentage of Marks allotted is more i.e. 70% as compared to Formative Assessment. The setting of the questions for External Exam will be such that each question should be mapped to at least one course outcome and all course outcomes should be covered in the pattern, which is shown in the template Figure 14 and Figure 15.

CONSOLIDATION OF COS BY ALL ASSESSMENT TOOLS

The percentage of COs attained using all assessment tools individually are consolidated by calculating average where weighted average taken into consideration as shown in Equation 5.

$$\text{Final } CO_1\% = \frac{(CO_1 \times CA)10 + (CO_1 \times FA)20 + (CO_1 \times SA)70}{100} \tag{5}$$

For the course considered as case study the percentage course outcome consolidation attainment obtained using the above-mentioned equations is shown in

Figure 12. Template of internal-2 question paper (20 marks)

Descriptive	15 Marks	
Question No	**Marks**	**COs**
Q1. (a)	3M	(CO1)
Q1. (b)	2M	(CO2)
Q2.	5M	(CO3)
Q3.	5M	(CO4)
Q4.	5M	(CO5)

Objective	5 Marks	
Question No	**Marks**	**COs**
Q1.	1M	(CO1)
Q2.	1M	(CO5)
Q3.	1M	(CO4)
Q4.	1M	(CO3)
Q5.	1M	(CO2)
Q6.	1M	(CO1)
Q7.	1M	(CO3)
Q8.	1M	(CO5)
Q9.	1M	(CO4)
Q10.	1M	(CO2)

Figure 16. The percentage attainment for the course can be analysed by constructing a pie chart (Figure 17), where CO2 and CO3 attained is less compared to other COs. This Analysis gives a clear idea about implementing the course and the strategies to be employed in upcoming semesters or years

Figure 13. Attainment template for internal exam-2

Questions are from second half of the syllabus																
		Descriptive Questions (Answer any three)					Objective Questions									
S.No	Roll.No	Q1 (a)	Q1 (b)	Q2	Q3	Q4	Q1	Q2	Q3	Q4	Q5	Q6	Q7	Q8	Q9	Q10
		CO1	CO2	CO3	CO4	CO5	CO1	CO5	CO4	CO3	CO2	CO1	CO3	CO5	CO4	CO2
		3M	2 M	5M	5M	5M	0.5	0.5	0.5	0.5	0.5	0.5	0.5	0.5	0.5	0.5
1																
2																
3																
4																
5																
	Total															
	No of students attempted (NSA)															
	Attempt %=(NSA/Total no of students)*100															
	Average (attainment)= Total marks/NSA															
	Attainment % = (average attainment/total marks for that question)*100															

ATTAINMENT OF POS

The POs planned for the Programme are defined at three levels 3, 2, 1 and blank which means 3=Fully matched; 2=Partially Matched; 1= Poorly Matched as shown in Figure 4. The Attained COs designed for 100% are deducted to significant levels 3, 2, 1 where it can describe whether the CO to PO mapping is fully, partially and poorly matched. The planned and attained level of CO to PO mapping is evaluated as shown in Figure 18. The Graph Figure 19 shows that PO4 is less than 60%, where PO7 & PO10 are less than 65%. These numbers indicate that out of all POs mentioned for programme, for the upcoming learners assessment tools need to be modified, changed or need new implementations for improving the POs.

Consolidation of PO attainments for all the courses of the programme is shown in Figure 20. From this the corrective action for any low level achieved PO with respect to course can be analysed for further improvement. The courses which plays major role for low attainment of specific PO, can be redefined or restructured for the benefit of learner in successive years.

CONCLUSION

In the Current work, an attempt is made with regard to level of PO attainment in correlation with COs duly considering one course as Case study. The method adapted for defining of POs based on the graduate attributes, further based on the formulated POs the procedure of defining COs and to establish relation between POs and Cos are explained with templates. The types of assessment strategies such as continuous, formative and summative by means of Flipped classroom, TPS, open book, internal and external examination and their establishment with respect to COs are presented

Figure 14. Template of external exam

Instructions:
Question paper comprises of Part-A and Part-B
Part-A (for 20 marks) must be answered at one place in the answer book.
Part-B (for 50 marks) consists of five questions with internal choice, answer all questions.

PART – A
(Answer ALL questions. All questions carry equal marks) 2 * 10 = 20 Marks

Question No.	Question	CO	Marks
1. a.		CO1	[2]
b.		CO1	[2]
c.		CO2	[2]
d.		CO2	[2]
e.		CO3	[2]
f.		CO3	[2]
g.		CO4	[2]
h.		CO4	[2]
i.		CO5	[2]
j.		CO5	[2]

PART – B
(Answer ALL questions. All questions carry equal marks) 10 * 5 = 50 Marks

Question No.	Question	CO	Marks
2.		CO1	[10]
	OR		
3.		CO5	[10]
4.		CO4	[10]
	OR		
5.		CO3	[10]
6.		CO2	[10]
	OR		
7.		CO1	[10]
8.		CO3	[10]
	OR		
9.		CO5	[10]
10.		CO4	[10]
	OR		
11.		CO2	[10]

Figure 15. Attainment template for external exam

S.No	Roll.No	Answer all Questions										Answer any Five Questions									
		Q1 (a)	Q1 (b)	Q1 (c)	Q1 (d)	Q1 (e)	Q1 (f)	Q1 (g)	Q1 (h)	Q1 (i)	Q1 (j)	Q2	Q3	Q4	Q5	Q6	Q7	Q8	Q9	Q10	Q11
		CO1	CO1	CO2	CO2	CO3	CO3	CO4	CO4	CO5	CO5	CO1	CO5	CO4	CO3	CO2	CO1	CO3	CO5	CO4	CO2
		2M	2M	2M	2M	2M	2M	2M	2M	2M	2M	10M	10M	10M	10M	10M	10M	10M	10M	10M	10M
1																					
2																					
3																					
4																					
5																					
	Total																				
	No of students attempted (NSA)																				
	Attempt %=(NSA/Total no of students)*100																				
	Average (attainment)= Total marks/NSA																				
	Attainment % = (average attainment/total marks for that question)*100																				

Figure 16. Consolidation of course outcomes with all assessment tools

Consolidated % of CO Attainment						
COs	Formative Assessment			Continous Assessment (FC, TPS, OB)	Summative Assessment	Total Assessment
	Internal Exam-1	Internal Exam-2	Avg of Internal Exams		End Exam	
CO1	74.32	60.23	67.3	72.3	87.30	81.80
CO2	75.82	75.82	75.8	65.2	59.45	63.30
CO3	55.07	55.07	55.1	75.6	73.45	69.99
CO4	30.04	94.42	62.23	85.25	80.39	77.25
CO5	85.3	72.54	78.92	91.23	77.15	78.91

with assumed model. The curriculum and teaching learning strategy modifications can be done based on the achieved outcomes of the course, further in order to create interest on the particular area/topic where ever low attainment is noticed, an alternative assessment tools adapted may be suggested for implementation. As by consolidation of all courses outcomes the attainment of Programme is evaluated, the remedial action with respect to low level attained POs has to be proposed to Board of evaluators with innovative approaches for improvement of the same.

FUTURE SCOPE

Research across the world has shown that ICT can lead to greater learning of students and better modes of teaching. A report from Japan's National Institute of Multimedia Education has already shown that enhancing the use of ICT in engineering with the

Figure 17. Analysis on attainment of COs using pie chart

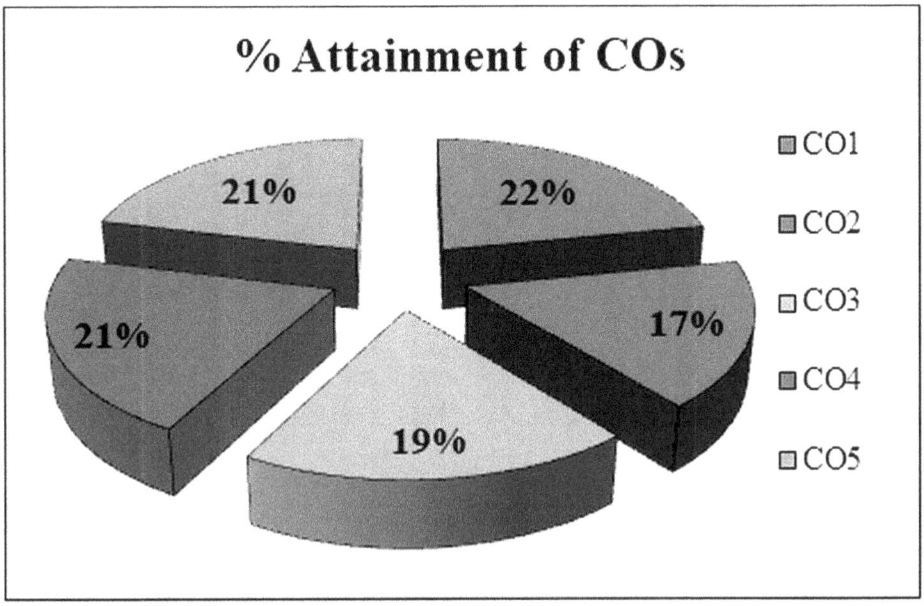

Figure 18. Attainment of POs in correlation with COs

Attained mappings of CO vs PO										
Course outcomes	CO attained %	PO1		PO2		PO3 to PO10	PO11		PO12	
		PL	AL	PL	AL		PL	AL	PL	AL
CO1	81.80	3	2.5	3	2.45		3	2.45	2	1.64
CO2	63.30	3	1.9	1	0.63		3	1.90	3	1.90
CO3	69.99	2	1.4	-	0.00		3	2.10	3	2.10
CO4	77.25	2	1.5	3	2.32		-	0.00	3	2.32
CO5	78.91	1	0.79	2	1.58		3	2.37	2	1.58
Total Attainment		11	8.09	9	6.98		12	8.82	13	9.53
PO attained %			73.52		77.58			73.50		73.31
		PL	Planned Level				AL	Attained Level		

inclusion of technology in to the curriculum has a considerable and significant effect on the performance of learners. Specifically, the findings showed that learners who are continually subjected to innovation through schooling have stronger understanding, lecture abilities, creative abilities, and are willing to take more effort into learning than their peers. The ICT tools can help a student in reaching the course outcomes very easily without stress within the student. The Access, availability and quality of

Figure 19. Analysis on attainment of POs using bar chart

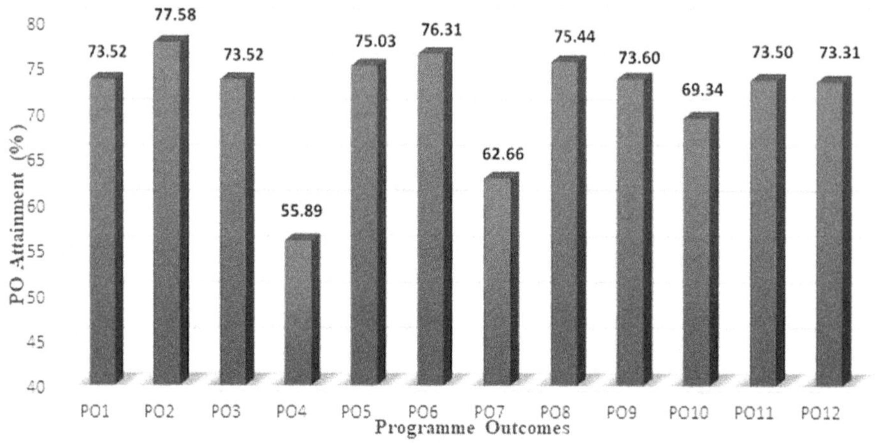

Figure 20. PO attainment for the programme

PO Attainment of Programme													
S.No	Course Name	PO1	PO2	PO3	PO4	PO5	PO6	PO7	PO8	PO9	PO10	PO11	PO12
1	Course 1	73.52	77.58	73.52	55.89	75.03	76.31	62.66	75.44	73.60	69.34	73.50	73.31
2	Course 2												
3	Course 3												
N	Course N												
Total													
Avg													

information is available to the learners through various ICT modes like MOOCS, NPTEL and SWAYAM through which the courses are taught over the web. Involving the students in the course through conducting quizzes, giving online assignments where duplicating can be avoided, creating chat box, discussion forum on a specific topic where students can involve in discussions and thought process can be improved. Taking feedbacks from learners for complex topics to see that everybody is acquiring the information. By implementing innovative methods, the learner's interest in the particular subject can be enhanced where the percentage attainment of course outcomes can be achieved without stressing the student.

REFERENCES

Akir, O., Eng, T. H., & Malie, S. (2012). Teaching and learning enhancement through outcome-based education structure and technology e-learning support. *Procedia: Social and Behavioral Sciences*, *62*, 87–92. doi:10.1016/j.sbspro.2012.09.015

Aziz, A. A. (2005). A Malaysian outcome-based engineering education model. *IACSIT International Journal of Engineering and Technology*, *2*(1), 14–21.

Deng, Z. T. (2002). *Developing assessment tools for outcome based engineering courses*. Academic Press.

Harden, R. M. (1999). AMEE Guide No. 14: Outcome-based education: Part 1-An introduction to outcome-based education. *Medical Teacher*, *21*(1), 7–14. doi:10.1080/01421599979969

Harden, R. M. (2002). Developments in outcome-based education. *Medical Teacher*, *24*(2), 117–120. doi:10.1080/01421590220120669 PMID:12098427

Jaafar, M. S. (2008). Assessment strategy for an outcome-based education. *International Conference on Engineering Education.*

Khalifa, M., & Lam, R. (2002). Web-based learning: Effects on learning process and outcome. *IEEE Transactions on Education*, *45*(4), 350–356. doi:10.1109/TE.2002.804395

Mansor, W. (2008). Preliminary results on the implementation of outcome-based education on the non-examinable computer engineering modules. In *2008 38th Annual Frontiers in Education Conference*. IEEE. 10.1109/FIE.2008.4720675

Rajaee, N. (2013). Issues and challenges in implementing outcome-based education in engineering education. *International Journal of Innovation Education and Research*, *1*(4), 1–9.

Ramchandra, Maitra, & MallikarjunaBabu. (2014). Method for estimation of attainment of program outcome through course outcome for outcome based education. In *2014 IEEE International Conference on MOOC, Innovation and Technology in Education (MITE)*. IEEE.

Sekhar, C. R. (2008). *Continuous improvement process based on outcome-based education*. Academic Press.

Spady, W. G. (1994). *Outcome-Based Education: Critical Issues and Answers*. American Association of School Administrators.

Tani, K., Dalzell, E., Ehambaranathan, N., Murugasu, S., & Steele, A. (2019). Evaluation of external factors affecting students' academic performance in higher education in New Zealand. *Journal of Pedagogical Research*, *3*(2), 1–10. doi:10.33902/JPR.2019252504

Taylor, L., Brogt, E., Cheer, U., Baird, N., Caldwell, J., & Wilson, D. (2017). Ethnicity and engagement in first-year New Zealand law programmes. *Journal Higher Education Research & Development*, *36*(5), 1047–1060. doi:10.1080/07294360.2016.1263935

Section 4
Ethics in Education

Chapter 14
Ethics in Engineering Profession:
Pedagogy and Practices

Jandhyala N. Murthy
Gokaraju Rangaraju Institute of Engineering and Technology, India

Lavanya C.
Gokaraju Rangaraju Institute of Engineering and Technology, India

Satyanarayana Kosaraju
https://orcid.org/0000-0001-5271-5743
Gokaraju Rangaraju Institute of Engineering and Technology, India

ABSTRACT

Engineering as a profession distinguishes itself by having and enforcing a code of ethics. Aberrations can lead to commercial considerations to dominate converting the profession into a business rather than promoting safety, health, and welfare of public and environment. Engineers, in their professional quest for an optimal solution, are forced into a dilemma due to clash of values or interests. The explosion of data and its usage is bringing in a lot of concern due to proliferation of unethical practices. Moral values and personal ethics at one end and professional and social ethics at the other end of the spectrum are of points of discussion in academics as well as in society. Engineering programmes strive to offer engineering ethics and professionalism either through direct courses or through embedded capsules in appropriate courses. Promotion of ethics integrating into the engineering profession at all levels could lead to a holistic alternative at universal level, which is self-satisfying, people-friendly, and eco-friendly.

DOI: 10.4018/978-1-7998-2245-5.ch014

INTRODUCTION

Wealth without Work,
Pleasure without Conscience,
Knowledge without Character,
Commerce (Business) without Morality (Ethics),
Science without Humanity,
Religion without Sacrifice,
Politics without Principles
- The Seven Social Sins – Mahatma Gandhi

Ms Senior has entered into her final semester of BTech Computer Science Engineering. She is required complete her final year major project by March of that year. She had competitive examinations like GATE in the month of February and booked a slot for GRE and TOEFEL in the month of March. As preparation for competitive examinations are in the way of project submission, She was tempted to purchase a project in the open market and submit it as her own. Is she right in her actions and what measures to be taken to curtail such ethical aberrations called plagiarism?

Mr Starter, soon after graduation, joined a start up company manufacturing airworthy washers. He was asked by his supervisor to file a tender before the closing hours of the day as it was the last date. He managed to reach the office just half an hour before closing time. For accepting tender, the official wanted his hand to be greased, which is against the personal ethics of Mr Starter. What should he do leave values or lose tender participation chance?

A young engineer, MrShuru, started his career working in Heat Treatment Shop in a car manufacturing company. One day his supervisor asked him to dump some life expired toxic chemical into the drainage. MrShuru feltit was against public interest as there arechances of getting into storm water drain contaminating the ground water as well as near by lake. He expressed the same to his supervisor, who bullied him stating that" you better do what I tell you. I have been here for a number of years, I know what is right. The amount of chemical is very little, it will soon get diluted in storm water before it reached lake. Even if few fish die, so what? MrShuru hesitated, but the boss continued "you better do what I tell you or look for another job?". What should MrShuru do?

Cases mentioned above are some ethical aberrations or temptationsfaced by engineers duringtheir academic pursuit and professional practice. This type of ethical cases can go far beyond issues of plagiarism, bribery and public safety and may involve neglect of duty, fraud, environmental protection, honesty in research and testing and conflicts of interest etc. During their undergraduate education, engineers receive training in basic and engineering sciences, problem - solving methodology, engineering design, etc. but generally receive little training in business practices, safety and ethics. Several notorious cases that have led to an awareness of the importance of ethics within the engineering profession as engineers realize how their technical work has far-reaching impacts on society, affecting public health and safety and influencing business practices and even politics. (Goswami, D., &Chakraborty, A. 2015).

Mr. Abhay, is an assistant professor with ten years of teaching experience in Department of Mechanical Engineering in a reputed autonomous Engineering College. He is ambitious and eagerly waiting for his promotion. But with change in policies, publications have increasingly become a measurable index for promotions. His experimental set up is hibernating due to fund crunch. He is aware that "no data, therefore no paper". A colleague advised him to fabricate data! MrAbhay is afraid of research misconduct. What should he do!

Fake science undermines trust in science and the capacity of individuals and society to make evidence-informed choices (Hopf, H., Krief, A., Mehta, G., &Matlin, S. A. 2019).As the pressure to publish increases, so has plagiarism. (Smith, R. 2006). Facing these challenges, it is especially important that the scientific world as a whole upholds the highest standards of ethical behaviour, honesty and transparency, aiming to sustain the gold standards of research integrity and validated information. Academic institutions have a key role to play.in promoting this behavioragainst the odds of personal ambitions and prevailing pressures.

The Sardar Sarovar Dam is a gravity dam on river Narmadanear Navagam, Gujarat in India. Four Indian states, Gujarat, Madhya Pradesh, Maharashtra and Rajasthan, receive water and electricity supplied from the dam. The foundation stone of the project was laid out by Prime Minister Jawaharlal Nehru on 5 April 1961. The project took form in 1979 as part of a development scheme funded by the World Bank through their International Bank for Reconstruction and Development, to increase irrigation and produce hydroelectricity, using a loan of US$ 200 million. The construction for dam began in 1987 but following a number of controversial cases before the Supreme Court of India (1999, 2000, 2003), the dam was finally inaugurated on 17 September 2017 by Prime Minister Narendra Modi. Considering the size of the project, it had great social and ecological impact, involving relocation *and rehabilitation of population including sizeable indigenous people, raising ethical issues. How does one arrive at a middle path solution leading to a win-win situation?*

Big Dams are to a Nation's "Development" what Nuclear Bombs are to its Military Arsenal. They are both weapons of mass destruction (Arundhati Roy 1999). A critical statement for engineers to dwell over to understand moral theories. Similar debating points arise when technology and its applications are discussed. They are like double edged swords. They have positive impact at the same time if misused or improperly harnessed can lead to the very destruction of society, like explosives, nuclear energy. Innovations are disruptive, but market-creating innovations can have tremendous impact on prosperity (Christensen, C. M. et al, 2019).An engineer may face such conflicting situations taking critical decisions at a fairly senior stage of ones life, but as a student need to be made aware of the ethical theories and solving techniques through appropriate pedagogical methods

Earth is unique as it is the only known planet housing life and humans arguably the most intelligent life on this earth. With his limitless intelligence and ideas, he has succeeded in harnessing the planet's resources. The tremendous progress he made in science and technology in turn yielded enhanced quality of life and development in all the domains. The development and progress is not uniform across the globe due to many reasons such as non-uniform availability of natural resources, unfavorable environmental conditions and some demographical such as cultural, racial or religious differences. In addition to growing inequalities, accidents and disasters of different magnitudes with avoidable losses to properties and lives stare at us pointing towards negligence, dereliction of duties, corruption and the very attitude of humans. Despite established codes and value systems, one often finds himself in a dilemma while taking decisions, trying not to traverse the *Lakshman Rekha*, which signifies the ethical limits of action, crossing may lead to undesirable consequences. Proper education is the only panacea and value system needs to be revisited.

The overall purpose of education is to enable a human being to live a fulfilling life with oneself and with family, society and nature (Gaur et al., 2011). No doubt focus on Science, Technology, Engineering and Mathematics (STEM) studies is increased all over the world considering their relevance to prosperity of any society. However, prosperity doesnot mean only becoming rich, morals are equally important. As quoted by Swami Vivekananda education should be for "Man Making and Character Building" (Barman & Bhattacharyya, 2012). Education aims at transforming an individual in terms of his attitudes, skills and knowledge (ASK).

Value system inputs form the basis for attitudes. Right from birth, individual get inputs on values and morals from home, neighborhood then schools, higher education institutes and from the society at large. At every stage, the value system, its awareness and courage and conviction to implement needed to be reinforced.

Moral values and personal ethics are at one end and professional and social ethics at the other end of the spectrum are of points of discussion in higher education as well as in society (Bairaktarova & Woodcock, 2015).What to teach, how to teach and who should teach ethics in engineering programmes shall remain the focus point in subsequent sections. Before appropriate pedagogy is discussed let us first review the finer points of engineering profession and the ethical environment surrounding it.

ENGINEER IN SOCIETY

Engineering is a profession in which knowledge of the mathematical and natural sciences gained by study, experience and practice is applied with judgment to develop ways to utilize economically the materials and forces of nature for the benefit of mankind. (ABET)

If one steps into an air craft or gets on a bridge, he or she does it with a confidence and trust in the designers and builders that they have balanced the science, technology and the resources to build a cost effective but aesthetically elegant structure for the utility and comfort of the users. This trust on the gadget or the process comes from the collective responsibility of operators, workers, designers, manufacturers, consultants, employees, managers, and administrators. An engineer in his professional career may assume any of these roles and in addition he may become researcher creating new knowledge through patents and publications or an entrepreneur / employer. In all the roles, a professional engineer should conduct his duty, which is right technically, legally and ethically, for the welfare of humanity, to improve quality of life, to protect and sustain life and environment. Examples of common judgments that engineers make are:

- When does a product or procedure become "safe"?
- How should the technical work of others be credited?
- What information must be disclosed?

The behaviour of engineers is guided by many types of rules – laws, government regulations, employer policies, professional consensus, peer expectations, cultural norms, religious standards, and consequences (Steve E. Watkins, 2015)These external rules and intrinsic values of engineer ultimately influences his moral efficacy, reasoning and judgement. The traits which can impede or promote a responsible behaviour are (Nair, 2006) as shown in Table 1

Table 1. Human traits

Roadblocks to Responsible Behaviour	Promoters
• Gluttony, Greed & Lust • Selfishness • Wrath & fear • Self-deception • Ignorance • Experts Blindness • Group Influence	• Honesty • Integrity • Courage • Empathy • Humility • Wisdom • Love

Value System

The onset of these positive values depends on the value system one grew with, what norms one is used to and what moral values one set himself for. Let us examine the subtle differences between these terminologies predominantly differentiate good and bad, positive or negative, in thought, word and deed. It is easier to discern human values by examining behaviour. Actions are based on choices, Choices are made by mind, Mind is guided by what it values. Thus, Values are guidelines for behaviour.

- **Norm** is an accepted standard or a way of behaving or doing things that most people agree with
- **Morals** are the degree to which something is right or wrong, good or bad, etc. They help in behaviour, instilling respect and enhancing relationship with others. Kohlbergs theorized how humans imbibe morals as they grow from elementary to high school with basis for behaviour changing from avoiding to punishment to sense of justice, whereas Gilligan extended the theory to include the differences exhibited by the women.
- **Ethics** are moral values guiding behaviour and actions. All moral values mentioned so far have tremendous impact on the profit of the business. If we allow economic systems to operate without moral foundation and without continued education, we will soon create an amoral, if not immoral society and business. (Covey, S. R., 1992).
    - **Personal Ethics** are based on individual's beliefs. Based on general morality, the religion, community and social-political affiliations of the individual. Parents, teachers and friends have a strong influence on the individual in defining and developing personal ethics.
    - **Professional Ethics** are moral code of conduct as applicable to persons belonging to a particular profession in the practice of that profession. Worldwide, many countries have professional bodies who issue codes of ethics applicable to their regions and for their profession. For example,

in USA -IEEE, NSPE, ASME, ASCE, in UK - Royal Academy of Engineering, in India - IE etc.
   - **Work Ethics** is value based on hard work and diligence. It is intrinsic, it should come from within. This involves attitude, behaviour, respect, communication and interaction. It involves honesty and accountability.
- **Laws** are binding codes of conduct; formally recognized and enforced company policies.

It may be appreciated that by the time an engineer begins his academic journey, his value systems upto personal ethics might be reaching maturity stage, where as he definitely requires awareness and practice on the following issues.

- Understanding Ethical Problem
- Ethical Dilemmas
- Solving the Ethical Problem

Let us review the content with an appropriate example so that course of action to teaching can be evolved.

ETHICAL RESPONSIBILITY FOR ENGINEERS

In March 2019 aviation regulators and airlines grounded the global fleet of 394 Boeing 737 MAX passenger airliners after two nearly-new aircraft crashed within four months and ten days, killing all 346 people aboard both flights. The accidents befell Lion Air Flight 610 on October 29, 2018 and Ethiopian Airlines Flight 302 on March 10, 2019. The 737 MAX's new Maneuvering Characteristics Augmentation System (MCAS) was suspected of sending each aircraft into a dive in response to erroneous data. Aviation engineers faulted Boeing's safety analysis that allowed repeated activation of the system beyond certified limits, and criticized the MCAS for using only one of the airplane's two angle of attack sensors, which created a single point of failure.

The crashes certainly impact the public's trust. In both cases, the questions to be answered are:

- *Was there a slip or gap in the design?*
- *Was the deficiency known but not informed?*
- *Were the pilots not informed on actions on MCAS malfunction?*

The enquiry is ON. But before the findings are out, it becomes a case of importance in ethics in addition to safety considerations for all engineers!

ETHICAL DILEMMAS IN ENGINEERING PROFESSION

In September 2016, Engineer James Robert Laing, leader of diesel competence for Volkswagen (VW) from 2008 through June 2016, pled guilty to a U.S. District Court grand jury's indictment of conspiracy to defraud the U.S. government, to commit wire fraud, and to violate the Clean Air Act. In his plea, Laing admitted that he was involved in VW s efforts to cover up the development and use of a "defeat device" to enable more than 500,000 of its vehicles from 2009 to 2015 to appear to pass U.S. emissions tests. In laboratory testing of the vehicles by the government, the emissions control system operated as it should. But the "defeat device" was disabled for driving in ordinary conditions. Regulators eventually discovered that in those ordinary conditions the cars emitted up to 40 times more smog-causing nitrogen oxide than the legal limits allow.

Trust and Reliability

Honesty is not just being truthful and not lying. Deception, misleading, exaggeration, willingly avoiding information etc are all different forms of dishonesty. Withholding negative aspects of a product, for example poor safety record and design deficiencies of a product is also deception.

Sometimes, withholding the information is considered as a form of dishonesty. But giving the information is against maintaining client-professional confidentiality and any such action can only mean lack of integrity. Integrity is strict adherence to ethical code of conduct and may be described as qualified honesty. Client professional agreements, Non-Disclosure Agreements (NDA), know-how transfer, Intellectual Property Rights are some instruments which demand highest form of confidentiality. Safety concerns in cyber space is on the rise, as increase in connectivity between everyday usage devices, automation, artificial intelligence, leading to growth and dependence on data. Loss of privacy and misuse of data are a cause of concern. Engineers as professionals must be reliable and dependable to their clients and to the society by following a path of honesty and integrity, acting wisely in the face of such conflicts.

Risks, Safety, and Liability

In late 1984, a pressure-relief valve on a tank used to store methyl isocyanate (MIC) at a Union Carbide plant in Bhopal, India, accidentally opened. MIC is a poisonous compound used in the manufacture of pesticides. When the valve opened, MIC was released from the tank, and a cloud of toxic gas formed over the area surrounding the plant. Unfortunately, this neighborhood was very densely populated. Some two thousand people were killed, and thousands more were injured as a result of the accident. Many of the injured have remained permanently disabled. The causes of the accident are not completely clear, but there appear to have been many contributing factors. Pipes in the plant were misconnected, and essential safety systems were either broken or had been taken off-line for maintenance. The effects of the leak were intensified by the presence of so many people living near the plant.

Among the many important issues this case brings up are questions of balancing risk to the local community with the economic benefits to the larger community of the state or nation. Undoubtedly, the presence of this chemical plant brought significant local economic benefit. However, the accident at the plant also brought disaster to the local community at an enormous cost in human lives and suffering. How can we decide if on balance the economic benefit brought by this plant outweighed the potential safety hazards?

Engineers have paramount responsibility to protect life and environment from hazards. Risk of hazard and safety in workplace are well recognized and there are statutory provisions in many countries to be complied with to minimize risk and maximize safety. Engineers are responsible for safety in design of products / structures, safety in work place, safety during construction/ manufacturing, safety in repair and maintenance, safety in cyber space due to interconnectivity of devices and dependence on digital data. Safety and risk are linked. Engineers, managers and owners of business are liable for legal actions for violations. Engineers must prepare for dealing effectively and responsibly with issues of risks, safety and liability.

Whistle Blowing

Vijay Pandhare, in 2012, was the Chief Engineer of the Water Resources Department and a member of the state level technical advisory committee, Maharashtra state in India. Pandhare is a whistleblower who wrote to the chief minister of the state of Maharashtra and its governor alleging irregularities and cost inflation in irrigation projects. The controversy raised by these communication caused the deputy chief minister of the stateAjitPawar to resign. In his letters he pointed out that Rs 120 billionspent on lift irrigation projects in the state are a total waste as around 99 per cent of the total 227 projects in Maharashtra are not working and 90 per cent never began functioning. Pandhare, in a letter to other engineers, alleged that corruption can be controlled by curtailing the powers. The irregularities were so serious Pandhare alleged that the government was "...playing with people's lives". (Daily Bhaskar, 2012)

As the example of VijayPandhare, in some situations, engineers find the actions of the employer to be so objectionable that they believe mere nonparticipation in the objectionable activity is insufficient. Rather, some form of outward protest, or whistleblowing is required

Conflict of Interest

Mr Job, a design engineer in a small company, usually recommends to his clients components made by a friend, even when the products made by other companies might be more suitable. Clients are justified if they complain that Mr Job is involved in conflict of interest, in that his professional judgment is unduly influenced by his personal relationship with his relative.

Conflict of interest is a conflict between an obligation to exercise good judgment and interest(s) that may compromise that judgment. This could range from self-interest, interests of or obligations to others such as friends, relations, employers and clients and competing interest of past and present clients. The subtlety between actual, apparent and potential conflicts need to be recognized.Actual conflict of interest is to be avoided by engineers and in case of the other two, if cannot be avoided, position should be declared to all concerned to clear any ambiguity.

Gifts and Bribes

Mr Rigid is an engineer in a large construction firm. Though he is not the final authority to decide, he has been tasked to recommend which steel fasteners should be used for the construction of a large apartment building. He carried out survey and testing and decides to recommend BEST steel fasteners for the job. On the day after Rigid's recommendation was made, a BEST representative visit him and gives him a voucher for an all- expense paid trip to their annual Technical Forum, which meets in Singapore. The trip will have considerable educational value but will also include day trips to Sentosaisland and other points of interest. If Rigid accepts, has he been bribed? Can it be termed as an acceptable gift?

Frequently, the boundary between a legitimate gift and a bribe is very subtle. Gifts of nominal value, such as coffee mugs or calendars with a vendor's logo and phone number on it, are just an advertising tool. Generally, there is no problem with accepting these types of items. Dining with a customer or a supplier is also an acceptable practice, especially if everyone pays his or her own way. It is important from the point of view of both suppliers and customers that good relations be maintained so that good service can be provided. Social interaction, such as eating together, often facilitates the type of close and successful interactions required by both sides. However, when meals or gifts are no longer of low cost and the expense of these items is not shared equally, the possibility for abuse becomes large.

Ethical Codes Through Professional Bodies

Like the Hippocratic Oath, which lays out a moral frame work of the conduct of doctors, engineering profession does not boast of any oath in the modern times. The oath called Obligation of Engineer was gradually changed to graduates pledge to social and environmental responsibility by majority of the colleges and universities all over the world. The Pledge states: "*I pledge to explore and take into account the social and environmental consequences of any job I consider and will try to improve these aspects of any organizations for which I work.*" In professional career they are guided by the professional bodies, each evolved as per the varying disciplines and place of influence and their codes of conduct.

Worldwide, many countries have professional bodies who issue codes of ethics applicable to their regions and for their profession. For example, in USA, IEEE, NSPE, ASME, ASCE, in UK Royal Academy of Engineering, in India IE etc. and their code of ethics are:

National Society of Professional Engineers (NSPE) issued code of Ethics for all licensed engineers, Engineers, in the fulfillment of their professional duties, shall:

- *Hold paramount the safety, health, and welfare of the public.*
- *Perform services only in areas of their competence.*
- *Issue public statements only in an objective and truthful manner.*
- *Act for each employer or client as faithful agents or trustees.*
- *Avoid deceptive acts.*
- *Conduct themselves honorably, responsibly, ethically, and lawfully so as to enhance the honor, reputation, and usefulness of the profession.*

The Royal Academy of Engineering, UK*, in collaboration with Engineering Council (UK) and a number of the leading professional engineering institutions, has created a Statement of Ethical Principles to which it believes all professional engineers and related bodies should subscribe, covering*

- *Accuracy and rigour*
- *Honesty and integrity*
- *Respect for life, law and the public good, and*
- *Responsible leadership: listening and informing*

Similarly, in India, the preamble of **Institution of Engineers (India)** for their code of ethics reads as, The Corporate Members of The Institution of Engineers (India) are committed to promote and practice the profession of engineering for the common good of the community bearing in mind the following concerns:

- *Concern for ethical standard*
- *Concern for social justice, social order and human rights*
- *Concern for protection of the environment*
- *Concern for sustainable development*
- *Public safety and tranquility.*

MORAL EFFICACY

While in profession, the situations may lead to ethical dilemmas, for which the moral theories may be applied to reason and to arrive at moral judgment. Having multiple theories to apply actually enriches the problem-solving process, allowing problems to be looked from different angles, since each theory stresses different aspects of a problem. There are four ethical theories that will be considered here (Charles B. Fleddermann, 2012), each differing according to what is held to be the most important moral concept.

- *Utilitarianism* seeks to produce the most utility, defined as a balance between good and bad consequences of an action, considering the consequences for everyone affected.
- *Duty ethics* contends that there are duties that should be performed regardless of whether these acts lead to the most good.
- *Rights ethics* emphasizes that we all have moral rights and any action that violates these rights is ethically unacceptable.

- V*irtue ethics* regards actions as right that manifest good character traits (virtues) and regards actions as bad that display bad character traits (vices)

DEALING WITH DILEMMAS

A micro ethical decision related to nanotechnology would be that a researcher ensures that all experiments be conducted with integrity and results reported honestly and completely. A macro ethical decision would be whether certain types of nanotechnologies be avoided until there is sufficient information regarding their risks.

In most cases, an engineer in his profession will be able to take decisions following the rules of the profession and morals. The ethical issues concerning day-to-day activities generally termed under micro-ethics can be easily dealt with, like, Stealing is immoral, so an engineer may not decide to embezzle company's funds. But some ethical problems specially termed under macro-ethics like issues relating to technology, information in present digital age, environment covering wider area and influence, are ambiguous or unclear involving conflicting morals. Typical case studies and their post-mortem findings along with professional code of ethics do surely offer guidance. But an engineer would still prefer clear cut ethical decision-making tools or methodologies to solve, reducing knee-jerk emotional reactions to a situation.

A first step in solving any ethical problem is to completely understand all of the issues involved and apply the general Methodology:

- Identification
  Recognize/acknowledge a state of conflict
  Identify source of conflict and stakeholders
- Analysis
  Investigation, facts…
  Alternatives, ranking…
- Action
  Deciding on and initiating a course of action/ resolution

Whenever there is a conflict, the following techniques may be used to arrive at resolution (Harris et al, 2017)

- **Line Drawing**: A line is drawn between known cases, one end being a "positive paradigm", morally acceptable example and the other end is a "negative paradigm", a known unacceptable example. The case under investigation is placed on the line considering the weight of all its features.

The relative position of the case with respect to the two extremes shall indicate the acceptability of the case.

- **Creative Middle Path**: When there is no unique solution acceptable to everybody, a creative middle path is taken satisfying everybody concerned, but requires imagination and convincing capability.
- **Flow Chart**: Graphically representing the decision process like a flow chart. Provides a visual picture of a situation and establishes sequences, helps to identify moral issues and consequences of actions.

A variation in solving grey ethical problems is the solution path followed for engineering design process problems offering multiple solutions. The ethical decision making process of Martin and Schinzinger was adapted for parallelism to the design process and presented to students as a step-wise technique for identification of the pertinent ethical issues, relevant moral theories, possible outcomes and a final decision (Bero, B., & Kuhlman, A. 2011).

Engineering Ethics has a large overlap from

- Scientific Ethics (Dishonesty in research – fabrication, faking, plagiarism, Disregard to Intellectual Property Rights)
- Business Ethics (corporate governance, insider trading, bribery, discrimination, corporate social responsibility)
- Computer Ethics (hacking, Privacy, embezzlement, piracy)
- Environmental Ethics (water and air pollution, the depletion of natural resources, loss of biodiversity, destruction of ecosystems, and global climate change)
- Legal Ethics (written laws)

ETHICS IN ENGINEERING PROGRAMMES

Importance of ethics and social responsibility grew from the last quarter of 20th century in engineering programs in USA. Accreditation Board for Engineering and Technology (ABET) student outcomes include "An ability to understand ethical and professional responsibilities and the impact of technical and/or scientific solutions in global, economic, environmental, and societal contexts." (ABET, 2019). Countries, seeking recognition of engineering qualifications and to assist mobility of professional engineers have become signatory to Washington Accord (International Alliance, 2019). India being signatory Washington Accord focuses on the same through its National Board of Accreditation (NBA, 2019). They are emphasizing on graduate attributes in terms of engineering professionalism, impact of engineering on society and the environment ethics, which are in turn mapped through corresponding Programme Outcomes. Engineering Programmes endeavour for this behavioural transformation in the graduates through courses such as Professional Ethics, Engineering Ethics, Human Values and Professional Ethics coupled with positive trait enhancers through good teaching-learning practices.

CHALLENGES IN TEACHING ENGINEERING ETHICS

Is It the Right Age?

The issues of whether ethics can be taught and whether it can be taught at the college level remain subject to debate in the academic community. When Arlow and Ulrich (1988) asked business school students and executives to rank factors influencing their ethical behavior, both groups reported that family training was most important and that university training was least important. Thus, some researchers have concluded that the development of moral reasoning occurs before a student enters college. Other researchers, basing their thinking on Kohlberg's cognitive development scheme, have concluded that moral reasoning is not well-developed until later in life and that the cognitive development necessary for making ethical decisions can be encouraged by practice with the process (Harris and Brown, 1990). Thus, some believe that training in ethical thinking may be valuable at any age. Of those who believe there is a place in the college curriculum for teaching ethics, some suggest that students can at least develop an ability to recognize ethical issues and to think through the consequences of alternative solutions (Andrews, 1989). Hosmer (1988) argues that the goal of dealing with ethical issues in the curriculum is not to change the values and beliefs of students but instead to teach systems of analysis to help students use their own values to weigh the potential benefits and harms of their actions to the organization, to society, and to individuals. (Adams, J. S. et al., 1998)

Attitude of Students

There is a difference in defining the concept of ethics by teachers and students. What is "acceptable" for students is "unacceptable" for teachers and vice versa. For example, helping one's friend to copy in the examination hall may be considered as ethically right by students, but not by the teachers. It is based on the huge gap between teachers" and students" understanding of the definition of the term "Engineering". For most engineering students, "Engineering" is only a means to get good salaried job. The teachers who teach Professional Ethics or Ethics in engineering education define "Engineering" as a means to ensure the safety, health and well-being of the society. Unless this ethical explanatory gap is closed, the major objective of preparing young engineers to be socially responsible and ethically sensitive will be worthless. It is also supported by the students' attitude that they do not need to study any humanities courses, because these disciplines are unimportant for designing, developing or manufacturing an engineering product. Moreover, there are engineering students who believe that they know what "ethics" is and what is "ethically right or wrong", and so there is no need to study an Ethics course. This type of attitude makes teaching human values and ethics to engineering students a tedious job. Often it is observed that students' position with regard to Ethics course is egoistic pragmatism while the teachers follow idealistic pragmatism. (Cheruvalath, R. 2015). This difference in outlook can better be explained by an often repeated statement by students of becoming rich by hook or crook in the shortest time, whereas the teachers support first part of the statement but try to moderate the second part to fair means and in reasonable periods.

Crowded Curriculum

Engineering curricula is constantly under review as the demands of industry are always in a state of flux due to ever changing societal needs and hence status of technologies. One can appreciate the change in expectations from graduating students over the period as industry is expecting them to be readily employable from the old terminology of industry ready. Demand for multi-disciplinary aspects along with the necessary soft skills has overburdened the number of courses. In this crowded scenario an additional course like ethics is a tricky task.

Taught by Humanities or Engineering Faculty

It is perceived that ethics handled by humanities staff specialized in psychology or beahavioural science may lack conviction to convince engineering students on safe designs as opposed to engineering faculty. But the later already burdened with domain subjects if tasked may have to carry this additional load with proper training.

International Ethical Standards

Ethics have strong bearing from religion and cultural practices. The diversity of the world cultures makes engineering students to learn and respect all cultures and not to carry preconceived assumptions and ideas about other societies and cultures. Attempts are underway to establish an international code of ethics for engineers. This requires an international committee of engineers where members will contribute their rich and local experiences. These standards have to be definitive with some room left for local accommodations. (Barakat, N., &Eng, P. 2004)

CURRICULUM MODELS

The two main strategies followed by engineering programs is to offer directly as *standalone courses* on engineering ethics and professionalism or indirectly provide through *embedded modules* in appropriate courses. Both have their advantages and limitations (Herkert, J. R. 2000), (May, D. R., & Luth, M. T.,2013). Standalone courses require space in the already crowded engineering curricula and trained faculty in engineering and/or philosophy. Embedded modules across the curriculum, increases the load and training requirement of existing engineering faculty of various related courses across the curriculum.

The notion of the ethics course is to supplement an engineer with professional responsibility, which includes – codes and canons of professional bodies, public safety and welfare, risk and the principle of informed consent, health and environment, conflict of interest, truthfulness, integrity and representation of data, whistle blowing, choice of a job, loyalty, accountability clients and customers, plagiarism and giving credit where due, quality control, confidentiality, trade secrets and industrial espionage, gift giving and bribes, employer/employee relations, discrimination and responsible conduct of research.

COURSE OUTCOMES

At the end of the course on ethics, the engineer should be both confident in content, in terms of ethical knowledge and judgment (Steve E Watkins, 2015) and courage to implement or exhibit an ethical behavior in real life. The course outcomes can be summarized on the following broad categories (Newberry B., 2004) as:

- Emotional engagement (the willingness and courage to make ethical decisions)
- Intellectual Engagement (knowing how to make ethical decision)
- Knowledge (knowing the standards of ethical practices)
- Perception Power (knowing the four natural orders & relation with environment)

PEDAGOGICAL TRENDS

Code of ethics, moral theories, value systems, case studies, vignettes with short abstracts, collaborative/ challenge games and role plays, debates and group discussions, presentations, online instructions, interactive multimedia packages, simulations (Alfred & Chung, 2012) are a multitude of instructional strategies which can be used for ethics. (Jehan Abu Hamad, et al., 2013).

Engineering ethics education is generally conducted by introducing the Code of Ethics of the profession and reviewing case studies. The discussion of case studies is brief and mainly done so students can apply the different cannons of the Code of Ethics. In the presentation of such cases, educators hope that the morals learned will leave a lasting impression on students. Some commonly used case studies are:

- Citicorp Centre Design
- Challenger Shuttle
- Gilbane Gold
- Goodrich
- Hyatt Regency Walkway Collapse
- Heinz Dilemma
- Mexican Plow
- Trench Failure

The other commonly used instructional techniques involve presenting short abstracts and more ambiguous vignettes, that are explicitly tied to the Code of Ethics. These vignettes present ethical dilemmas in more ambiguous ways, but educators hope that when students are introduced to and become familiar with the Code of Ethics they will easily recognize a breach of ethics. The cases can be hypothetical too, typically inspired by events or to suit a specialization or a specific ethical issue.

ONLINE RESOURCES

Online Ethics Center (OEC) is a resource maintained by the Center for Engineering Ethics and Society (CEES) with a mission to provide engineers, scientists, faculty and students with resources for understanding and addressing ethically significant issues that arise in scientific and engineering practice and from the developments of science and engineering (https://www.onlineethics.org/, 2019) Another example is A&M's resource (https://ethics.tamu.edu, 2019).

ENVIRONMENTAL ENABLING FACTORS

In addition to classroom engagement on ethics, teaching-learning environment with its systems and enforcement rules can also indirectly influence and mould the personality of the students, aligning with the required programme outcome in terms of self, society and nature.

Three students were working on a senior cap stone engineering design project, involving design, build, and test of a meter that would be mounted on the dashboard of automobiles. The professor directing the project was so impressed that he found a source of funds to buy the flow meter. He also encouraged the three students to draft an article describing their design for a technical journal. Few weeks later, the professor was surprised to receive a congratulatory letter from the editor of a prominent journal, announcing that the journal was accepting for publication the excellent article that the professor had coauthored with his three senior design students. The professor knew that the flow meter had not yet even arrived, nor had he seen any draft version of the paper. On enquiry, students informed tha as it was his idea to write the paper and he was the faculty advisor, they communicated without informing him. Furthermore, they really could not wait for the flow-measuring instrument to arrive because they were all graduating in a few weeks and planned to begin new jobs, they simulated the data and communicated the paper.

As a matter of fact, the students' assumption of the simulated behaviour was incorrect. They also made false assumptions about the response of the professor to their actions. The result was that the paper was withdrawn from the journal and the students sent letters of apology to the journal. Copies of the letters were placed in their files, the students received an F in the senior design course, and their graduation was delayed by six months. (Harris,Jr.,C.E., et al, 2017).

Apart from formal ethics course or modules, every academic institution should promote students to build their ethical habits through academic honesty during course submissions, assignments, projects, communication of credentials through CVs. Harsh measures to be taken against fabrication of data or falsifying results. Proper report writing rules to be insisted upon, emphasizing how credit to be given for other contributors, maintaining confidentiality and avoiding plagiarism.

Involve students in Institutional Social Responsibility Activities such as disaster relief, working with under privileged or old age homes etc, so that engineer understands his duty towards society and be responsible towards it when he graduates. Corporate Social Responsibility (CSR) activities in industry will continue to remind him and help his conscience to be directed towards "WE" rather than "I".

Computer Ethics

Pervasiveness of social media and proliferation of IoT devices are promoting the adoption of cyberspace for our socioeconomic activities. With Industry 4.0 round the corner, automation, inter connectivity and data dependency are on the rise and so are loss of privacy and misuse of data. Cyber attacks may paralyze the human activities to a stand still. Exposure to ethics concerning the cyber space usage, etiquette and safety measures is essential and should find place in the engineering ethics. Unlimited free access to the internet is a privilege generally given to the students.It is often observed that there are students who utilize these facilities for more non-academic purposes than for academic and they rarely try to examine if whatever they are doing is ethically right or wrong. For instance, misusing technology (say hacking) for fun is unethical for some students and for others it is not. They have their own ethical justification for that, saying if the intention is good then the act is ethical, otherwise not. For instance, the fun often takes in accessing the email account of juniors by some senior students and sending mails from those accounts without having the formers permission. These aberrations are to be corrected with strict enforcement of laws in the environment, so that the ethical conduct could be displayed in all aspects of engineering.

Environment Ethics

All Engineering programmes offer Environmental Engineering as a compulsory course (UGC). Ethics for sustainability are amply covered there. The biodiversity loss or excess/ mis-use of the five elements of the nature can greatly be reduced if proper positive traits are instilled in the graduating engineers. Apart from classroom interaction, students get benefitted if involved in ISR activities through movements like NSS (National Social Service Scheme) and maintenance of college campus environment.

I used to think that top environmental problems were biodiversity loss, ecosystem collapse and climate change. I thought that thirty years of good science could address these problems. I was wrong. The top environmental problems are selfishness, greed and apathy, and to deal with these we need a cultural and spiritual transformation. And we scientists don't know how to do that.

Gus Speth, a US advisor on climate change

Society is surely expecting engineers with knowledge and skills soundly built on strong positive traits for its sustainable prosperity.

CLOSURE REMARKS

The two basic aspirations of a human being have always been continuous happiness and prosperity. But these must be extended to universal level ensuring sustainability and for future generations and no exploitation of people or planet. An engineer is expected to utilize the science and technological progress to do maximum good more people, use the planetary resources to the need. This is possible if engineer conducts himself, upholding ethical principles following the universal statement "*loka samasta sukhino bhavantu*', meaning "May all in the universe live peacefully"! Positive traits like honesty, integrity, and empathy need to be nurtured and implications of security concerns of physical and virtual worlds need to be understood. Moral reasoning, judgment and moral courage can be strengthened through relevant case studies of ethical dilemmas along with introspection into codes of ethics promulgated by appropriate professional bodies. Pedagogy on ethics in engineering during engineering programmes through standalone courses on engineering ethics or embedded capsules of ethics in appropriate courses can deliver the required outcomes. The value system in terms of ethics need to be revisited at all academic levels, to avoid the trap of knowledge without character, commerce (business), without morality (ethics) and science without humanity. Engineering Faculty has to raise to meet the challenges in integrating engineering ethics into curriculum already crowded with domain specific courses and attitudes hesitant to revisit the value system at this stage of life. Promotion of ethics integrating into the engineering profession at all levels could lead to a holistic alternative at universal level of which is self-satisfying, people-friendly and eco-friendly.

REFERENCES

ABET. (2019). Available http://www.abet.org

Adams, J. S., Harris, C., & Carley, S. S. (1998). Challenges in teaching business ethics: Using role set analysis of early career dilemmas. *Journal of Business Ethics, 17*(12), 1325–1335. doi:10.1023/A:1005784303157

Alfred, M., & Chung, C. A. (2012). Design, Development, and Evaluation of a Second Generation Interactive Simulator for Engineering Ethics Education (SEEE2). *Science and Engineering Ethics, 18*(4), 689–697. doi:10.100711948-011-9284-0 PMID:21633864

Andrews, K. R. (1989). Ethics in Practice. *Harvard Business Review, 89*(5), 99–105. PMID:10295476

Arlow, P., & Ulrich, T. A. (1988). A longitudinal survey of business school graduates' assessments of business ethics. *Journal of Business Ethics*, *7*(4), 295–302. doi:10.1007/BF00381835

Bairaktarova, D., & Woodcock, A. (2015). Engineering ethics education: Aligning practice and outcomes. *IEEE Communications Magazine*, *53*(11), 18–22. doi:10.1109/MCOM.2015.7321965

Barakat, N., & Eng, P. (2004). Issues and Challenges of Teaching Engineering Ethics. *Proceedings of the CSME 2004 Forum.*

Barman, P., & Bhattacharyya, D. (2012). Vivekananda's Thoughts on Man-Making through Moral Values and Character Development and Its Present Relevancy in School Education. *International Journal of Multidisciplinary Educational Research*, *1*(2).

Bero, B., & Kuhlman, A. (2011). Teaching ethics to engineers: Ethical decision making parallels the engineering design process. *Science and Engineering Ethics*, *17*(3), 597–605. doi:10.100711948-010-9213-7 PMID:20524076

Daily Bhaskar. (2012, Sept. 27). *Maharashtra's Irrigation Scam: Meet the whistleblower Vijay Pandhare.* Author.

Cheruvalath, R. (2015). Teaching ethics to engineering students in India: Issues and challenges. In *Contemporary ethical issues in engineering* (pp. 121–132). IGI Global. doi:10.4018/978-1-4666-8130-9.ch009

Christensen, C. M., Ojomo, E., & Dillon, K. (2019). *The prosperity paradox: How innovation can lift nations out of poverty*. New York: HarperBusiness.

Covey, S. R. (1992). *Principle centered leadership*. Simon and Schuster.

Fleddermann. (2012). *Engineering Ethics*. Prentice Hall.

Gaur, R. R., Sangal, R., & Bagaria, G. P. (2011). *A Foundation Course in Human Values and Professional Ethics*. New Delhi, India: Excel Books.

Goswami, D., & Chakraborty, A. (2015, November). Sensitizing engineers: A brief study of the role of ethics in engineering education. In *2015 5th Nirma University International Conference on Engineering (NUiCONE)* (pp. 1-4). IEEE. 10.1109/NUICONE.2015.7449589

Hamad, Hasanain, Abdulwahed, & Al-Ammari. (2013). *Ethics in Engineering Education: A Literature Review*. IEEE.

Harris, C., & Brown, W. (1990). Developmental constraints on ethical behavior in business. *Journal of Business Ethics*, *9*(11), 855–862. doi:10.1007/BF00382909

Herkert, J. R. (2000). Engineering ethics education in the USA: Content, pedagogy and curriculum. *European Journal of Engineering Education, 25*(4), 303–313. doi:10.1080/03043790050200340

Hopf, H., Krief, A., Mehta, G., & Matlin, S. A. (2019). Fake science and the knowledge crisis: Ignorance can be fatal. *Royal Society Open Science, 6*(5), 190161. doi:10.1098/rsos.190161 PMID:31218057

Hosmer, L. T. (1988). Adding ethics to the business curriculum. *Business Horizons, 31*(4), 9–15. doi:10.1016/0007-6813(88)90062-6

Harris Jr, Pritchard, & Rabins. (2017). *Engineering ethics: Concepts and cases*. Cengage Learning.

May, D. R., & Luth, M. T. (2013). The effectiveness of ethics education: A quasi-experimental field study. *Science and Engineering Ethics, 19*(2), 545–568. doi:10.100711948-011-9349-0 PMID:22212360

Nair, C. G. K. (2006). *Engineering Ethics*. Bangalore: Hari Sri Publishing Co.

NBA. (2019). Available https://www.nbaind.org/

Newberry, B. (2004). The dilemma of ethics in engineering education. *Science and Engineering Ethics, 10*(2), 343–351. doi:10.100711948-004-0030-8 PMID:15152860

Online Ethics Center. (n.d.). Retrieved from https://www.onlineethics.org/

Roy, A. (1999). *The Cost of Living*. Modern Library.

Smith, R. (2006). Research misconduct: The poisoning of the well. *Journal of the Royal Society of Medicine, 99*(5), 232–237. doi:10.1177/014107680609900514 PMID:16672756

Texas A&M. (n.d.). Retrieved from https://ethics.tamu.edu

UGC. (2019). Available https://www.ugc.ac.in/

Watkins. (2015). Teaching Engineering Ethics. American Society for Engineering Education.

Compilation of References

AACSB. (2016). *A collective vision for business education*. AACSB International.

ABET. (2019). Available http://www.abet.org

Abeysekera, L., & Dawson, P. (2015). Motivation and cognitive load in the flipped classroom: Definition, rationale and a call for research. *Higher Education Research & Development*, *34*(1), 1–14. doi:10.1080/07294360.2014.934336

Achten, H. (2003). New design methods for computer aided architectural design methodology teaching. *International Journal of Architectural Computing*, *1*(1).

Adams, J. S., Harris, C., & Carley, S. S. (1998). Challenges in teaching business ethics: Using role set analysis of early career dilemmas. *Journal of Business Ethics*, *17*(12), 1325–1335. doi:10.1023/A:1005784303157

Adbelaal, M., Ismail, M., Khater, M., & Zaki, M. (2018). *Digital Business Transformation and Strategy: What Do We Know So Far?* Working Paper, Cambridge Service Alliance.

Adetokunbo, A., Adenowo, A., & Patel, A. (2014). A metamodel for designing an intelligent tutoring system authoring tool. *Computer and Information Science*, *7*(2), 82–98.

Adukaite, A., Van Zyl, I., & Cantoni, L. (2016). The role of digital technology in tourism education: A case study of South African secondary schools. *Journal of Hospitality, Leisure, Sport and Tourism Education*, *19*, 54–65. doi:10.1016/j.jhlste.2016.08.003

Aehnelt, M., & Bader, S. (2014). Tracking assembly processes and providing assistance in smart factories. *Proceedings of the 6th International Conference on Agents and Artificial Intelligence (ICAART)*.

Aftermarket Business. (1994). *'Sell' ... you're on 'candid camera'*. Author.

Agndal, H., & Nordin, F. (2009). Consequences of outsourcing for organizational capabilities: Some experiences from best practice. *Benchmarking: An International Journal*, *16*(3), 316–334. doi:10.1108/14635770910961353

Ahearne, M., Jelinek, R., & Jones, E. (2007). Examining the effect of salesperson service behavior in a competitive context. *Journal of the Academy of Marketing Science*, *35*(4), 603–616. doi:10.100711747-006-0013-1

Akcapinar, G., & Hasnine, M., Nehal, M., Majumdar, R., Flanagan, B., & Ogata, H. (2019). Developing an early-warning system for spotting at-risk students by using eBook interaction logs. *Smart Learning Environments*, *6*(4), 1–18.

Akin, O. (1989). *Computational design instruction: towards a pedagogy CAAD Futures, Digital Proceedings*. Boston: MIT Press.

Akir, O., Eng, T. H., & Malie, S. (2012). Teaching and learning enhancement through outcome-based education structure and technology e-learning support. *Procedia: Social and Behavioral Sciences*, *62*, 87–92. doi:10.1016/j.sbspro.2012.09.015

Aleven, V., McLaren, B., Sewall, J., van Velsen, M., Popescu, O., Demi, S., ... Koedinger, K. (2016). Example-tracing tutors: Intelligent tutor development for non-programmers. *International Journal of Artificial Intelligence in Education, 26*(1), 224–269. doi:10.100740593-015-0088-2

Alexander, B., Ashford-Rowe, K., Barajas-Murphy, N., Dobbin, G., Knott, J., McCormack, M., . . . Weber, N. (2019). EDUCAUSE horizon report 2019 higher education edition. *EDUCAUSE*. Retrieved from https://library.educause.edu/resources/2019/4/2019-horizon-report

Alfred, M., & Chung, C. A. (2012). Design, Development, and Evaluation of a Second Generation Interactive Simulator for Engineering Ethics Education (SEEE2). *Science and Engineering Ethics*, *18*(4), 689–697. doi:10.100711948-011-9284-0 PMID:21633864

Al-Hemyari & Alsarmi. (2015). Standards, benchmarks, and qualitative indicators to enhance the institutions' activities and performance: Surveys and data analysis. *International Journal of Knowledge-Based Organization, 5*(4), 37-61.

Alkhatlan, A., & Kalita, J. (2018). *Intelligent tutoring systems: A comprehensive historical survey with recent developments.* Cornell University. Retrieved from: http://arxiv.org/abs/1812.09628

Allais, S. (2007). Education Service Delivery: The Disastrous Case of Outcomes-Based Qualifications Frameworks. Progress in Development Studies, 7(1), 65–78. doi:10.1177/146499340600700106 doi:10.1177/146499340600700106

Allen, K., Sheve, J., & Nieter, V. (2010). *Understanding Learning Styles: Making a Difference for Diverse Learners*. Shell Education.

Al-Qawasmi, J. (2005). Digital media in architectural design education: reflections on the estudiopedagog. *Art, Des. Commun. Higher Educ., 4*(3).

Anderson, L. W., & Krathwohl, D. R. (2001). *A taxonomy for learning, teaching, and assessing: A revision of Bloom's taxonomy of educational objectives.* Allyn and Bacon.

Anderson, D. N., Groves, D. L., Lengfelder, J., & Timothy, D. (2001). A research approach to training: A case study of mystery guest methodology. *International Journal of Contemporary Hospitality Management, 13*(2), 93–102. doi:10.1108/09596110110381906

Andersson, K. (2012). Using Model-Based Design in Engineering Design Education. *ASME 2012 International Design Engineering Technical Conference & Computers and Information in Engineering Conference*. 10.1115/DETC2012-71081

Andrews, K. R. (1989). Ethics in Practice. *Harvard Business Review, 89*(5), 99–105. PMID:10295476

Angelil, M., & Uziyel, L. (2003). Inchoate: An Experiment in Architectural Education. Swiss Federal Institute of Technology (ETHZ).

Antsaklis, P. (2014). Goals and Challenges in Cyber-Physical Systems Research Editorial of the Editor in Chief. *IEEE Transactions on Automatic Control, 59*(12), 3117–3119. doi:10.1109/TAC.2014.2363897

Arlow, P., & Ulrich, T. A. (1988). A longitudinal survey of business school graduates' assessments of business ethics. *Journal of Business Ethics, 7*(4), 295–302. doi:10.1007/BF00381835

Armstrong, J., & Arvantis, C. (1996). *MC: Evolution of Large-Scale Programme within Financial Institutions*. ESOMAR/EFMA Conference on Banking and Insurance: Optimising Change, Barcelona, Spain.

Austin, T. (2014). *Goals 2000 - The Clinton Administration Education Program*. Academic Press.

Avlonitis, G., Hart, S., & Tzokas, N. (2000). An analysis of product deletion scenarios. *Journal of Product Innovation Management, 17*(1), 41–56. doi:10.1111/1540-5885.1710041

Azcona, D., Hsiao, I., & Smeaton, A. (2019). Detecting students-at risk in computer programming classes with learning analytics from students' digital footprints. *User Modeling and User-Adapted Interaction*, 1–30. doi:10.100711257-019-09234-7

Aziz, A. A. (2005). A Malaysian outcome-based engineering education model. *IACSIT International Journal of Engineering and Technology, 2*(1), 14–21.

Bagheri, M. (2015). Intelligent and Adaptive Tutoring Systems: How to Integrate Learners. *International Journal of Education, 7*(2), 1–16. doi:10.5296/ije.v7i2.7079

Bairaktarova, D., & Woodcock, A. (2015). Engineering ethics education: Aligning practice and outcomes. *IEEE Communications Magazine, 53*(11), 18–22. doi:10.1109/MCOM.2015.7321965

Baird, D., & Pane, J. (2019). Translating standardized effects of education programs into more interpretable metrics. *Educational Researcher, 48*(4), 217–229. doi:10.3102/0013189X19848729

Baker, R. (2016). Stupid tutoring systems, intelligent humans. *International Journal of Artificial Intelligence in Education, 26*(2), 600–614. doi:10.100740593-016-0105-0

Bandura, A. (1977). Self-efficacy: Towards a unifying theory of behavioural change. *Psychological Review, 84*(2), 191–215. doi:10.1037/0033-295X.84.2.191 PMID:847061

Barakat, N., & Eng, P. (2004). Issues and Challenges of Teaching Engineering Ethics. *Proceedings of the CSME 2004 Forum.*

Barbieri, P., Ciabuschi, F., Fratocchi, F., & Vignoli, M. (2018). What do we know about manufacturing reshoring? *Journal of Global Operations and Strategic Sourcing, 11*(1), 79–122. doi:10.1108/JGOSS-02-2017-0004

Barksdale-Ladd, M. A., & Thomas, K. F. (2000). What's at stake in high-stakes testing: Teachers and parents speak out. *Journal of Teacher Education, 51*(5), 384–397. doi:10.1177/0022487100051005006

Barman, P., & Bhattacharyya, D. (2012). Vivekananda's Thoughts on Man-Making through Moral Values and Character Development and Its Present Relevancy in School Education. *International Journal of Multidisciplinary Educational Research, 1*(2).

Barney, J. B. (1991). Firm resources and sustained competitive advantage. *Journal of Management, 17*(1), 99–110. doi:10.1177/014920639101700108

Barney, J. B. (2001). Resource-based theories of competitive advantage: A ten-year retrospective on the resource-based view. *Journal of Management, 27*(6), 643–650. doi:10.1177/014920630102700602

Bartolomé, A., & Cebrian-de-la-Serna, M. (2017). Personal Learning Environments: A study among Higher Education students' designs. *International Journal of Education and Development Using Information and Communication Technology, 13*(2), 21–41.

Bateson, J. E. G. (1992). *Managing Services Marketing* (2nd ed.). Orlando, FL: Dryden Press.

Bauernhansl, T., ten Hompel, M., & Vogel-Heuser, B. (2014). *Industrie 4.0 in Produktion, Automatisierung undLogistik — Anwendung, Technologien, Migration.* Wiesbaden: Springer. doi:10.1007/978-3-658-04682-8

Beausoleil, A. (2018). Revisiting Rogers: The diffusion of his innovation development process as a normative framework for innovation managers, students and scholars. *Journal of Innovation Management, 6*(4), 73–97. doi:10.24840/2183-0606_006.004_0006

Becker, M. C., & Zirpoli, F. (2017). How to Avoid Innovation Competence Loss in R&D Outsourcing. *California Management Review, 59*(2), 24–44. doi:10.1177/0008125617697941

Beiting J. (2001). Taking the mystery out of patient satisfaction. *Quirks Marketing Research Review*, 651.

Berger, U., Hartmann, A., & Schmid, D. (2013). *Additive manufacturing Fertigungsverfahren: Rapid prototyping, Rapid tooling, Rapid manufacturing*. Europe Lehrmittel.

Bero, B., & Kuhlman, A. (2011). Teaching ethics to engineers: Ethical decision making parallels the engineering design process. *Science and Engineering Ethics*, *17*(3), 597–605. doi:10.100711948-010-9213-7 PMID:20524076

Berry, L. L., & Gresham, L. G. (2001). Relationship retailing: Transforming customers into clients. *Business Horizons*, *296*, 43–47.

Berry, L. L., Parasuraman, A., & Zeithaml, V. (1988). The Service -Quality Puzzle. *Business Horizons*, *31*, 35–43. doi:10.1016/0007-6813(88)90053-5

Bersin, J. (2004). How Did We Get Here? The History of Blended Learning. In *The Blended Learning Book: Best Practices, Proven Methodologies, and Lessons Learned.* Wiley.

Bevan, D., & Kipka, C. (2012). Experiential learning and management education. *Journal of Management Development*, *31*(3), 193–197. doi:10.1108/02621711211208943

Beyrouti, N. (2017). Digital technology management and educational innovation: The marketabilityand employability of the higher education degrees. *Journal of Developing Areas*, *51*(1), 392–400. doi:10.1353/jda.2017.0022

Bezjak, J. (2010). Contemporary engineer pedagogic's project research - using multimedia at technology classes in technical and vocational schools. *Procedia: Social and Behavioral Sciences*, *2*(2), 407–411. doi:10.1016/j.sbspro.2010.03.035

Bhuiyan, T., Peng, W., & Mahmud, I. (2015). Measuring learning motivation of students in supply chain management games setting: A case study of Innov8.0 game. *Problems and Perspectives in Management*, *13*(4), 81–90.

Bickel, V. B., & Alexa, M. (2013). Computational Aspects of Fabrication: Modeling, Design, and 3D Printing. *IEEE Computer Graphics and Applications*, *33*(6), 24–25. doi:10.1109/MCG.2013.89 PMID:24921096

Bienkowski, M., Feng, M., & Means, B. (2012). Enhancing teaching and learning through educational data mining & learning analytics: An issue brief. U.S. Dept. of Education, Center for Technology in Learning.

Bloom, B. (1984). The 2-Sigma problem: The search for methods of group instruction as effective as one-on-one tutoring. *Educational Researcher*, *13*(6), 4–16. doi:10.3102/0013189X013006004

Boatright, J. (2000). *Ethics and the Conduct of Business* (3rd ed.). Prentice Hall.

Bonk, C. J., & Graham, C. R. (2006). *The handbook of blended learning environments: Global perspectives, local designs*. San Francisco: Jossey-Bass/Pfeiffer.

Bonner, D., Gilbert, S., Dorneich, M., Winer, E., Sinatra, A., Slavina, A., … Holub, J. (2016). The challenges of building intelligent tutoring systems for teams. *Proceeding of the Human Factors and Ergonomics Social Annual Meetings*, *60*(1), 1981-1985. 10.1177/1541931213601451

Boyd, M. (1995). A case for incentives. *Incentive*, *169*(2), 14.

Brauner, P., & Ziefle, M. (2015). Human Factors in Production Systems. Springer International Publishing. doi:10.1007/978-3-319-12304-2_14

Braun, I., Pull, K., Alewell, D., Störmer, S., & Thommes, K. (2011). HR outsourcing and service quality: Theoretical framework and empirical evidence. *Personnel Review*, *40*(3), 364–382. doi:10.1108/00483481111118667

Breen, J. (2004). Changing roles for (multi) media tools in design: assessing developments and applications of (multi)media techniques in design education, practice and research. *22nd eCAADe Conference Proceedings: Architecture in the Network Society,* 530–539.

Brettel, M., Friederichsen, N., Keller, M., & Rosenberg, M. (2014). How virtualization, decentralization and network building change the manufacturing landscape: An Industry 4.0 Perspective, International Journal of Science. *Engineering and Technology*, *8*(1), 37–44.

Brew, A., & Cahir, J. (2014). Achieving sustainability in learning and teaching initiatives. *The International Journal for Academic Development*, *19*(4), 341–352. doi:10.1080/1360144X.2013.848360

Brinkhurst, M., Rose, P., Maurice, G., & Ackerman, J. D. (2011). Achieving campus sustainability: Top-down, bottom-up, or neither? *International Journal of Sustainability in Higher Education*, *12*(4), 338–354. doi:10.1108/14676371111168269

Brookhart, S. M. (2013). How To Create And Use Rubrics For Formative Assessment And Grading. Alexandria, VA: ASCD.

Brown, V. A., & Lambert, J. A. (2013). *Collective Learning for Transformational Change: A Guide to Collaborative Action*. New York, NY: Routledge.

Bruce, B. C., & Hogan, M. C. (1998). The disappearance of technology: Toward an ecological model of literacy. In D. Reinking, M. McKenna, L. Labbo, & R. Kieffer (Eds.), Handbook of literacy and technology: Transformations in a post-typographic world (pp. 269–281). Academic Press.

Bruno, K. (1988). Today's mystery guest is ... a food spy, hired by management, to check service. *Nation's Restaurant News*, *22*(11), F24.

Bryson, J. (1991). MC: Uses and abuses. *Quirks Marketing Research Review,* 456.

Buil, M., Aznar, P., Galiana, J., & Rocafort-Marco, A. (2016). An explanatory study of MBA students with regards to sustainability and ethics commitment. *Sustainability*, *8*(3), 280–293. doi:10.3390u8030280

Burnside, A. (1994). In-store spies snuff out poor service. *Marketing, 32*(3).

Bustinza, O. F., Molina, L. M., & Gutierrez-Guiterrez, L. J. (2010). Outsourcing as seen from the perspective of knowledge management. *Supply Chain Management*, *46*(3), 23–39. doi:10.1111/j.1745-493X.2010.03196.x

Butterfield, K., Trevinäo, L. K., & Weaver, G. (2000). Moral Awareness in Business Organisations: Influences of Issue-Related and Social Context Factors. *Human Relations, 53*(7), 981-1018.

Buxton E. (1999). MC reports: Are quality and timeliness competing priorities? *Quirks Marketing Research Review,* 454.

Buxton, E. (2000). Similar but different. *Quirks Marketing Research Review*, 20000102.

Cannella, G. S., & Reiff, J. C. (1994). Individual constructivist teacher education: Teachers as empowered learners. *Teacher Education Quarterly*, 27–38.

Cantrell, S., & Visser, L. (2011). Factors influencing the integration of technology to facilitate transfer of learning processes in South African, Western Cape province schools. *Quarterly Review of Distance Education, 12*(4), 275–285.

Carr. (2012). Project-based learning: increasing social studies interest and engagement. *Rising Tide, 5.*

Cashin, W. E. (1995). Student Ratings of Teaching: The Research Revisited. IDEA Paper No. 32. Kansas State Univ.

Chalmers, C. (2018). Robotics and Computational thinking in primary school, Int. *J. Child-Computer Interaction, 17*, 93–100. doi:10.1016/j.ijcci.2018.06.005

Chanal, V., Martin, A., Breda, G., & Bergantz, C. (2017). From research to innovation, how to create value for an IT consulting and engineering service firm through its internal research activities. *International Journal of Knowledge-Based Organization, 7*(3).

Chen, C. (2004). Searching for intellectual turning points: Progressive knowledge domain visualization. *Proceedings of the National Academy of Sciences of the USA, 101*(1), 5303-53.

Chen, H. Z. (2005). *The application of MC in measuring the service quality of hotel telephone reservation: A case study of a resort hotel. Master theses*. Taipei: National Dong Hwa University.

Cheruvalath, R. (2015). Teaching ethics to engineering students in India: Issues and challenges. In *Contemporary ethical issues in engineering* (pp. 121–132). IGI Global. doi:10.4018/978-1-4666-8130-9.ch009

Cheung, A., & Slavin, R. (2016). How methodological features affect effect sizes in education. *Educational Researcher, 45*(5), 283–292. doi:10.3102/0013189X16656615

Chiu, P., & Huang, Y. (2016). The development of a decision support system for mobile learning: A case study in Taiwan. *Innovations in Education and Teaching International, 53*(5), 532–544. doi:10.1080/14703297.2015.1020328

Christensen, C. M., Ojomo, E., & Dillon, K. (2019). *The prosperity paradox: How innovation can lift nations out of poverty*. New York: HarperBusiness.

Claver, E., González, R., Gascó, J., & Llopis, J. (2002). Information systems outsourcing: Reasons, reservations and success factors. *Logistics Information Management, 15*(4), 294–308. doi:10.1108/09576050210436138

Clemente, J., Ramírezb, J., & Antoniob, A. (2014). Applying a student modeling with non-monotonic diagnosis to intelligent virtual environment for training/instruction. *Expert Systems with Applications, 41*(2), 508–520. doi:10.1016/j.eswa.2013.07.077

Cobb, R. (1995). Magical mystery lure. *Marketing,* 26.

Cohen, K., & Katz, R. (2015). Teaching Mechanical Design Practice in Academia. *Procedia CIRP, 36*, 177–181. doi:10.1016/j.procir.2015.01.043

Commission Staff Working Document. (2017). *Europe's Digital Progress Report – 2017.* SWD (2017) 160 final. Available from https://ec.europa.eu/digital-single-market/en/news/europes-digital-progress-report-2017

Commission Staff Working Document. (2017). *Europe's Digital Progress Report (EDPR) 2017 – Country Profile Romania.* SWD (2017) 160 final. Part 29/62. Available from https://ec.europa.eu/digital-single-market/en/news/europes-digital-progress-report-2017

Corral de Zubielqui, G., Jones, J., & Statsenko, L. (2015). Managing Innovation Networks for Knowledge Mobility and Appropriability: A Complexity Perspective. *Entrepreneurship Research Journal, 6*(1), 75–109.

Costigan, R., & Brink, K. (2015). Another perspective on MBA program alignment: An investigation of learning goals. *Academy of Management Learning & Education, 14*(2), 260–276. doi:10.5465/amle.2013.0315

Courtenay-Hall, P., & Rogers, L. (2002). Gaps in mind: Problems in environmental Knowledge-behavior modeling research. *Environmental Education Research, 8*(3), 283–297. doi:10.1080/13504620220145438

Covey, S. R. (1992). *Principle centered leadership*. Simon and Schuster.

Cramp, B. (1994). Industrious espionage and witnesses for the defence. *Marketing, 18*(17-18).

Currie, V. L., Michell, W., & Abanishe, O. (2008). Knowledge process outsourcing in financial services: The vendor perspective. *European Management Journal, 26*(2), 94–104. doi:10.1016/j.emj.2007.11.002

Customers Satisfaction Council. (1995). *Customer satisfaction assessment guide*. Schaumburg, IL: Motorola University Press.

Dafermos, V. (2005). *Social statistics using SPSS*. ZITI Publishing.

Daft, R. L. (2007). *Understanding the Theory and Design of Organizations.* Mason, OH: Thompson South-Western Publishing.

Daily Bhaskar. (2012, Sept. 27). *Maharashtra's Irrigation Scam: Meet the whistleblower Vijay Pandhare.* Author.

Damsa, C. (2019). Learning with digital technologies in higher education. *Journal of Educational Sciences & Psychology, 9*(2), 5–9.

Dasic, P., Dasic, J., Crvenkovic, B., & Serifi, V. (2016). A review of intelligent tutoring systems in e-learning. *Annals of the Oradea University – Fascicle of Management and Technological Engineering, 15*, 85-90.

Davies, R. S., & West, R. E. (2014). Technology Integration in Schools. In *Handbook of Research on Educational Communications and Technology* (pp. 841–853). New York, NY: Springer New York. doi:10.1007/978-1-4614-3185-5_68

Dawande, M., Pinedo, M., & Sriskandarajah, C. (2009). Multiple Part-Type Production in Robotic Cells: Equivalence of Two Real-World Models. *Manufacturing & Service Operations Management: M & SOM, 11*(2), 210–228. doi:10.1287/msom.1070.0208

Dawes, J., Sharp, B. (2000). The reliability and validity of objective measures of customer Service: MC. *Australian Journal of Market Research, 8*(1), 29-46.

Dawson, J., & Hillier, J. (1995). Competitor MC: Methodological considerations and implications for MRS code of conduct. *Journal of the Market Research Society, 37*(4), 417–427. doi:10.1177/147078539503700406

Day, C. (2004). *A Passion for Teaching".* Psychology Press. doi:10.4324/9780203464342

Day, I., van Blankenstein, F., Westenberg, M., & Admiraal, W. (2017). A review of the characteristics of intermediate assessment and their relationship with student grades. *Assessment & Evaluation in Higher Education, 43*(6), 908–929. doi:10.1080/02602938.2017.1417974

De Beer, N., (2013). *Additive manufacturing – Turning Mind into Mater.* Sierra College for Applied Competitive Technologies.

Deng, Z. T. (2002). *Developing assessment tools for outcome based engineering courses.* Academic Press.

Dermeval, D., Paiva, R., Bittencourt, I. I., Vassileva, J., & Borges, D. (2018). Authoring tools for designing intelligent tutoring systems: A systematic review of the literature. *International Journal of Artificial Intelligence in Education, 28*(3), 336–384. doi:10.100740593-017-0157-9

Di Gregorio, D., Musteen, M., & Thomas, D. E. (2009). Offshore outsourcing as a source of international competitiveness for SMEs. *Journal of International Business Studies, 40*(4), 969–988. doi:10.1057/jibs.2008.90

Dimitriadis, S., & Stevens, E. (2008). Integrated customer relationship management for service activities: An internal/external gap model. *Managing Service Quality, 18*(5), 496–511. doi:10.1108/09604520810898857

Dobson, J., & Bauerly, R. (2019). Ten years after: An update on globalization and the future of business education. *Proceedings of the American Society of Business & Behavioral Sciences*, *26*, 182–188.

Doherty, B., Meehan, J., & Richards, A. (2015). The business case and barriers for responsible management education in business schools. *Journal of Management Development*, *34*(1), 34–60. doi:10.1108/JMD-06-2013-0082

Donnelly, K. (2007). Australia's Adoption of Outcomes Based Education – A Critique. Issues in Educational Research.

Dorman, K. G. (1994). MC results can shape your future. *Bank Marketing*, *26*(8), 17–21.

Dormido, S., Dormido, S., Dormido, R., S'anchez, J., & Duro, N. (2005). The role of interactivity in control learning. *International Journal of Engineering Education*, *21*(6), 1122–1133.

Douglas, A., & Davies, J. (2007). *The Impact of Mystery Customers on Employees Disruptive behaviour*. Higher Education View project Lean Six Sigma in East Africa View Project.

Douglas, J. (2015). Mystery shoppers: An evaluation of their use in monitoring performance. *The TQM Journal*, *27*(6), 705–715. doi:10.1108/TQM-04-2015-0052

Durst, S., & Edvardsson, I. R. (2012). Knowledge management in SMEs: A literature review. *Journal of Knowledge Management*, *16*(6), 879–903. doi:10.1108/13673271211276173

Dutka, A. (1995). *AMA handbook of customer satisfaction: A complete guide to research, planning and implementation*. Lincolnwood, IL: NTC Business Books.

Dwek, R. (1996). Magic of MC. *Marketing*, *17*, 41–44.

Earl, L. M. (2003). Assessment as Learning: Using Classroom Assessment to Maximize Student Learning. Thousand Oaks, CA: Corwin Press, Inc.

Edvardsson, I. R. (2011). The impact of outsourcing strategies on companies' intellectual capital. *Knowledge Management Research and Practice*, *9*(4), 286–292. doi:10.1057/kmrp.2011.33

Edvardsson, I. R., & Durst, S. (2014). Outsourcing of knowledge processes: A literature review. *Journal of Knowledge Management*, *18*(4), 795–811. doi:10.1108/JKM-01-2014-0033

Edvardsson, I. R., & Teitsdóttir, U. D. (2015). Outsourcing and financial crisis: Evidence from Icelandic service SMEs. *Employee Relations*, *37*(1), 30–47. doi:10.1108/ER-11-2013-0168

Eggen, P. D., & Kauchak, D. P. (2006). *Strategies and Models for Teachers Teaching Content and Thinking Skills* (5th ed.). Pearson Education Inc.

Eisman, R. (1993). Going undercover. *Incentive*, *167*(5), 70–71.

Ekizoglu, N., & Ozcinar, Z. (2010). The relationship between the teacher candidates' computer and internet based anxiety and perceived self-efficacy. *Procedia: Social and Behavioral Sciences*, *2*(2), 5881–5890. doi:10.1016/j.sbspro.2010.03.962

El Fadil, J., & St-Pierre, J. (2016). Analysis and management of risks associated with outsourcing in China: The experience of seven Canadian firms. *Strategic Outsourcing*, *9*(2), 218–242. doi:10.1108/SO-08-2015-0018

Erstad, M. (1998). MC programmes and human resource management. *International Journal of Contemporary Hospitality Management*, *10*(1), 34–38. doi:10.1108/09596119810199318

Ertmer, P. A. (2005). Teacher pedagogical beliefs: The final frontier in our quest for technology integration. *Educational Technology Research and Development*, *53*(4), 25–39. doi:10.1007/BF02504683

Ertu, U., Atkinson, P., & Tancock, D. (2006). Applying 3D scanning and modelling in transport design education. *Computer-Aided Design and Applications*, *3*(1-4), 41–48. doi:10.1080/16864360.2006.10738440

Eschenbrenner, B., & Fui-Hoon Nah, F. (2019). Learning through mobile devices: Leveraging affordances as facilitators of engagement. *International Journal of Mobile Learning and Organization*, *13*(2), 152–170. doi:10.1504/IJMLO.2019.098193

European Commission. (2013). *Commission presents new Rethinking Education strategy.*. European Commission.

European Commission. (2016). *Communication from the European Commission to the European Parliament, the Council, the European Economic and Social Committee and the Committee of the Regions. A new skills agenda for Europe – Working together to strengthen human capital, employability and competitiveness.* COM (2016) 381 final. Available from https://eur-lex.europa.eu/legal-content/EN/TXT/?uri=CELEX%3A52016DC0381

European Commission. (2017). *Communication from the European Commission to the European Parliament, the Council, the European Economic and Social Committee and the Committee of the Regions. A renewed agenda for higher education.* COM (2017) 247 final. Available from https://eur-lex.europa.eu/legal-content/EN/TXT/?uri=COM%3A2017%3A247%3AFIN

Evanschitzky, H., Iyer, G., & Caemmerer, B. (2008). Dimensions of satisfaction in retail settings: A research note. *Journal of Relationship Marketing*, *7*(3), 275–285. doi:10.1080/15332660802409621

Fahimrad, M., & Kotamjani, S. (2018). A review on application of artificial intelligence in teaching and learning in educational contexts. *International Journal of Learning and Development*, *8*(4), 106–118. doi:10.5296/ijld.v8i4.14057

Fang, D. Y., Zhang, A. M., & Qiu, Y. (2012). Research and practice on new measuring and drafting mode of mechanical drawing based on reverse engineering. *International Conference on Measurement, Instrumentation and Automation. Applied Mechanics and Materials*, *239*(240), 645–648. doi:10.4028/www.scientific.net/AMM.239-240.645

Farooq, M., Kawoosa, H., & Muttoo, M. (2017). CAL: Computer assisted learning. *International Journal of Computer Science and Mobile Computing*, *6*(6), 254–258.

Fathema, N., Shannon, D., & Ross, M. (n.d.). Expanding the technology acceptance model (TAM) to examine faculty use of learning management systems (LMSs) in higher education institutions. *MERLOT Journal of Online Learning and Teaching, 11*(2), 210-232.

Fed'ak, V., & Bauer, P. (2005). E-learning in education of electrical drives and power electronics: Opportunities and challenges. *International Conference on Electrical Drives and Power Electronics*.

Felah, B., & Noreddine, H. (2017). Pedagogical robotics-A way to experiment and innovate in education teaching in Morocco, *Int. J. Education and Learning Systems*, *2*, 71–75.

Felder. (1988). Learning and Teaching Styles: In Engineering Education. Energy Education, 78(7), 674–681.

Felijakowski, K., & Kosinski, R. (2014). Opinion formation and self-organization in a social network in an intelligent agent system. *Acta Physica Polonica B*, *45*(11), 2123–2134. doi:10.5506/APhysPolB.45.2123

Ferguson, C. (2009). An effect size primer: A guide for clinicians and researchers. *Professional Psychology, Research and Practice*, *40*(5), 532–538. doi:10.1037/a0015808

Finn, A. (2001). MC benchmarking of durable-goods chains and stores. *Journal of Service Research*, *3*(4), 310–320. doi:10.1177/109467050134004

Finn, A., & Kayandé, U. (1997). Reliability Assessment and and Optimization of Marketing Measures. *JMR, Journal of Marketing Research*, *34*(2), 262–275. doi:10.1177/002224379703400206

Finn, A., & Kayande, U. (1999). Unmasking a phantom: A psychometric assessment of MC. *Journal of Retailing*, *75*(2), 195–217. doi:10.1016/S0022-4359(99)00004-4

Fisher, L., Brinthaupt, T., Gardner, J., & Raffo, D. (2014). Conditional release of course materials: Assessing best practice recommendations. *Journal of Online Learning and Teaching / MERLOT*, *10*(2), 228–239.

Flavin, M. (2016). Technology-enhanced learning and higher education. *Oxford Review of Economic Policy*, *3*(4), 632–645. doi:10.1093/oxrep/grw028

Fleacă, E., Fleacă, B., & Maiduc, S. (2017). Modeling Stakeholders Relationships to Strengthen the Entrepreneurial Behavior of Higher Education Institutions. *Proc. Eng*, *181*, 935–942. doi:10.1016/j.proeng.2017.02.490

Fleacă, E., Fleacă, B., & Maiduc, S. (2018). Aligning Strategy with Sustainable Development Goals (SDGs): Process Scoping Diagram for Entrepreneurial Higher Education Institutions (HEIs). *Sustainability*, *10*(4), 1032. doi:10.3390u10041032

Fleaca, E., & Stanciu, R. D. (2019). Digital age learning and Business Engineering Education-A pilot study on Student's E-skills. *Procedia Manufacturing*, *32*, 1051–1105. doi:10.1016/j.promfg.2019.02.320

Fleddermann. (2012). *Engineering Ethics*. Prentice Hall.

Fletcher, J., & Sottilare, R. (2018). *A perspective of intelligent tutoring systems design and use, in Assessment of intelligent tutoring systems technologies and opportunities*. North Atlantic Treaty Organization, STO Technical Report, AC/323(HFM-237)TP/764, 2, 1-11.

Ford, P., & Dean, L. (2013) Additive manufacturing in product design education: out with the old and in with the new? *Proceedings of E and PDE 2013, the 15th International conference on engineering and product Design education,* 611-616.

Ford, S., & Minshell, T. (2019). Invited review article: Where and how 3D printing is used in teaching and education. *Additive Manufacturing*, *25*, 131–150. doi:10.1016/j.addma.2018.10.028

Frankfort-Nachmias, C., & Nachmias, D. (1992). *Research methods in social sciences*. London: St. Martin's Press Inc.

Freeman, J., & Zachary, W. (2018). Intelligent Tutoring for Team Training: Lessons Learned from US Military Research. In Building Intelligent Tutoring Systems for Teams. Emerald Publishing Limited.

Furtado, T. (1998). Ethical Dilemmas RegardingCompetitive Intelligence. In *Perspectives in Business Ethics*. London: Irwin McGraw Hill.

Gaur, R. R., Sangal, R., & Bagaria, G. P. (2011). *A Foundation Course in Human Values and Professional Ethics*. New Delhi, India: Excel Books.

Ghemawat, P. (2017). Strategies for higher education in the digital age. *California Management Review*, *59*(4), 56–78. doi:10.1177/0008125617717706

Ghodeswar, B., & Vaidyanathan, J. (2008). Business process outsourcing: An approach to gain access to world-class capabilities. *Business Process Management Journal*, *14*(1), 23–28. doi:10.1108/14637150810849382

Gibson, I., Rosen, D. W., & Stucker, B. (2010). *Additive manufacturing technologies: Rapid prototyping to direct digital Manufacturing*. Springer. doi:10.1007/978-1-4419-1120-9

Gofton, K. (1995). *Service without a smile; customer service, marketing guides; customer loyalty. Marketing, 2(3).*

Goi, C. (2019). The use of business simulation games in teaching and learning. *Journal of Education for Business*, *94*(5), 342–349. doi:10.1080/08832323.2018.1536028

Gonzalez, M. (2014). Virtual teams and international business teaching and learning: The case of the global enterprise experience. *Journal of Teaching in International Business*, *25*(3), 200–213. doi:10.1080/08975930.2014.925738

González-Marcos, A., Alba-Elíasa, F., Navaridas-Naldab, F., & Ordieres-Merέc, J. (2016). Student evaluation of a virtual experience for project management learning. *Computers & Education*, *102*, 172–187. doi:10.1016/j.compedu.2016.08.005

Gorla, N., & Lau, M. B. (2010). Will negative experience impact future IT outsourcing? *Journal of Computer Information Systems*, *50*(3), 91–101.

Gosselt, J., van Hoof, J., De Jong, M., & Prinsen, S. (2007). MC and alcohol sales: Do supermarkets and liquor stores sell alcohol to underage customers? *The Journal of Adolescent Health*, *41*(3), 302–308. doi:10.1016/j.jadohealth.2007.04.007 PMID:17707301

Goswami, D., & Chakraborty, A. (2015, November). Sensitizing engineers: A brief study of the role of ethics in engineering education. In *2015 5th Nirma University International Conference on Engineering (NUiCONE)* (pp. 1-4). IEEE. 10.1109/NUICONE.2015.7449589

Gottschalk, P., & Solli-Sæther, H. (2007). Knowledge transfer in IT outsourcing relationships: Three international case studies. *International Journal of Innovation and Learning*, *4*(2), 103–111. doi:10.1504/IJIL.2007.011687

Graesser, A., Hu, X., Nye, B., VanLehn, K., Kumar, R., Heffernan, C., ... Woolf, B. (2018). Intelligent tutoring systems. In *International Handbook of the Learning Sciences* (pp. 246–255). Routledge. doi:10.4324/9781315617572-24

Graham, C. R. (2011). Theoretical considerations for understanding technological pedagogical content knowledge (TPACK). *Computers & Education*, *57*(3), 1953–1960. doi:10.1016/j.compedu.2011.04.010

Graham, C., Woodfield, W., & Harrison, B. (2013). A framework for institutional adoption and implementation of blended learning in higher education. *The Internet and Higher Education*, *18*, 4–14. doi:10.1016/j.iheduc.2012.09.003

Grant, R.M. (1996). Towards a knowledge-based theory of the firm. *Strategic Management Journal*, *17*, 109-122.

Grattan Institute Report No. 2017-01, February 2017.

Green, R. (1994). *The Ethical Manager: A New Method for Business Ethics*. Englewood Cliffs, NJ: Macmillan.

Grigoroudis, E., & Siskos, Y. (2002). Preference disaggregation for measuring and analyzing customer satisfaction: The MUSA method. *European Journal of Operational Research*, *143*(1), 148–170. doi:10.1016/S0377-2217(01)00332-0

Grigoroudis, E., & Spyridaki, O. (2003). Derived vs. stated importance in customer satisfaction surveys. *Operational research. International Journal (Toronto, Ont.)*, *3*(3), 229–247.

Groover, M. P. (2007). *Automation, Production Systems, and Computer-Integrated Manufacturing*. Prentice Hall Press.

Grove, S. J., & Fisk, R. P. (1992). Observational data collection methods for services marketing: An overview. *Journal of the Academy of Marketing Science*, *20*(3), 217–224. doi:10.1007/BF02723408

Guney, D. (2015). The importance of computer-aided courses in architectural education. *Procedia: Social and Behavioral Sciences, 176*, 757–765. doi:10.1016/j.sbspro.2015.01.537

Gupta, A., Herath, S. K., & Mikouiza, N. C. (2005). Outsourcing in higher education: An empirical examination. *International Journal of Educational Management, 19*(4), 396–412. doi:10.1108/09513540510607734

Guskey, T. (2007). Closing achievement gaps: Revisiting Benjamin S. Bloom's earning for mastery. *Journal of Advanced Academics, 19*(1), 8–31. doi:10.4219/jaa-2007-704

Hall, S., Grant, G., Arora, D., Karaksha, A., McFarland, A., Lohning, A., Anoopkumar-Dukie, S., (2017). A pilot study assessing the value of 3D printed molecular modelling tools for pharmacy student education. *Currents in Pharmacy Teaching and Learning*.

Hamad, Hasanain, Abdulwahed, & Al-Ammari. (2013). *Ethics in Engineering Education: A Literature Review*. IEEE.

Hanafizadeh, P., & Ravasan, A. Z. (2018). An empirical analysis on outsourcing decision: The case of e-banking services. *Journal of Enterprise Information Management, 31*(1), 146–172. doi:10.1108/JEIM-11-2016-0182

Hanke, E. (1993). Through the looking glass. *Credit Union Management, 42*, 45-6.

Harden, R. M. (1999). AMEE Guide No. 14: Outcome-based education: Part 1-An introduction to outcome-based education. *Medical Teacher, 21*(1), 7–14. doi:10.1080/01421599979969

Harden, R. M. (2002). Developments in outcome-based education. *Medical Teacher, 24*(2), 117–120. doi:10.1080/01421590220120669 PMID:12098427

Harris Jr, Pritchard, & Rabins. (2017). *Engineering ethics: Concepts and cases*. Cengage Learning.

Harris, C., & Brown, W. (1990). Developmental constraints on ethical behavior in business. *Journal of Business Ethics, 9*(11), 855–862. doi:10.1007/BF00382909

Harvey, E., & Reitz, E. (2017). Enhanced stem subject outcomes from the use of intelligent tutors. Interservice/Industry Training. *Simulation, and Education Conference., 17239*, 1–10.

Hawe, E., & Dixon, H. (2016). Assessment for learning: A catalyst for student self-regulation. *Assessment & Evaluation in Higher Education*.

Heath, C. (2016). Why diversity matters: designing the right design team. *New Design Magazine, 6*(132). Retrieved from http://newdesignmagazine.co.uk/

Herkert, J. R. (2000). Engineering ethics education in the USA: Content, pedagogy and curriculum. *European Journal of Engineering Education, 25*(4), 303–313. doi:10.1080/03043790050200340

Herrington, J., & Arnold, D. (2013). Undergraduate business education: It's time to think outside the box. *Journal of Education for Business, 88*(2), 202–209. doi:10.1080/08832323.2012.672934

Hesselink, M., & van der Wiele, A. (2003). *MC: In depth measurement of customer satisfaction.* Research Paper ERS-2003-020-ORG Revision, Erasmus Research Institute of Management (ERIM). Retrieved September, 2017, from http://repub.eur.nl/res/pub/281/

Hesselink, M., Iwaarden, J.D. van & Wiele, A. van der (2004). *MC: A tool to develop insight into customer service provision.* ERIM report series research in management, ERS-2004-082-ORG.

Hill, N. (1996). *Handbook of customer satisfaction measurement.* Aldershot, UK: Gower Press.

Hills, T. (2015). Crowdsourcing content creation in the classroom. *Journal of Computing in Higher Education, 27*(1), 47–67. doi:10.100712528-015-9089-2

Hislop, D. (2013). *Knowledge Management in Organizations: A Critical Introduction* (3rd ed.). Oxford, UK: Oxford University Press.

Hoffman, G. (1993). Customers can hone MC. *Bank Marketing, 25*(8), 36–37.

Holliday, K. K. (1994). Mutual funds. *Bank Marketing, 26*(7), 23–31.

Homburg, C., & Rudolph, B. (2001). Customer satisfaction in industrial markets: Dimensional and multiple role issues. *Journal of Business Research, 52*(1), 15–33. doi:10.1016/S0148-2963(99)00101-0

Homes, W., Bialik, M., & Fadel, C. (2019). *Artificial intelligence in education: Promises and implications for teaching and learning.* Boston: Center for Curriculum Redesign.

Hopf, H., Krief, A., Mehta, G., & Matlin, S. A. (2019). Fake science and the knowledge crisis: Ignorance can be fatal. *Royal Society Open Science, 6*(5), 190161. doi:10.1098/rsos.190161 PMID:31218057

Hosmer, L. T. (1988). Adding ethics to the business curriculum. *Business Horizons, 31*(4), 9–15. doi:10.1016/0007-6813(88)90062-6

Hotchkiss, D. A. (1995). What do your customers really think? *Bank Marketing, 27*(3), 13–20.

Howells, J., Gagliardi, D., & Malik, K. (2008). The growth and management of R&D outsourcing: Evidence from UK pharmaceuticals. *R & D Management, 38*(2), 205–219. doi:10.1111/j.1467-9310.2008.00508.x

Hsiung, C. (2012). The effectiveness of cooperative learning. *Journal of Engineering Education, 101*(1), 119–137. doi:10.1002/j.2168-9830.2012.tb00044.x

Hughes, T. P., Carpenter, S., Rockström, J., Scheffer, M., & Walker, B. (2013). Multiscale regime shifts and planetary boundaries. *Trends in Ecology & Evolution, 28*(7), 389–395. doi:10.1016/j.tree.2013.05.019 PMID:23769417

Huhta, A. (2010). Diagnostic and Formative Assessment. In The Handbook of Educational Linguistics (pp. 469–482). Oxford, UK: Blackwell.

Hurley, W. G., & Lee, C. K. (2005). Development, implementation, and assessment of a web-based power electronics laboratory. *IEEE Transactions on Education*, *48*(4), 567–573. doi:10.1109/TE.2005.856147

IBM SPSS Statistics. (2015). Version 22. Retrieved from https://www.ibm.com/products/spss-statistics

Iqbal, Z., & Dad, A. M. (2013). Outsourcing: A review of trends. Winners & losers and future directions. *International Journal of Business and Social Science*, *4*(8), 91–107.

Jaafar, M. S. (2008). Assessment strategy for an outcome-based education. *International Conference on Engineering Education.*

Jacob, W. (2015). Interdisciplinary trends in higher education. *Palgrave Communications*, *1*(15001), 1–5.

Janghorbani, S., Modi, A., Buhmann, J., & Kapadia, M. (2019). Domain Authoring Assistant for Intelligent Virtual Agents. *Proceedings of the 18th International Conference on Autonomous Agents and Multi-gent Systems*, 1-9.

Jankal, R., & Jankalová, M. (2011). Mystery shopping - the tool of employee communication skills evaluation. Business. *Theory into Practice*, *12*(1), 45–49. doi:10.3846/btp.2011.05

Jashapara, A. (2004). *Knowledge management: An integrated approach.* Harlow: Prentice Hall.

Jawahar, J. (2015). A study on role of intelligent agent in education. *International Journal of Applied Engineering Research*, *10*(76), 497–501.

Jesson, J. K., Matheson, L., & Lacey, F. M. (2011). *Doing Your Literature Review: Traditional and Systematic Techniques.* Los Angeles, CA: Sage.

Jiménez, S., Juárez-Ramírez, R., Castillo, V., & Ramírez-Noriega, A. (2018). Integrating affective learning into intelligent tutoring systems. *Universal Access in the Information Society*, *17*(4), 679–692. doi:10.100710209-017-0524-1

Johnson, D., & Samora, D. (2016). The potential transformation of higher education through computer-based adaptive learning systems. *Global Education Journal*, *16*(1), 1–17.

Joyce, T., Evans, I., Pallan, W., & Hopkins, C. (2013). A Hands-on Project-based Mechanical Engineering Design Module Focusing on Sustainability. *Engineering Education*, *8*(1), 65–80. doi:10.11120/ened.2013.00008

Jumadinova, J., Dasgupta, P., & Soh, L. (2014). Strategic capability learning for improving multi-agent collaboration in ad hoc environments. *IEEE Transactions on Systems, Man, and Cybernetics. Systems*, *44*(8), 1003–1014. doi:10.1109/TSMC.2013.2285527

Kafchehi, P., Hasani, K., & Gholami, A. (2016). The Relationship between Innovation orientation and strategic typology in business firms. *International Journal of Knowledge-Based Organization*, *6*(2), 1-20.

Kakabadse, A., & Kakabadse, N. (2000). Critical review – outsourcing: A paradigm shift. *Journal of Management Development*, *19*(8), 670–728. doi:10.1108/02621710010377508

Kakabadse, A., & Kakabadse, N. (2005). Outsourcing: Current and future trends. *Thunderbird International Business Review*, *47*(2), 183–204. doi:10.1002/tie.20048

Kakumanu, P., & Portanova, A. (2006). Outsourcing: Its benefits, drawbacks and other related issues. *The Journal of American Academy of Business, Cambridge*, *9*(2), 1–8.

Kane, G. C., Palmer, D., Phillips, A. N., & Kiron, D. (2015). Is your business ready for a digital future? *MIT Sloan Management Review*, *56*(4), 37–44.

Kane, G., Palmer, D., Phillips, A., & Kiron, D. (2017). Winning the Digital War for Talent. *MIT Sloan Management Review*, *58*(2), 17–19.

Kanyarusoke, K. (2019). Mechanical Engineering Design: Going Over the Analysis-Synthesis Mountain to Seed Creativity. Engineering Design and Innovation Methods. doi:10.5772/intechopen.85174

Karady, G. G., & Heydt, G. T. (2000). Increasing student interest and comprehension in power engineering at the graduate and undergraduate levels. *IEEE Transactions on Power Systems*, *15*(1), 16–21. doi:10.1109/59.852095

Karakas, F., Manisaligilb, A., & Sarigolluc, E. (2015). Management learning at the speed of life: Designing reflective, creative, and collaborative spaces for millennials. *International Journal of Management Education*, *13*(3), 237–248. doi:10.1016/j.ijme.2015.07.001

Kara, L. (2015). A critical look at the digital technologies in Architectural education: When, where, and how? *Procedia: Social and Behavioral Sciences*, *176*, 526–530. doi:10.1016/j.sbspro.2015.01.506

Kaufman, D. (1996). Constructivist-based experiential learning in teacher education. *Action in Teacher Education*, *18*(2), 40–50. doi:10.1080/01626620.1996.10462832

Kehagias, J., Rigopoulou, I., & Vassilikopoulou, A. (2011). Linked mystery shopping inventory to customer-seller encounters. *Journal of Customer Behaviour*, *10*(1), 7–34. doi:10.1362/147539211X570492

Kennedy, K. (2011). Conceptualising Quality Improvement in Higher Education: Policy, Theory and Practice for Outcomes Based Learning In Hong Kong. Journal of Higher Education Policy and Management, 33(3), 205–218. doi:10.1080/1360080X.2011.564995 doi:10.1080/1360080X.2011.564995

Khalifa, M., & Lam, R. (2002). Web-based learning: Effects on learning process and outcome. *IEEE Transactions on Education*, *45*(4), 350–356. doi:10.1109/TE.2002.804395

Kim, M., Park, J. E., Dubinsky, A. J., & Chaiy, S. (2012). Frequency of CRM implementation activities: A customer-centric view. *Journal of Services Marketing*, *26*(2), 83–93. doi:10.1108/08876041211215248

Kimonen, E., & Nevalainen, R. (2000). Active learning in the process of educational change. *Teaching and Teacher Education*, *21*(6), 623–635. doi:10.1016/j.tate.2005.05.003

Kluge, J., Wolfram, S. & Licht, T. (2001). *Knowledge Unplugged. The McKinsey & Company Global Survey on Knowledge Management*. Houndsmills: Palgrave.

Kochhar-Bryant, C., Bassett, D. S., & Webb, K. W. (2009). *Transition to post secondary education for students with disabilities*. Thousand Oaks, CA: Corwin Press.

Koedinger, K., Corbett, A., & Perfetti, C. (2012). The knowledge-learning-instruction framework: Bridging the science-practice chasm to enhance robust student learning. *Cognitive Science*, *36*(5), 757–798. doi:10.1111/j.1551-6709.2012.01245.x PMID:22486653

Koehier, M., Mishra, P., & Cain, W. (2013). What is Technological Pedagogical Content Knowledge (TPACK)? *Journal of Education*, *193*(3), 13–19. doi:10.1177/002205741319300303

Kolberg, D., & Zühlke, D. (2015). Lean automation enabled by industry 4.0 technologies. *IFAC-PapersOnLine*, *48*(3), 1870–1875. doi:10.1016/j.ifacol.2015.06.359

Korhan, A. (2015). Effect of knowledge sharing and supply chain management on organizational performance. *International Journal of Knowledge-Based Organizations*, *5*(1), 19–32.

Kotabe, M., Mol, M. J., & Ketkar, S. (2008). An evolutionary stage model of outsourcing and competence destruction: A triad comparison of the consumer electronics industry. *Management International Review*, *48*(1), 65–93. doi:10.100711575-008-0004-1

Kovatsch, M., Mayer, S., & Ostermaier, B. (2012). Moving application logic from the firmware to the cloud: Towards the thin server architecture for the internet of things. *Proc. Innovative Mobile and Internet Services and Ubiquitous Computing (IMIS), 2012 Sixth International Conference, IEEE*, 751-756.

Kremic, T., Tukel, I. G., & Rom, O. W. (2006). Outsourcing decision support: A survey of benefits, risks, and decision factors. *Supply Chain Management*, *11*(6), 467–482. doi:10.1108/13598540610703864

Kriesi, C., Steinert, M., Meboldt, M., & Balters, S. (2014) Physiological Data acquisition for deeper insights into Prototyping, In *Proceedings of NordDesign 2014 Conference*. Aalto Design Factory and Design Society.

Kulik, J., & Fletcher, J. (2016). Effectiveness of intelligent tutoring systems: A meta-analytic review. *Review of Educational Research*, *86*(1), 42–78. doi:10.3102/0034654315581420

Kumar, S., Deivasigamani, A. J., & Omer, W. W. (2010). Knowledge for sale – the benefits and effects of off-shoring knowledge-based jobs in engineering, design, and R&D – a case study. *Knowledge Management Research and Practice*, *8*(4), 351–368. doi:10.1057/kmrp.2010.24

Lacity, M., Solomon, S., Yan, A., & Willcocks, L. (2011). Business process outsourcing studies: A critical review and research directions. *Journal of Information Technology*, *26*(4), 221–258. doi:10.1057/jit.2011.25

Lam, W., & Chua, A. Y. K. (2009). Knowledge outsourcing: An alternative strategy for knowledge management. *Journal of Knowledge Management, 13*(3), 28–43. doi:10.1108/13673270910962851

Lanz, M., Pieters, R., & Ghabcheloo, R. (2019). Learning environment for robotics education and industry-academia collaboration. *Procedia Manufacturing, 31*, 79–94. doi:10.1016/j.promfg.2019.03.013

Larsen, R., Blocher, E., Brewer, P., Cokins, G., Sorensen, J., Stout, D., ... Wouters, M. (2014). Focusing accounting curricula on students' long-run careers: Recommendations for an integrated competency-based framework for accounting education. *Issues in Accounting Education, 29*(2), 295–317. doi:10.2308/iace-50673

Lazovic, V., Montenegro, J., & Durickovic, T. (2014). The digital economy in developing Countries -challenges and opportunities. *37th International Convention on Information and Communication Technology, Electronics and Microelectronics (MIPRO)*, 580-1585. 10.1109/MIPRO.2014.6859817

Leeds, B. (1992). MC offers clues to quality service. *Bank Marketing, 24*(11), 24–26.

Leeds, B. (1995). MC: From novelty to necessity. *Bank Marketing, 27*(6), 17–23.

Lee, H., & Mehta, M. (2015). E-Mentoring through a network of practice on Facebook. *International Journal of Knowledge-Based Organizations, 5*(1), 34–45. doi:10.4018/ijkbo.2015010103

Lee, J., Huynh, M. Q., & Hirschheim, R. (2008). An integrative model of trust on IT outsourcing: Examining a bilateral perspective. *Information Systems Frontiers, 10*(2), 145–163. doi:10.100710796-008-9066-7

Lee, J., Kao, H.-A., & Yang, S. (2014). Service innovation and smart analytics for industry 4.0 and big data environment. *Procedia CIRP, 16*, 3–8. doi:10.1016/j.procir.2014.02.001

Lee, J., Lapira, E., Yang, S., & Kao, H. A. (2013). Predictive manufacturing system trends of next generation production systems. *Proceedings of the 11th IFAC workshop on intelligent manufacturing systems*, 150–6. 10.3182/20130522-3-BR-4036.00107

Lee, S.-H. (1999). Taxonomy of HRM policies of knowledge-based firms: A resource-based explanation. *International Journal of Innovation Management, 3*(4), 379–395. doi:10.1142/S1363919699000190

Leso, T., & Peck, K. L. (1992). Computer anxiety and different types of computer courses. *Journal of Educational Computing Research, 8*(4), 469–478. doi:10.2190/Q1TJ-8JCU-LDAP-84H8

Li, N., Matsuda, N., Cohen, W., & Koedinger, K. (2015). Integrating representation learning and skill learning in a human-like intelligent agent. *Artificial Intelligence, 219*, 67–91. doi:10.1016/j.artint.2014.11.002

Li, X., Wei, J., Ding, S., & Wang, Q. (2017). Teaching Mode of Mechanical Design Course based on Simulation Analysis Technology. *International Journal of Emerging Technologies in Learning, 12*(7), 112–123. doi:10.3991/ijet.v12i07.7221

Lohmann, H. (1954). The Technique and its Teaching-A research sub -program for a scientific engineering pedagogy. *Scientific Journal of the TH Dresden*, *3*(4), 602–629.

Loveless, A. (2003). *Role of ICT*. A&C Black.

Lowman, M., Trott, P., Hoecht, A., & Sellam, Z. (2012). Innovation risks of outsourcing in pharmaceutical new product development. *Technovation*, *32*(2), 99–109. doi:10.1016/j.technovation.2011.11.004

Lubeck, J., Cheng, B., Myszkowski, G., Doedijns, G., Drew, D., & Snow, S. (2015). *Future Trends in Business Education*. Association to Advance Collegiate Schools of Business (AACSB), the Executive MBA Council (EMBAC), and the Executive Education Consortium (UNICON), Executive Core.

Luckin, R., Holmes, H., Griffiths, M., & Forcier, L. (2016). *Intelligence unleashed: An argument for AI in education. UCL Knowledge Lab*. London: Pearson.

Lu, O., Huang, A., Huang, J., Lin, A., Ogata, H., & Yang, S. (2018). Applying learning analytics for the early prediction of students' academic performance in blended learning. *Journal of Educational Technology & Society*, *21*(2), 220–232.

Luvison, D., & Bendixen, M. (2010). The behavioral consequences of outsourcing: Looking through the lens of paradox. *The Journal of Applied Management and Entrepreneurship*, *15*(4), 28–52.

Mahmoodzadeh, E., Jalalinia, S., & Nekui Yazdi, F. (2009). A business process outsourcing framework based on business process management and knowledge management. *Business Process Management Journal*, *15*(6), 845–864. doi:10.1108/14637150911003748

Maldonado, E., & Seehusen, V. (2018). Data mining student choices: A new approach to business curriculum planning. *Journal of Education for Business*, *93*(5), 196–203. doi:10.1080/08832323.2018.1450212

Mansor, W. (2008). Preliminary results on the implementation of outcome-based education on the non-examinable computer engineering modules. In *2008 38th Annual Frontiers in Education Conference*. IEEE. 10.1109/FIE.2008.4720675

Martin, C., & Polly, D. (2017). Handbook of Research on Teacher Education and Professional Development. IGI Global. doi:10.4018/978-1-5225-1067-3

Martínez-Noya, A., & García-Canal, E. (2015). The framing of knowledge transfers to shared R&D suppliers and its impact on innovation performance: A regulatory focus perspective. *R & D Management*, *46*(2), 354–368. doi:10.1111/radm.12191

Martyn, M. (2003). The hybrid online model: Good practice. *EDUCAUSE Quarterly*, 18–23.

Mavromatidis, L. (2018), Coupling architectural synthesis to applied thermal engineering, constructed thermodynamics and fractal analysis: An original pedagogic method to incorporate sustainability into architectural education during the initial conceptual stages. *Sustainable Cities and Society*, *39*, 689-707.

Ma, W., Adesope, O., Nesbit, J., & Liu, Q. (2014). Intelligent tutoring systems and learning outcomes: A meta-analysis. *Journal of Educational Psychology, 106*(4), 901–918. doi:10.1037/a0037123

May, D. R., & Luth, M. T. (2013). The effectiveness of ethics education: A quasi-experimental field study. *Science and Engineering Ethics, 19*(2), 545–568. doi:10.100711948-011-9349-0 PMID:22212360

McNerney, D. J. (1996). The link to customer satisfaction. *HRFocus, 73*(9), 1–4.

McWhirter, N. D., & Shealy, T. (2018). Teaching decision-making for sustainable Infrastructure: A wind energy case study module. *International Journal of Sustainability in Higher Education, 19*(5), 893–911. doi:10.1108/IJSHE-10-2017-0183

Mehrotra, A., & Elias, H. (2017). Employers' Feedback on Business Graduates. *US-China Education Review, 7*(4), 190–199.

Mehta, A., Armenakis, A., Mehta, N., & Irani, F. (2006). Challenges and opportunities of business process outsourcing in India. *Journal of Labor Research, 27*(3), 324–338. doi:10.100712122-006-1026-8

Melonçon, L., Rosselot-Merritt, J., & St.Amant, K. (2019). A Field-Wide Metasynthesis of Pedagogical Research in Technical and Professional Communication. *Journal of Technical Writing and Communication*. doi:10.1177/0047281619853258

Mendo, S., León-del-Barco, B., Felipe-Castaño, E., Polo-del-Río, M., & Iglesias-Gallego, D. (2018). Cooperative team learning and the development of social skills in higher education: The variables involved. *Frontiers in Psychology, 9*, 15–36. PMID:29445346

Mergendoller, J. (2006). *Project Based Learning Handbook* (2nd ed.). Novatto, CA: Buck Institute for Education.

Michaels, R. E., & Day, R. L. (1985). Measuring orientation of salespeople: A replication with industrial buyers. *JMR, Journal of Marketing Research, 22*(4), 443–446. doi:10.1177/002224378502200409

Michelson, M. (2001). The difference between mystery shopping and marketing research. *Quirks Marketing Research Review*, 653. Retrieved from www.quirks.com

Miles, L. (1993). Rise of the MC. *Marketing, 29*, 19–20.

Mimirinis. (2018). Qualitative differences in academics. *Springer Nature, 30*(3), 207-210.

Miozzo, M., & Grimshaw, D. (2005). Modularity and innovation in knowledge-intensive business services: IT outsourcing in Germany and the UK. *Research Policy, 34*(9), 1419–1439. doi:10.1016/j.respol.2005.06.005

Mishra, P., & Koehler, M. J. (2006). Technological Pedagogical Content Knowledge: A Framework for teacher knowledge. *Teachers College Record, 108*(6), 1017–1054. doi:10.1111/j.1467-9620.2006.00684.x

Mishra, P., Spiro, R. J., & Feltovich, P. J. (1996). Technology, representation, and cognition: The prefiguring of knowledge in cognitive flexibility hypertexts. In H. van Oostendorp & A. de Mul (Eds.), *Cognitive aspects of electronic text processing* (pp. 287–305). Norwood, NJ: Ablex.

Mohapatra, S. (2015). Business school education and technology: A case study. *Education and Information Technologies, 20*(2), 335–346. doi:10.100710639-013-9287-3

Mohayidin, M. G., Suandi, T., Mustapha, G., Konting, M. M., Kamaruddin, N., Man, N. A., . . . Abdullah, S. N. (2008). Implementation of Outcome-Based Education In Universiti Putra Malaysia: A Focus On Students Learning Outcomes. International Education Studies, 1(4). doi:10.5539/ies.v1n4p147 doi:10.5539/ies.v1n4p147

Mokyr. (1998). *The Second Industrial Revolution, 1870-1914, The Lever of Riches*. Academic Press.

Monterrubio, L.E., & Sirinterlikci, A. (2015). A Hands-on Approach in Teaching Machine Design. *122nd American Society for Engineering Education.*

Moore, J. (2005). Seven recommendations for creating sustainability education at the university level: A guide for change agents. *International Journal of Sustainability in Higher Education, 6*(4), 326–339. doi:10.1108/14676370510623829

Morral, K. (1994). MCs serve tests service and compliance. *Bank Marketing, 26*(2), 13.

Morrison, E. (1994). Role definitions and organizational citizenship behaviour: The importance of the employee's perspective. *Academy of Management Journal, 37*, 1543–1567.

Morrison, L. J., Colman, A. M., & Preston, C. C. (1997). Mystety Customer Research: Cognitive Processes Affecting Accuracy. *Journal of the Market Research Society, 39*(2), 349–361. doi:10.1177/147078539703900203

Mostert-van der Sar, M., Mulder, I., Remijn, L., & Troxler, P. (2013). Fablabs in design education. *Proceedings of E and PDE 2013, the 15th international conference on engineering and product design education*, 629-634.

MRS. (2009). *Market Research Society*. Retrieved October 2017, from www.mrs.org.uk

Muhson, A., Lestari, B., Supriyanto, S., & Baroroh, K. (2017). The Development of Practical Item Analysis Program for Indonesian Teachers. *International Journal of Instruction, 10*(2), 199–210. doi:10.12973/iji.2017.10213a

Murray, J. Y., Kotabe, M., & Westjohn, S. A. (2009). Global sourcing strategy and performance of knowledge-intensive business services: A two-stage strategic fit model. *Journal of International Marketing, 17*(4), 90–105. doi:10.1509/jimk.17.4.90

Nair, C. G. K. (2006). *Engineering Ethics*. Bangalore: Hari Sri Publishing Co.

Najar, A., Mitrovic, J., & McLaren, B. (2016). Learning with intelligent tutors and worked examples: Selecting learning activities adaptively leads to better learning outcomes than a fixed curriculum. *User Modeling and User-Adapted Interaction, 26*(5), 459–491. doi:10.100711257-016-9181-y

Naumann, E., & Giel, K. (1995). Customer satisfaction measurement and management: Using the voice of the customer. Thomson Executive Press.

NBA. (2019). Available https://www.nbaind.org/

Newberry, B. (2004). The dilemma of ethics in engineering education. *Science and Engineering Ethics, 10*(2), 343–351. doi:10.100711948-004-0030-8 PMID:15152860

Nicol, D. J., & Dick, D. M. (2006). Formative assessment and self regulated learning: A model and seven principles of good feedback practice. *Studies in Higher Education, 31*(2), 199–218. doi:10.1080/03075070600572090

Nikabadi, M.S., Hoseini, S.A.M., & Zangouri, S. (2014). The effect of total quality management on innovation performance with mediator role of organizational learning: (Case study: ABBAND mazandaran). *International Journal of Knowledge-Based Organization, 4*(4), 51-66.

Nikolova, I. (2001). Teacher Development in ICT: Vision and Implementation. In *Information and Communication Technologies in Education, The School of the Future. IFIP TC3/WG3.1 International Conference on The Bookmark of the School of the Future.* Springer Science & Business Media. 10.1007/978-0-387-35403-3_6

Niraula, N., Rus, V., & Banjade, R. (2015). DeepTutor: An effective, online intelligent tutoring system that promotes deep learning. *Proceedings of the Twenty-Ninth AAAI Conference on Artificial Intelligence.*

Novakova-Marcincinova, L., Fecova, V., Novak-Marcincin, J., Janak, M., & Barna, J. (2012). Effektive Utilization of Rapid prototyping technology. *Materials Science Forum, 713*, 61–66. doi:10.4028/www.scientific.net/MSF.713.61

O'Brien, K. (2013). Global environmental change III: Closing the gap between knowledge and action. *Progress in Human Geography, 37*(4), 587–596. doi:10.1177/0309132512469589

O'Dell, C., Hubert, C., Lopez, K., Odem, P., & Raybourn, C. (2003). Successful KM implementations: a study of best-practice organizations. In C. W. Holsapple (Ed.), *Handbook on Knowledge Management* (Vol. 2, pp. 411–441). Berlin: Springer. doi:10.1007/978-3-540-24748-7_19

Ogbu. (2012). Benchmark on the Teaching Effectiveness of Basic Electricity Teachers in South Eastern Nigeria. European Scientific Journal, 8(22), 299-326.

Olsen, J., Belenky, D., Aleven, V., Rummel, N., Sewall, J., & Ringenberg, M. (2014). Authoring tools for collaborative intelligent tutoring system environments. *Lecture Notes in Computer Science, 847*, 523–528. doi:10.1007/978-3-319-07221-0_66

Online Ethics Center. (n.d.). Retrieved from https://www.onlineethics.org/

Oonk, W., Verloop, N., & Gravemeijer, K. P. (2015). Enriching Practical Knowledge: Exploring Student Teachers' Competence in Integrating Theory and Practice of Mathematics Teaching. *Journal for Research in Mathematics Education*, *46*(5), 559–598. doi:10.5951/jresematheduc.46.5.0559

Owston, R. (2013). Blended learning policy and implementation. *The Internet and Higher Education*, *18*, 1–3. doi:10.1016/j.iheduc.2013.03.002

Pacheco, E., Lips, M., & Yoong, P. (2018). Transition 2.0: Digital technologies, higher education, and vision impairment. *The Internet and Higher Education*, *37*, 1–10. doi:10.1016/j.iheduc.2017.11.001

Paine, L. S. (1991). Corporate Policy and the Ethics of Competitor Intelligence Gathering. *Journal of Business Ethics, 10*(6), 423-436.

Pallant, J. (2001). *SPSS Survival manual: A step by step guide to data analysis using SPSS for Windows Version 10*. Allen & Unwin.

Parker, L. (1988). MC for clues to improve service. *The Bottom Line (New York, N.Y.)*, *5*(1), 49–50.

Peffers, K., Tuunanen, T., Rothenberger, M. A., & Chatterjee, S. (2007). A design science research methodology for information systems research. *Journal of Management Information Systems*, *24*(3), 45–77. doi:10.2753/MIS0742-1222240302

Perera, L., Nguyen, H., & Watty, K. (2015). Formative feedback through summative tutorial-based assessments: The relationship to student performance. *Accounting Education*, *23*(5), 424–442. doi:10.1080/09639284.2014.947093

Philips, F., Hsieh, C., Ingene, C., & Golden, L. (2016). Business schools in crisis. *Journal of Open Innovation*, *2*(1), 1–21. doi:10.118640852-016-0037-9

Pistilli, M., & Arnold, K. (2012). *Course signals at Purdue: Using learning analytics to increase student success*. 2nd International Conference on Learning Analytics and Knowledge, Vancouver, Canada.

Polanyi, M. (2009). *The Tacit Dimension*. Chicago: University of Chicago Press.

Popham, W. J. (2008). *Classroom Assessment: What Teachers Need to Know* (5th ed.). Boston: Allyn and Bacon.

Posada, J., Toro, C., Barandiaran, I., Oyarzun, D., Stricker, D., De Amicis, R., ... Vallarino, I. (2015). Visual computing as a key enabling technology for industrie 4.0 and industrial internet. *Computer Graphics and Applications, IEEE*, *35*(2), 26–40. doi:10.1109/MCG.2015.45 PMID:25807506

Prasad, H., & Prasad, D. (2005). Towards professionalism in education. *University News*, *43*(18), 323–337.

Presbitero, A., Roxas, H., & Chadee, D. (2017). Sustaining innovation of information technology service providers: Focus on the role of organisational collectivism. *International Journal of Physical Distribution & Logistics Management*, *47*(2-3), 156–174. doi:10.1108/IJPDLM-07-2015-0171

Priego, R., Orive, D., & Marcos, M. (2014). Maintaining the availability of the Control System in Industrial Automation. In *Agenten im Umfeld von Industrie 4.0* (pp. 15-19). Sierke Verlag.

Promethean, R. (2019). Retrieved from https://Resourced.Prometheanworld.Com/Types-Of-Summative-Formative-Assessment/

Qaqish, R., & Hanna, R. (Eds.). (1997). The Impact of CAL strategies on CAD: A case study of the effective use of computers in the teaching of architectural design. CAADRIA '97 Conference.

Qi, C., & Chau, P. Y. K. (2013). Investigating the roles of interpersonal and interorganizational trust in IT outsourcing success. *Information Technology & People*, *26*(2), 120–145. doi:10.1108/ITP-09-2012-0088

Quinn, J. B. (1999). Strategic outsourcing: Leveraging knowledge capabilities. *Sloan Management Review*, *40*(4), 9–21.

Quintas, P. (2002). Managing knowledge in a new century. In S. Littler, P. Quintas, & T. Ray (Eds.), *Managing Knowledge: An essential reader* (pp. 1–14). London: The Open University of Sage Publications.

Rajaee, N. (2013). Issues and challenges in implementing outcome-based education in engineering education. *International Journal of Innovation Education and Research*, *1*(4), 1–9.

Raju, D., & Schumacker, R. (2016). Comparing data mining models in academic analytics. *International Journal of Knowledge-Based Organizations*, *6*(2), 38–53. doi:10.4018/IJKBO.2016040103

Ramchandra, Maitra, & MallikarjunaBabu. (2014). Method for estimation of attainment of program outcome through course outcome for outcome based education. In *2014 IEEE International Conference on MOOC, Innovation and Technology in Education (MITE)*. IEEE.

Ramírez-Noriega, A., Juárez-Ramíreza, R., & Martínez-Ramírezb, Y. (2017). Evaluation module based on Bayesian networks to intelligent tutoring systems. *International Journal of Information Management*, *37*(1), 1488–1498. doi:10.1016/j.ijinfomgt.2016.05.007

Ranger, B. J., & Mantzavinou, A. (2018). Design Thinking in Development Engineering Education: A Case Study on Creating Prosthetic and Assistive Technologies for the Developing World. *Development Engineering*, *3*, 166–174. doi:10.1016/j.deveng.2018.06.001

Raths, L. E., Harmin, M., & Simon, S. B. (1966). *Values and Teaching*. Columbus, OH: Charles E. Merriu Publishing Company and A. Bell & Howel Company.

Reggia, E., Calabro, K. M., & Albrecht, J. (2015). *A scalable instructional method to introduce first year engineering students to design and manufacturing processes by coupling 3D printing with CAD assignments.* 122nd ASEE Annual conference and exposition, Seattle, WA.

Remesal, A. (2011). Primary and secondary teachers' conceptions of assessment: A qualitative study. *Teaching and Teacher Education*, *27*(2), 472–482. doi:10.1016/j.tate.2010.09.017

Rickel, J., & Johnson, W. (1999). Animated agents for procedural training in virtual reality: Perception, cognition, and motor control. *Applied Artificial Intelligence, 13*(4-5), 343–382. doi:10.1080/088395199117315

RightScale. (2016). *State of the cloud report, Hybrid Cloud Adoption Ramps as cloud users and cloud providers mature*. Retrieved from www.rightscale.com/lp/2016-state-of-the-cloud-report

Rodrigues, M. A., Kormann, M., & Davison, L. (2011). A case study of 3D technologies in Higher education: Scanning the metal work collection of museums Sheffield and its implication to teaching and learning. *ITHET 2011: IEEE 10th International conference on Information Technology based Higher education and Training,* 1-6.

Rogers, E. (2003). *Diffusion of innovations* (5th ed.). New York: Free Press.

Roll, I., & Wylie, R. (2016). Evolution and revolution in artificial intelligence in education. *International Journal of Artificial Intelligence in Education, 26*(2), 582–599. doi:10.100740593-016-0110-3

Ronchetti. (2010). Using video lectures to make teaching more interactive. *International Journal of Emerging Technologies in Learning*.

Roy, A. (1999). *The Cost of Living*. Modern Library.

Royalty, A. (2018). Design-based Pedagogy: Investigating an emerging approach to teaching design to non-designers. *Mechanism and Machine Theory, 125*, 137–145. doi:10.1016/j.mechmachtheory.2017.12.014

Rubel, C. (1995). CSR works best when it also satisfies company needs. *Marketing News TM, 23*, 38.

Rundquist, J., & Halila, F. (2010). Outsourcing of NPD activities: A best practice approach. *European Journal of Innovation Management, 13*(1), 5–23. doi:10.1108/14601061011013203

S'anchez, J., Dormido, S., & Esquembre, F. (2004). The learning of control concepts using interactive tools. *Computer Applications in Engineering Education, 13*(1), 84–98. doi:10.1002/cae.20033

Saengbanchong, V., Wiratchai, N., & Bowarnkitiwong, S. (2014). Validating the Technological Pedagogical Content Knowledge appropriate for instructing Students (TPACK-S) of pre-service teachers. *Procedia: Social and Behavioral Sciences, 116*, 524–530. doi:10.1016/j.sbspro.2014.01.252

Sales, A., & Pane, J. (2019). The role of mastery learning in an intelligent tutoring system: Principal stratification on a latent variable. *The Annals of Applied Statistics, 13*(1), 420–443. doi:10.1214/18-AOAS1196

Sawilowsky, S. (2009). New effect size rules of thumb. *Journal of Modern Applied Statistical Methods; JMASM, 8*(2), 467–474. doi:10.22237/jmasm/1257035100

Scheifele, S., Friedrich, J., Lechler, A., & Verl, A. (2014). Flexible, Self -configuring Control System for a Modular Production System. *2nd International Conference on System -Integrated Intelligence, 15*, 398–405.

Schenk, P. (2005). Reflections on the teaching of drawing in the digital age: Attitudes of senior academics in the United Kingdom to the place of drawing tuition on the design curriculum in higher education. *Art. Design & Communication in Higher Education, 4*(3), 189–203. doi:10.1386/adch.4.3.189/1

Scherer, R., Tondeur, J., Siddiq, F., & Baran, E. (2018). The importance of attitudes toward technology for pre-service teachers' technological, pedagogical, and content knowledge: Comparing structural equation modeling approaches. *Computers in Human Behavior, 80*, 67–80. doi:10.1016/j.chb.2017.11.003

Schlechtendahl, J., Kretschmer, F., Lechler, A., & Verl, A. (2014). Communication Mechanisms for Cloud based Machine Controls. *CIRP Conference on Manufacturing Systems, 17*, 830–834

Schlossberg, H. (1991). There's no mystery in how to retain customers. *Marketing News, 25*(3), 10.

Schlosser, F., Templer, A., & Ghanam, D. (2006). How human resource outsourcing affects organizational learning in the knowledge economy. *Journal of Labor Research, 27*(3), 291–303. doi:10.100712122-006-1024-x

Schlossherg, H. (1992, September). Latest image study finds mixed results for researchers. *Marketing News, 8-9*. doi:10.1108/02634500010327917

Schmid, R., & Petko, D. (2019). Does the use of educational technology in personalized learning environments correlate with self reported digital skills and beliefs of secondary-school students? *Computers & Education, 136*, 75–86. doi:10.1016/j.compedu.2019.03.006

Schuh, G., Potente, T., Wesch-Potente, C., Weber, A.R., & Prote, J.-P. (2014). Collaboration Mechanism s to Increase Productivity in the Context of Industrie 4.0. *CIRP Robust Manufacturing Conference, 19*, 51–56.

Scott, G., Tilbury, D., Sharp, L., & Deane, E. (2012). Turnaround leadership for sustainability in higher education. Final report for the Australian Government Office for Learning and Teaching, Commonwealth of Australia.

Seif, M. A. (1997). Design Case Studies: a practical approach for teaching machine design. In *Proceedings Frontiers in Education 1997 27th Annual Conference. Teaching and Learning in an Era of Change*. IEEE Explore 10.1109/FIE.1997.632754

Sekhar, C. R. (2008). *Continuous improvement process based on outcome-based education.* Academic Press.

Selwyn, N. (2007). The use of computer technology in university teaching and learning: A critical perspective. *Journal of Computer Assisted Learning, 23*(2), 83–94. doi:10.1111/j.1365-2729.2006.00204.x

Shahiri, A., Husain, W., & Rashid, N. (2015). A review on predicting student's performance using data mining techniques. *Procedia Computer Science*, *72*, 414–422. doi:10.1016/j.procs.2015.12.157

Shah, M. M., Hassan, R., & Embi, R. (2012). Technology acceptance and computer anxiety. *Proceedings of the 2012 international conference on innovation management and technology research (ICIMTR),* 306–309. 10.1109/ICIMTR.2012.6236408

Sharma, A., & Loh, P. (2009). Emerging trends in sourcing of business services. *Business Process Management Journal*, *15*(2), 149–165. doi:10.1108/14637150910949425

Shi, J., Wan, J., Yan, H., & Suo, H. (2011), A survey of cyber-physical systems. *International conference on wireless communications and signal processing (WCP)*, 1–6.

Shinaberger, L. (2017). Components of a flipped classroom influencing student success in an undergraduate business statistics course. *Journal of Statistics Education: An International Journal on the Teaching and Learning of Statistics*, *25*(3), 122–130. doi:10.1080/10691898.2017.1381056

Sigala, M. (2008). Service quality and customer relationship management: Interrelationships and implications. *Managing Service Quality*, *18*(5), msq.2008.10818eaa.001. doi:10.1108/msq.2008.10818eaa.001

Simon, H. (1969). *Sciences of the artificial.* Cambridge, MA: MIT Press.

Simon, J. C., Poston, R. S., & Kettinger, B. (2009). Creating better governance of offshore services. *Information Systems Management*, *26*(2), 110–122. doi:10.1080/10580530902794778

Simpson, A. (2017). The misdirection of public policy: Comparing and combining standardized effect sizes. *Journal of Education Policy*, *32*(4), 450–466. doi:10.1080/02680939.2017.1280183

Sims, R. R., & Sims, S. J. (1995). *The Importance of Learning Styles: Understanding the Implications for Learning, Course Design, and Education.* Greenwood Publishing Group.

Singh, R., Sandhu, H.S., Metri, B.A., & Kaur, R. (2014). Understanding organized retail supply chain performance indicators for business growth. *International Journal of Knowledge-Based Organization, 4*(2), 64-76.

Sipos, Y., Battisti, B., & Grimm, K. (2008). Achieving transformative sustainability learning: Engaging head, hands and heart. *International Journal of Sustainability in Higher Education*, *9*(1), 68–86. doi:10.1108/14676370810842193

Smith, R. (2006). Research misconduct: The poisoning of the well. *Journal of the Royal Society of Medicine*, *99*(5), 232–237. doi:10.1177/014107680609900514 PMID:16672756

Soetanto, R., Childs, M., Poh, P., Austin, S., & Hao, J. (2012). Global multidisciplinary learning in construction education: Lessons from virtual collaboration of building design teams. *Civil Engineering Dimension, 14*(3), 173–181.

Soliman, S., Taha, D., & El Sayad, Z. (2019). *Architectural education in the digital age Computer applications: Between academia and practice. Alexandria Engineering Journal.*

Sottilare, R., Burke, S., Salas, E., Sinatra, A., Johnston, J., & Gilbert, S. (2018). Designing adaptive instruction for teams: A meta-analysis. *International Journal of Artificial Intelligence in Education*, *28*(2), 225–264. doi:10.100740593-017-0146-z

Spady, W. G. (1994). *Outcome-Based Education: Critical Issues and Answers*. American Association of School Administrators.

Spady. (1994). Outcome-Based Education: Critical Issues and Answers. American Association of School Administrators.

Spain, R., Priest, H., & Murphy, J. (2012). Current trends in adaptive training with military applications: An introduction. *Military Psychology*, *24*(2), 87–95. doi:10.1080/08995605.2012.676984

Spender, J. (2016). How management education's past shapes its present. *BizEd*. Retrieved from: http://www.bizedmagazine.com/

Spiro, R. J., & Jehng, J.-Ch. (1990). Cognitive flexibility and hypertext: Theory and technology for the nonlinear and mutlidimensional traversal of complex subject matter. In D. Nix & R. Spiro (Eds.), Cognition, education, and multimedia: Exploring ideas in high technology (pp. 163–204). Academic Press.

Spiro, R. L., & Weitz, B. A. (1990). Adaptive Selling: Conceptualization, Measurement, and Nomological Validity. *JMR, Journal of Marketing Research*, *27*, 61–69. doi:10.1177/002224379002700106

Spooner, L. (1985, July). MCs uncover keys to sales and service. *Savings Institutions,* 92.

Stachowicz-Stanusch, A., & Amann, W. (2018). Artificial intelligence at universities in Poland. *Organization and Management Scientific Quarterly*, *2*(42), 62–82.

Steenbergen-Hu, S., & Cooper, H. (2014). A Meta-Analysis of the effectiveness of intelligent tutoring systems on college students' Academic Learning. *Journal of Educational Psychology*, *106*(2), 331–347. doi:10.1037/a0034752

Stefanelli, J. M. (1994). Using MCs to maintain hospitality company service standards. *Hospitality and Tourism Educator,* *6*(1), 17-18.

Stone, R., Brooks, R., Brynjolfsson, E., Calo, R., & Etzioni, O. (2016). *Artificial intelligence and life in 2030*. Stanford University. Retrieved from: https://ai100.stanford.edu/2016-report

Stonebraker, I., Brookem, M., Shaw, R., & Moss, J. (2016). Student see versus student do: A comparative study of two online tutorials. *TechTrends*, *60*(2), 176–182. doi:10.100711528-016-0026-7

Stosic, L. (2015). The importance of educational technology in teaching. International Journal of Cognitive Research in Science. *Engineering and Education*, *3*, 111–114.

Stovall, S. A. (1993). Keeping tabs on customer service. *Bank Marketing*, *25*(6), 29–33.

Suhadi, S., Bahru, J., Mohamed, H., & Abdullah, Z. (2016). Enhancing student's higher order thinking skills (HOTS) through the Socratic Method approach with technology. *International Journal of Knowledge-Based Organizations*, *6*(4), 14–28. doi:10.4018/IJKBO.2016100102

Sullivan, P., & McCartney, H. (2017). Integrating 3D printing into an early childhood teacher preparation course: Reflections on practice. *Journal of Early Childhood Teacher Education*, *38*(1), 39–51. doi:10.1080/10901027.2016.1274694

Sumak, B., Hericko, M., & Pusnik, M. (2011). A meta-analysis of e-learning technology acceptance: The role of user types and e learning technology types. *Computers in Human Behavior*, *27*(6), 2067–2077. doi:10.1016/j.chb.2011.08.005

Sweet, K. (2016). Resurrecting the master builder: A pedagogical strategy for robotic construction. *Automation in Construction*, *72*, 33–38. doi:10.1016/j.autcon.2016.07.001

Tam, F. Y., Moon, K. L., Ng, S. F., & Hui, C. L. (2007). Production sourcing strategies and buyer-supplier relationships: A study of the differences between small and large enterprises in the Hong Kong clothing industry. *Journal of Fashion Marketing and Management: An International Journal*, *11*(2), 297–306. doi:10.1108/13612020710751446

Tani, K., Dalzell, E., Ehambaranathan, N., Murugasu, S., & Steele, A. (2019). Evaluation of external factors affecting students' academic performance in higher education in New Zealand. *Journal of Pedagogical Research*, *3*(2), 1–10. doi:10.33902/JPR.2019252504

Targowski, A. (2009). The architecture of service systems as the framework for the definition of service science scope. *International Journal of Information Systems in the Service Sector*, *1*(1), 54–77. doi:10.4018/jisss.2009010104

Taylor, L., Brogt, E., Cheer, U., Baird, N., Caldwell, J., & Wilson, D. (2017). Ethnicity and engagement in first-year New Zealand law programmes. *Journal Higher Education Research & Development*, *36*(5), 1047–1060. doi:10.1080/07294360.2016.1263935

Teaching Oxford Learner's Dictionaries. (n.d.). Retrieved from oxfordlearnersdictionaries.com/definition/english/teaching

Teo, T. (2011). *Technology acceptance in education. Research and issues*. Rotterdam: Sense Publishers. doi:10.1007/978-94-6091-487-4

Tepper, G.C. (1994). The merits of self-testing. *Mortgage Banking, 54*(8), 76.

Texas A&M. (n.d.). Retrieved from https://ethics.tamu.edu

Tinto, V. (1993). *Leaving college: Rethinking the causes and cures of student attrition* (2nd ed.). Chicago: University of Chicago Press.

Tițan, E., Burciu, A., Manea, D., & Ardelean, A. (2014). From traditional to digital: The labour market demands and education expectations in an EU context. *Procedia Economics and Finance*, *10*, 269–274. doi:10.1016/S2212-5671(14)00302-5

Topp, G. (2011, Oct. 6). Flipped classrooms take advantage of technology. *USA Today.*

Tsiakas, K., Abujelala, M., & Makedon, F. (2018). Task engagement as personalization feedback for socially-assistive robots and cognitive training. *Technologies*, *49*, 1–17.

UGC. (2019). Available https://www.ugc.ac.in/

Undeland, M., Panaitescu, R. C., Jain, A. K., Jose, P., & Begalke, T. (2003). Restructuring of first courses in power electronics and electric drives that integrates digital control. IEEE Transactions on Power Electronics, 18(1), 429–437.

Van der Leeuw, S., Wiek, A., Harlow, J., & Buizer, J. (2012). How much time do we have? Urgency and rhetoric in sustainability science. *Sustainability Science*, *7*(S1), 115–120. doi:10.100711625-011-0153-1

van der Wiele, A., Hesselink, M., & Van Iwaarden, J. (2005). Mystery shopping: A tool to develop insight into customer service provision. *Total Quality Management & Business Excellence*, *16*(4), 529–541. doi:10.1080/14783360500078433

van Gorp, D., Jagersma, K. P., & Livshits, A. (2007). Offshore behaviour of service firms: Policy implications for firms and nations. *Journal of Information Technology Case and Application Research*, *9*(1), 7–19. doi:10.1080/15228053.2007.10856104

Van Zyl, I. J. (2013). *Technology encounters and the symbolic narrative. Localising the 'technology for development' experience in South African education settings* (Unpublished PhD dissertation). Universitàdella Svizzeraitaliana.

Vandeyar, T. (2013). Practice as policy in ICT for education: Catalysing communities of practice in education in South Africa. *Technology in Society*, *35*(4), 248–257. doi:10.1016/j.techsoc.2013.10.002

Vandeyar, T. (2015). Policy intermediaries and the reform of e- Education in South Africa. *British Journal of Educational Technology*, *46*(2), 344–359. doi:10.1111/bjet.12130

Vavra, T. G. (1997). *Improving your measurement of customer satisfaction: A guide to creating, conducting, analyzing, and reporting customer satisfaction measurement programs*. Milwaukee, WI: ASQC Quality Press.

Vietor, R.H.K., & Veytsman, A. (2005). *American outsourcing.* Harvard Business School Case 705-037.

von Tunzelmann. (1996). *Engineering and Innovation in the Industrial Revolutions.* STEEP Discussion Paper No 30, Science Policy Research Unit.

Vongkulluksn, V. W., Xie, K., & Bowman, M. A. (2018). The role of value on teachers' internalization of external barriers and externalization of personal beliefs for classroom technology integration. *Computers & Education*, *118*, 70–81. doi:10.1016/j.compedu.2017.11.009

Vrasidas, Charalambos, Zembylas, Michalinos, & Glass. (2009). *ICT for Education, Development, and Social Justice*. IAP.

Wang, D., Han, H., Zhan, Z., Xu, J., Liu, Q., & Rena, G. (2014). A problem solving oriented intelligent tutoring system to improve students' acquisition of basic computer skills. *Computers & Education, 81*, 102–112. doi:10.1016/j.compedu.2014.10.003

Washington Accord. (2012). International Engineering Alliance.

Washington Accord. (2017). International Engineering Alliance. Retrieved 20 April 2017, http://www.Ieagreements.org/Accords/Washington/Signatories/

Watkins. (2015). Teaching Engineering Ethics. American Society for Engineering Education.

Weltman, M. W., & Whiteside, M. (2010). Comparing the Effectiveness of Traditional and ActiveLearning Methods in Business Statistics: Convergence to the Mean. *Journal of Statistics Education: An International Journal on the Teaching and Learning of Statistics*, *18*(1). doi:10.1080/10691898.2010.11889480

Weston, M. E., & Bain, A. (2010). The end of techno-critique: the naked truth about 1:1laptop initiatives and educational change. *The Journal of Technology, Learning, and Assessment*, *9*(6).

Weyer, S., Schmitt, M., Ohmer, M., & Gorecky, D. (2015). Towards Industry 4.0-Standardization as the crucial challenge for highly modular, multi-vendor production systems. *IFAC-PapersOnLine*, *48*(3), 579–584. doi:10.1016/j.ifacol.2015.06.143

White, C. (2004). *Customer satisfaction research and MC: Why do both?* Retrieved September 2017, from http://www.marketresearchworld.net

Widden, M., & Gunn, K. (2010). Design build test of model aerofoils for engineering education using FDM. *Virtual and Physical Prototyping*, *5*(4), 189–194. doi:10.1080/17452759.2010.528841

Wiley, D., Bliss, T. J., & McEwen, M. (2014). Open Educational Resources: A Review of the Literature. In Handbook of Research on Educational Communications and Technology (pp. 781–789). New York, NY: Springer. doi:10.1007/978-1-4614-3185-5_63

William. (2001). *To Teach: The Journey of a Teacher*. Teachers College Press.

Williams, C., & Durst, S. (2018). Exploring the Transition Phase in Offshore Outsourcing: Decision Making amidst Knowledge at Risk. *Journal of Business Research*. doi:10.1016/j.jbusres.2018.01.013

Williamson, O. (1979). Transaction-cost economics: The governance of contractual relations. *The Journal of Law & Economics*, *22*(2), 233–261. doi:10.1086/466942

Wilson, A. M. (1998). The Role of MC in the Measurement of Service Performance. *Managing Service Quality*, *8*(6), 414–420. doi:10.1108/09604529810235123

Wilson, A. M. (1998). The use of MC in the measurement of service delivery. *Service Industries Journal*, *18*(3), 148–163. doi:10.1080/02642069800000037

Wilson, A. M. (2000). The use of performance information in the management of service delivery. *Marketing Intelligence & Planning*, *18*(3), 127–134.

Wilson, A. M., & Gutmann, J. (1998). Public transport: The role of MC in investment decisions. *Journal of the Market Research Society*, *40*(4), 285–293. doi:10.1177/147078539804000401

Wilson, G. (2001). Technical Communication and Late Capitalism: Considering a Postmodern Technical Communication Pedagogy. *Journal of Business and Technical Communication*, *15*(1), 72–99. doi:10.1177/105065190101500104

Winch, J., & Cahn, E. (2015). Improving student performance in a management science course with supplemental tutorial videos. *Journal of Education for Business*, *90*(7), 402–409. doi:10.1080/08832323.2015.1081865

Wohlers, T. (2012). *Wohlers Report 2012: Additive Manufacturing 3D printing state of the industry. Annual worldwide progress report*. Wohlers Associates Inc.

Wood, J., Campbell, M., Wood, K., & Jensena, D. (2005). Enhancing the Teaching of Machine Design by Creating a Basic Hands-On Environment with Mechanical 'breadboards'. *International Journal of Mechanical Engineering Education*, *33*(1), 1–25. doi:10.7227/IJMEE.33.1.1

Woodruff, R. B., & Gardial, S. F. (1996). *Know your customer: New approaches to understanding customer value and satisfaction*. Oxford, UK: Blackwell.

Wu, Y. J., Yuan, C. H., & Pan, C. I. (2018). Entrepreneurship Education: An Experimental Study with Information and Communication Technology. *Sustainability*, *10*(3), 691. doi:10.3390u10030691

Xia, Y., & Tang, T. L.-P. (2011). Sustainability in supply chain management: Suggestions for the auto industry. *Management Decision*, *49*(4), 495–512. doi:10.1108/00251741111126459

Xie, K., & Luthy, N. (2017). *Textbooks in the digital world*. Retrieved October 24, 2019, from https://theconversation.com/textbooks-in-the-digital-world-78299

Yakhlef, A. (2009). Outsourcing as a mode of organizational learning. *Strategic Outsourcing*, *2*(1), 37–53. doi:10.1108/17538290910935882

Yang, C. C. (2002). *Customer needs and satisfaction survey*. Taipei: SGS Taiwan Limited.

Yanli, Z., & Yao, Z. (2015). Challenges in organizational control: The economic and management perspectives. *International Journal of Knowledge-Based Organizations*, *5*(1), 33–44.

Young, S. (2008). Outsourcing in public health: A case study of contract failure and its aftermath. *Journal of Health Organization and Management*, *22*(5), 446–464. doi:10.1108/14777260810898688 PMID:18959298

Zeithaml, V. A., & Bitner, M. J. (2000). *Services Marketing*. New York: McGraw-Hill.

Zeng. (2016). Study on the Third Industrial Revolution and Paradigm Transformation of China's Manufacturing Industry—Based on Theoretical Analysis of Scale Economy and Scope Economy. *American Journal of Industrial and Business Management*, 73-82.

Zhang, Y., Fei, Q., Quddus, M., & Davis, C. (2014). An examination of the impact of early intervention on learning outcomes of at-risk students. *Research in Higher Education*, *26*, 1–12.

Zhu, Z.T., Yu, M.H., & Riezebos, P. (2016). A research framework of smart education. *Smart Learning Environments,* 3-4.

About the Contributors

Kaushik Kumar, B.Tech (Mechanical Engineering, REC (Now NIT), Warangal), MBA (Marketing, IGNOU), and Ph.D (Engineering, Jadavpur University), is presently an Associate Professor in the Department of Mechanical Engineering, Birla Institute of Technology, Mesra, Ranchi, India. He has 14 years of Teaching & Research and over 11 years of industrial experience in a manufacturing unit of Global repute. His areas of teaching and research interest are Quality Management Systems, Optimization, Non-conventional machining, CAD / CAM, Rapid Prototyping and Composites. He has 9 Patents,15 Book, 6 Edited Book 35 Book Chapters, 120 international Journal publications, 18 International and 8 National Conference publications to his credit. He is on the editorial board and review panel of 7 International and 1 National Journals of repute. He has been felicitated with many awards and honours.

J. Paulo Davim received his Ph.D. degree in Mechanical Engineering in 1997, M.Sc. degree in Mechanical Engineering (materials and manufacturing processes) in 1991, Mechanical Engineering degree (5 years) in 1986, from the University of Porto (FEUP), the Aggregate title (Full Habilitation) from the University of Coimbra in 2005 and the D.Sc. from London Metropolitan University in 2013. He is Senior Chartered Engineer by the Portuguese Institution of Engineers with an MBA and Specialist titles in Engineering and Industrial Management as well as in Metrology. He is also Eur Ing by FEANI-Brussels and Fellow (FIET) by IET-London. Currently, he is Professor at the Department of Mechanical Engineering of the University of Aveiro, Portugal. He has more than 30 years of teaching and research experience in Manufacturing, Materials, Mechanical and Industrial Engineering, with special emphasis in Machining & Tribology. He has also interest in Management, Engineering Education and Higher Education for Sustainability. He has guided large numbers of postdoc, Ph.D. and master's students as well as has coordinated and participated in several financed research projects. He has received several scientific awards. He has worked as evaluator of projects for ERC-European Research Council and other international research agencies as well as examiner of

Ph.D. thesis for many universities in different countries. He is the Editor in Chief of several international journals, Guest Editor of journals, books Editor, book Series Editor and Scientific Advisory for many international journals and conferences. Presently, he is an Editorial Board member of 30 international journals and acts as reviewer for more than 100 prestigious Web of Science journals. In addition, he has also published as editor (and co-editor) more than 125 books and as author (and co-author) more than 10 books, 80 book chapters and 400 articles in journals and conferences (more than 250 articles in journals indexed in Web of Science core collection/h-index 54+/9500+ citations, SCOPUS/h-index 58+/11000+ citations, Google Scholar/h-index 75+/18500+).

* * *

Esther Akinlabi is a Full Professor at the Department of Mechanical Engineering Science, Faculty of Engineering and the Built Environment, University of Johannesburg. Her research interest is in the field of modern and advanced manufacturing processes – Friction stir welding and additive manufacturing. Her research in the field of laser-based additive manufacturing includes laser material processing and surface engineering. She also conducts research in the field of renewable energy, and biogas production from waste. She is a rated National Research Foundation (NRF) researcher and has demonstrated excellence in all fields of endeavors. Her mentorship and research experience are enviable as she guides her team of postgraduate students through the research journey. She is a recipient of several research grants and has received many awards of recognition to her credit. She is a member of the prestigious South African Young Academy of Science and registered with the Engineering Council of South Africa. Prof Akinlabi has filed two patents, edited one book, published four books and authored/co-authored over 400 peers reviewed publications.

Anitha Lakshmi Akkireddy, Assistant Professor of Mechanical Engineering, GRIET, Hyderabad and pursuing a Ph.D. from Andhra University, Visakhapatnam and has 11 years of academic and research experience in Indian Universities. Her Ph. D work is on the Extended formability of Austenitic stainless steel 304 in Hot Forming Region. She had earned a Bachelor of Engineering in Mechanical Engineering from VITAM College of Engineering and Masters of Engineering with CAD/CAM specialization in Mechanical Engineering from GITAM Engineering College, Affiliated to Andhra University. Her areas of research interests include Material Testing & Characterization, Experimental, Analytical and Finite Element Investigations of sheet metal forming processes in which she has publications. She is continuing research on the Tensile testing and formability characteristics of Aluminum alloy

8011 from cold forming to hot forming region, Forming Limit Diagrams (FLDs) at various temperatures and optimization. She is a recipient of Best Researcher in sheet metal Forming of the year from Research Ratna Award. 2019. She is a life member of the Indian Institute of Engineers.

Ankita Awasthi has done her B.Tech, Mechanical Engineering and M.Tech, Computer Aided Design from Abdul Kalam Technical University, Lucknow. She has 10 years of experience in academics and industry. Ms. Awasthi holds expertise in Designing, Composite and Finite element analysis. She has authored many research papers which are published in reputed international journals. She has presented her work in many international conferences and forums. Ms. Awasthi is associated with AUTODESK as an official Autodesk student's expert. She was the faculty advisor for BAJA- SAE. She has authored two books on Engineering Thermodynamics and Material Science. She is currently working as an Assistant professor in College of Engineering and Technology, Greater Noida.

Akinsanya Baruwa is a Ph.D. candidate in the department of mechanical engineering science, University of Johannesburg, South Africa. He has interests in surface engineering, energy, and engineering materials. He also has experience in tutoring in various undergraduate modules at the University of Johannesburg. He has published several peer-reviewed articles.

Tanya Buddi, Assistant Professor of Mechanical Engineering, completed her Ph.D from KL University and has 8 years of academic and research experience. Her Ph.D work is on Development of Bio-adhesive in Plywood Manufacturing, part of which was carried out at RGUKT, Basar and Victoria University, Melbourne, Australia. She had earned Bachelor of Engineering from JNTU Kakinada in Mechanical Engineering, and Master of Technology also from JNTU Kakinada in Machine Design. Her research interests include Wood-based Bio-Composites, Nano Composites, Sheet Metal Forming in which she has 7 publications, in various journals and conferences. She has executed a funded project in the area of Wood-based Bio-Composites funded by UGC in the year 2014 for 2.0 lakhs.

Lavanya C., Professor of GRIET is graduated from Jawaharlal Nehru Technological University (JNTU), Hyderabad in Civil Engineering (2002). She has done her Masters from Jawaharlal Nehru Technological University (JNTU), Kakinada in Soil Mechanics (2005) and her Ph.D from Jawaharlal Nehru Technological University (JNTUH), Hyderabad in Civil Engineering (2015). She has over ten years of teaching and research experience. Her research interest is on expansive soils and ground improvement techniques in which she has more than 16 research publications in international and national journals and conferences. She is a life member for Indian Geotechnical Society (IGS).

Archana Dixit is an Assistant Professor in GLA University Mathura U.P, since 2008. Her teaching experience spans over twelve years in various Engineering institutes and Degree colleges as well. She has completed her M.Phil. and doctorate degree from Dr B. R. Ambedkar University Agra. Her expertise is in the field of Mathematics. She has published many research papers in reviewed and refereed National and International journals.

Jitendra Kumar Dixit has 10+ years of teaching and research experience in the field of management. His area of research is Consumer Psychology, Brand Management and Econometrics. He has authored 15+ research papers published in reputed international journals indexed by SCI/Scopus. He has presented his work in many international conferences and forums. He is currently working as an Assistant Professor in Institute of Business Management, GLA University, Mathura.

Susanne Durst is a Full Professor of Business Administration at the School of Business at University of Skövde and a Visiting Professor at University West (both in Sweden). She is also the leader of the research group knowledge, innovation, and marketing (KIM) at the University of Skövde. Her research interests include small business management, SME succession/transfers, knowledge management, knowledge risk management, and corporate governance. She has been conducting several national and international research projects. Her work has been awarded different awards and has been published in international peer-reviewed journals. Before joining academia, she worked with private enterprises.

Ingi Runar Edvardsson is a Professor at the School of Business, University of Iceland. He received his PhD in Sociology from the University of Lund, Sweden. His research and publication includes: knowledge and human resource management, outsourcing, regional universities and Nordic labour markets. His articles have appeared in Employee Relations, Journal of Knowledge Management, Knowledge Management Research & Practice, International Journal of Entrepreneurship and Small Business, International Journal of Knowledge-Based Organizations, International Journal of Knowledge-Based Development, Journal of Innovation Management, Marine Policy, Measuring Business Excellence and Scandinavian Journal of Education Research. He has also published several book chapters.

Durga Prasad Garapati received the B.Tech. Degree from Pondicherry University, Pondicherry, and M.Tech. from VIT, Deemed to be an University, Vellore, and completed Ph.D. in Karunya Institute of Technology and Sciences Deemed to be an University, Karunya nagar, Coimbatore, TN, India. He is working presently as Associate Professor in Shri Vishnu Engineering College for Women, Bhimavaram, India. He has presented technical papers in various National and International Journals and Conferences. His area of interest includes Power Electronics, Industrial Drives, and Instrumentation.

Owen P. Hall, Jr. holds the Corwin D. Denney Academic Chair and is a tenured full professor of Decision Sciences, at Pepperdine University Graziadio School of Business. He is a Julian Virtue Professorship and a Rothschild Applied Research Fellow. Dr. Hall has more than 35 years of academic and industry experience in computer decision systems and information technology. He has authored numerous technical papers and several books on computer-based management decision systems. Dr. Hall received the Harriet and Charles Luckman Distinguished Teaching Fellow in 1993, the Howard A. White Teaching Excellence Award in 2009 and 2017, and the Sloan-C Effective Teaching Practice Award in 2013. His current area of research includes the application of artificial intelligent agents to integrated learning systems. Dr. Hall is the vice chairman of the INFORMS University Analytics Program Committee. He is a registered professional engineer, State of California. Dr. Hall received his Ph.D. from the University of Southern California and undertook post-doctoral studies at the Center for Futures Research.

S. Julius Fusic obtained his B.E degree in Electrical and Electronics Engineering from PSNA Engineering College, Anna University, Chennai in 2010 and M.E degree in Power Electronics and Drives from MAM College of Engineering in 2014. Currently he is an Assistant Professor in the Department of Mechatronics Engineering at Thiagarajar College of Engineering Madurai. His research interests include Modeling and analysis of Power Electronic Converter circuits, autonomous system navigation and control, smart educational methods. He is a mentor for FDP 101x and 201x organised by IIT Bombay.

U. S. Jyothi, B.Tech (Mechanical Engineering, JNTU, Anantapur), M.Tech (Thermal Engineering, JNTU, Hyderabad), and Ph.D (Hydrogen fuel, JNTUH, Hyderabad), is presently as Professor in the Department of Mechanical Engineering, GRIET, Hyderabad, India. She has a professional experience of 12 years in Andhra Pradesh Road Transport Corporation (PSU) at different designations such as Asst. Mechanical Engineer, Stores & Purchase officer and Deputy Chief mechanical Engineer. As a Dy. CME contributed for holding the responsibility of achieving all targeted technical parameters of the fleet in Region with proper coordination with unit managers and manufacturers. She has a teaching and research experience of 8 years in GRIET and areas of interest are thermodynamics, heat transfer and Internal combustion Engines. Under UGC minor project grant, she did investigations on bio diesel by changing the piston bowl geometry and with additives.

Hridayjit Kalita is presently pursuing a PhD in Mechanical Engineering from BIT, Mesra. Kalita completed their Masters in Design of Mechanical Equipments from BIT, Mesra and did their B.Tech in Mechanical engineering from SRM University, Kattankulathur.

Satyanarayana Kosaraju, Associate Professor of Mechanical Engineering, completed his Ph.D from National Institute of Technology Warangal and has over Five years of academic and research experience in Indian universities. His Ph.D work was on Machinability characteristics: Modeling, Simulation, analysis and optimization of Ti-6Al-4V. Prior to PhD, he had earned Bachelors of Engineering in Mechanical Engineering from JayaPrakash Narayan college of Engineering, and Masters of Engineering with Product Design Development specialization in Mechanical Engineering from National Institute of Technology Warangal. Dr. Satyanarayana research interests include Material Testing & Characterization, single and multi-objective optimization techniques, Experimental, Analytical and Finite Element Investigations of various machining and for mining processes in which he has more than 30 publications, in various journals and conferences. He is currently working on uniaxial and biaxial deformation behavior of Titanium 21S alloy up to 4000C, anisotropic yield criteria for titanium 21S alloy, Forming Limit Diagrams (FLDs) at various temperatures and optimization. Dr Satyanarayana is recipient of Bharth Vidhya Rathan Award. He is a member of ASME, IEANG and life member of ISTE.

Thangavel M. is an Assistant Professor presently working in the Department of Information Technology at Thiagarajar college of Engineering, Madurai. He presently holds 5 years of Teaching and Research experience in Thiagarajar college of Engineering, Madurai and 2 years of Teaching and Research experience in Madras Institute of Technology, Anna University - Chennai. He graduated as a B.E. Computer Science and Engineering from M.A.M College of Engineering, Trichy (Anna University - Chennai) and as an M.E. Computer Science and Engineering from J.J. College of Engineering and Technology, Trichy (Anna University - Chennai) and Pursuing his PhD from Madras Institute of Technology, Chennai under Anna University - Chennai. He is a Gold Medalist in UG and Anna University - First Rank Holder with Gold Medal in PG. His specialization is Cloud Computing, and

Information Security. His Areas of Interest include DNA Cryptography, Ethical Hacking, Compiler Design, Computer Networks, Data Structures and High Performance Computing. He has published 6 articles in International Journals, 10 book chapters in International Publishers, 15 in the proceedings of International Conferences and 3 in the proceedings of national conferences /seminars. He has attended 40 Workshops / FDPs/Conferences in various Higher Learning Institutes like IIT, Anna University. He has organized 21 Workshops / FDPs /Contests/Industry based courses over the past 5 years of experience. He has been a delegate for Cyber Week 2017 organized by Tel Aviv University, Israel. He has been recognized by IIT Bombay; SAP CSR as SAP Award of Excellence with cash reward of Rs.5000/- for the best Participation in IITBombayX: FDPICT001x Use of ICT in Education for Online and Blended Learning. He shows interest in student counseling, in motivating for better placements and in helping them design value-based life-style.

Jandhyala N. Murthy, Professor and Director of GRIET studied PUC(MPC) from Maharajah's College of Vizianagaram (1969-70) and BTech (Mechanical) from IIT Madras (1970-75), MS (Thermal Power)(1982-84)and PhD (Thermal Power) (1985-88) from Cranfield Institute of Technology, UK. He served in the Maintenance Branch of the Indian Air Force as an AE(M) officer for over 25 years since 1975, held various appointments looking after operations, overhaul, training at unit, station and command levels. He was an Instructor at Air Force Technical College, Bangalore and had a tenure at Combustor division, Gas Turbine Research and Development Establishment, Bangalore. He led an IAF Technical Training Team to Botswana, Africa, providing training and maintenance support to Botswana Defence Force. He is an alumnus of Defence Services Staff College, Wellington. He took voluntary retirement from the IAF as a Wing Commander in 2001 and immediately joined as a professor in the Department of Mechanical Engineering at Gokaraju Rangaraju Institute of Engineering and Technology (Autonomous), Hyderabad, an accredited and permanently affiliated college under JNTUH. He assumed the responsibility as the Director of GRIET since 31 January 2018, after a successful tenure as the Principal of GRIET since March 2004. His areas of interest span the Thermal Engineering domain, Gas Turbine Combustion Chambers, simulation and education.

Fredrick Mwema is an assistant lecturer at Dedan Kimathi University of Technology, DeKUT, Kenya where has been working from the level of teaching assistant since 2011. He has taught several modules in the department of mechanical engineering including machine design, materials science, the strength of materials and engineering materials and so many others. He has an interest in engineering education and the quality of training of undergraduate and graduate students in engineering. He is currently a doctoral candidate at the University of Johannesburg,

Auckland Park, Kingsway Campus, South Africa. His Ph.D. research focuses on the deposition of thin films using magnetron sputtering technology. He holds a Bachelor of Science and Master of Science degrees in Mechanical Engineering from Jomo Kenyatta University of Agriculture & Technology (JKUAT), Nairobi, Kenya. Mr. Mwema has published more than twenty-five (25) articles in peer-reviewed journals, conferences and book chapters. His research interests include thin films, material characterization, advanced and modern manufacturing. He is a graduate engineer with the Engineers Board of Kenya (EBK) and a member of the International Society for Stereology & Image Analysis (ISSIA).

Anandh N. obtained his B.E degree in Electrical and Electronics Engineering from Ranipet Engineering College, Anna University, Chennai in 2006 and M.Tech degree in Power Electronics, Electrical Machines and Drives from Indian Institute of Technology Delhi in 2014. Currently he is an Assistant Professor in the Department of Electrical and Electronics Engineering at Manipal Institute of Technology, Manipal Academy of Higher Education (Deemed to be University), Manipal. His research interests include Modeling and analysis of Power Electronic Converter circuits, Soft-Switching techniques, Power Quality, and harmonic analysis. He is a life member of Indian Society for Technical Education and The Indian Science Congress Association.

Dharmendra Patel received Bachelor Degree in Industrial Chemistry- BSc. (Industrial Chemistry) from North Gujarat University, Gujarat, India and Master Degree in Computer Applications(MCA) from North Gujarat University, Gujarat, India. He completed his Ph.D., in the field of Web Usage Mining. Currently, he is working as an associate professor at Charotar University of Science and Technology, Changa, Gujarat, India. He has published many research papers in the international and national journals/conferences of repute. He is associated with many international journals as a reviewer board or editorial board member.

Padmaja S. M. received M.Tech. & Ph.D. Degree from JNTUH, Hyderabad, BE from Andhra University. She has 17 years teaching & Research Experience and held many positions & presently she is working as Professor & HOD/EEE in Shri Vishnu Engineering College for Women, Bhimavaram, India. She has published many technical papers in various National and International Journals and Conferences. Her area of interest includes Power Electronics, FACTS etc.

Jogindra Nath Sahu is a serving metallurgist starting his technical education Industrial training (ITI) in Fitter tradesman and finished his educational journey by completing Ph.D in materials and metallurgical engineering from MANIT (BHOPAL).. Author have wide experience in teaching from 6th standard to post graduation (PG) during his 30 years' professionals' career. The whole education journey is completed by learning while earning. This will give a strong massage excellent motivation to the readers to face the challenges in their career. There are endless case studies created by the author out of which highlighted for readers to know for produce a bench mark in their work and skill. A simple living and high thinking personality crated thousands of dispels all most all technical and non-technical field and presently a trainer for ITI, Diploma and engineering apprentices in BHEL, Bhopal. Self-disciplined person having wish to improve and lead other to improve with this hobby many motivational classes and modules are conducted by the author and by the same time many conferences as well as training session are attended during the entire journey of learning while earning process. Author has put the matters in this chapter as per his experience and work on hand skill in the teaching and learning system for a noble cause to give motivating and some thought process in this challenging time.

Kuldeep Kumar Saxena has done his B.Tech, Mechanical Engineering and M.Tech, Material Science and Engineering from MNNIT Allahabad. He has 9+ years of experience in academics, Research and industry. Dr. Saxena holds expertise in hot deformation behaviour of materials, microstructural characterization of materials, and micro manufacturing. He has served as Senior Research Fellow (SRF) for 2 years and 8 months on a project sponsored by Board of Research in Nuclear Sciences (BRNS), a research unit of Bhabha Atomic Research Centre, Trombay, Mumbai. He has authored 20+ research papers which are published in reputed international journals indexed by SCI/ Scopus. He has presented his work in many international conferences and forums. He is currently working as an Associate professor in Institute of Engineering and Technology, GLA University, Mathura. He is an active member of The Indian Institute of Metals (IIM) and secretary of The Indian Institute of Metals Mathura Chapter.

Index

S

T

U

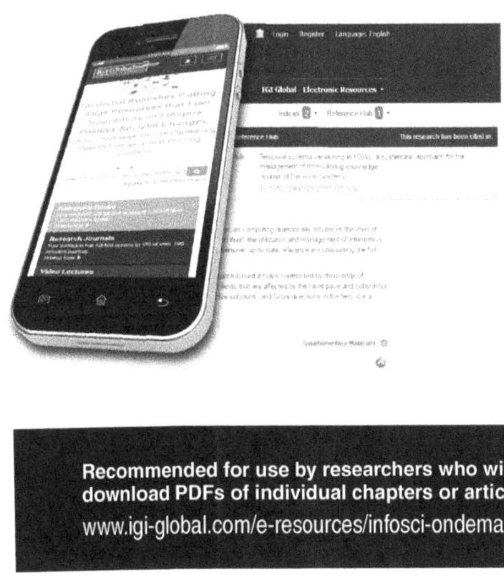

www.ingramcontent.com/pod-product-compliance
Lightning Source LLC
LaVergne TN
LVHW080604300926
849162LV00005B/1045
* 9 7 8 1 7 9 9 8 2 2 4 5 5 *